Natural Fiber Based Composites II

Natural Fiber Based Composites II

Editor

Philippe Evon

Basel • Beijing • Wuhan • Barcelona • Belgrade • Novi Sad • Cluj • Manchester

Editor
Philippe Evon
LCA Laboratory & AGROMAT
Toulouse INP & ENSIACET
Tarbes
France

Editorial Office
MDPI
St. Alban-Anlage 66
4052 Basel, Switzerland

This is a reprint of articles from the Special Issue published online in the open access journal *Coatings* (ISSN 2079-6412) (available at: www.mdpi.com/journal/coatings/special_issues/fiber_basedII).

For citation purposes, cite each article independently as indicated on the article page online and as indicated below:

Lastname, A.A.; Lastname, B.B. Article Title. *Journal Name* **Year**, *Volume Number*, Page Range.

ISBN 978-3-0365-9031-8 (Hbk)
ISBN 978-3-0365-9030-1 (PDF)
doi.org/10.3390/books978-3-0365-9030-1

© 2023 by the authors. Articles in this book are Open Access and distributed under the Creative Commons Attribution (CC BY) license. The book as a whole is distributed by MDPI under the terms and conditions of the Creative Commons Attribution-NonCommercial-NoDerivs (CC BY-NC-ND) license.

Contents

About the Editor vii

Philippe Evon
Special Issue "Natural Fiber Based Composites II"
Reprinted from: *Coatings* 2023, *13*, 1694, doi:10.3390/coatings13101694 1

Aisyah Humaira Alias, Mohd Nurazzi Norizan, Fatimah Athiyah Sabaruddin, Muhammad Rizal Muhammad Asyraf, Mohd Nor Faiz Norrrahim and Ahmad Rushdan Ilyas et al.
Hybridization of MMT/Lignocellulosic Fiber Reinforced Polymer Nanocomposites for Structural Applications: A Review
Reprinted from: *Coatings* 2021, *11*, 1355, doi:10.3390/coatings11111355 4

Benjamin Barthod-Malat, Maxime Hauguel, Karim Behlouli, Michel Grisel and Géraldine Savary
Influence of the Compression Molding Temperature on VOCs and Odors Produced from Natural Fiber Composite Materials
Reprinted from: *Coatings* 2023, *13*, 371, doi:10.3390/coatings13020371 46

Jiaxin Zhao, Xiaoxiao Wu, Xushuo Yuan, Xinjie Yang, Haiyang Guo and Wentao Yao et al.
Nanocellulose and Cellulose Making with Bio-Enzymes from Different Particle Sizes of Neosinocalamus Affinis
Reprinted from: *Coatings* 2022, *12*, 1734, doi:10.3390/coatings12111734 67

Jifei Chen, Qiansun Zhao, Guifeng Wu, Xiaotian Su, Wengang Chen and Guanben Du
Design and Analysis of a 5-Degree of Freedom (DOF) Hybrid Three-Nozzle 3D Printer for Wood Fiber Gel Material
Reprinted from: *Coatings* 2022, *12*, 1061, doi:10.3390/coatings12081061 83

Tomasz Kozior, Al Mamun, Marah Trabelsi and Lilia Sabantina
Comparative Analysis of Polymer Composites Produced by FFF and PJM 3D Printing and Electrospinning Technologies for Possible Filter Applications
Reprinted from: *Coatings* 2022, *12*, 48, doi:10.3390/coatings12010048 95

Elio Padoan, Enzo Montoneri, Giorgio Bordiglia, Valter Boero, Marco Ginepro and Philippe Evon et al.
Waste Biopolymers for Eco-Friendly Agriculture and Safe Food Production
Reprinted from: *Coatings* 2022, *12*, 239, doi:10.3390/coatings12020239 113

Baodong Liu, Xinjie Huang, Shuo Wang, Dongmei Wang and Hongge Guo
Performance of Polyvinyl Alcohol/Bagasse Fibre Foamed Composites as Cushion Packaging Materials
Reprinted from: *Coatings* 2021, *11*, 1094, doi:10.3390/coatings11091094 125

Aicha Bouaziz, Dorra Dridi, Sondes Gargoubi, Souad Chelbi, Chedly Boudokhane and Abderraouf Kenani et al.
Analysis of the Coloring and Antibacterial Effects of Natural Dye: Pomegranate Peel
Reprinted from: *Coatings* 2021, *11*, 1277, doi:10.3390/coatings11111277 139

S. El-Sayed Saeed, Tahani M. Al-Harbi, Ahmed N. Alhakimi and M. M. Abd El-Hady
Synthesis and Characterization of Metal Complexes Based on Aniline Derivative Schiff Base for Antimicrobial Applications and UV Protection of a Modified Cotton Fabric
Reprinted from: *Coatings* 2022, *12*, 1181, doi:10.3390/coatings12081181 148

Jing Gao, Liang-Shi Hao, Bing-Bing Ning, Yuan-Kang Zhu, Ju-Bo Guan and Hui-Wen Ren et al.
Biopaper Based on Ultralong Hydroxyapatite Nanowires and Cellulose Fibers Promotes Skin Wound Healing by Inducing Angiogenesis
Reprinted from: *Coatings* **2022**, *12*, 479, doi:10.3390/coatings12040479 . . . . **166**

Jie Zhou, Xiaosan Song, Boyang Shui and Sanfan Wang
Preparation of Graphene Oxide Composites and Assessment of Their Adsorption Properties for Lanthanum (III)
Reprinted from: *Coatings* **2021**, *11*, 1040, doi:10.3390/coatings11091040 . . . . **185**

About the Editor

Philippe Evon

Dr. Philippe Evon is a Research Engineer at the Laboratoire de Chimie Agro-industrielle (LCA), Toulouse INP, France. He specializes in the valorization of wastes and co-products from biomass to produce extracts and to design agromaterials. He is mainly developing studies for using biomass as a raw material for:

- The production of bioactive extracts through fractionation processes using "green" solvents and twin-screw extrusion technology as a continuous extraction technique;

- The manufacture of agromaterials by combining single- or twin-screw extrusion technologies with molding processes (e.g., injection molding or thermopressing). He is the Manager of the LCA's Industrial Technological Hall "AGROMAT", dedicated to agromaterials, which is located in Tarbes (south-west of France).

Editorial

Special Issue "Natural Fiber Based Composites II"

Philippe Evon

Laboratoire de Chimie Agro-Industrielle, École Nationale Supérieure des Ingénieurs en Arts Chimiques et Technologiques, Université de Toulouse, INP, 4 Allée Emile Monso, 31030 Toulouse, France; philippe.evon@toulouse-inp.fr

In the last twenty years, the use of cellulosic and lignocellulosic agricultural by-products for composite applications has been of great interest, especially for reinforcing matrices. Fibers of renewable origin have many advantages. They are abundant and cheap, have a reduced impact on the environment, and are independent from fossil resources. Their ability to mechanically reinforce thermoplastic matrices is well known, as is their natural heat-insulation ability. The matrices can themselves be of renewable origin (e.g., proteins, thermoplastic starch, poly(lactic acid), polyhydroxyalkanoates, etc.), thus contributing to the development of 100% bio-based composites with a controlled end of life.

Continuing the Special Issue, *"Natural Fiber Based Composites"* [1], this Special Issue provides an inventory of the latest research in the area of composites reinforced with natural and wood fibers, focusing particularly on the preparation and molding processes of such materials (e.g., extrusion, injection-molding, hot pressing, 3D printing, etc.) and their characterizations. It contains one review and ten research reports authored by researchers from three continents and eleven countries, namely, China, Egypt, France, Germany, Italy, Malaysia, Poland, Russia, Saudi Arabia, Tunisia, and Yemen.

An overview of the hybridization of MMT/lignocellulosic fiber-reinforced polymer nanocomposites for structural applications is provided in a comprehensive review paper [2]. The use of montmorillonite nanoparticles (MMTs) improves the mechanical properties of composites while ensuring their thermal stability. The dispersion of MMTs, particularly in water-soluble polymers, is facilitated by its intrinsic properties. This review article provides an exhaustive list of the latest advances in the field, from methods of obtaining these innovative materials to their mechanical, thermal, and fire-retardant properties. Such materials can now be used as reinforcing agents in a variety of high-end industrial applications.

Some of the research articles in this Special Issue suggest using very different raw materials to obtain bio-based composite materials from different molding techniques, which may have future applications in many fields, e.g., automotive, packaging, load-bearing composites, tissue engineering and biomedicine, filtering, agriculture, and fabrics.

- Non-woven preforms made from natural and thermoplastic fibers are frequently used as vehicle interior parts. Lightweight and recyclable, they can nevertheless release volatile organic compounds (VOCs) and odors into the vehicle interior. In [3], VOCs and odors released by flax/PP non-woven composites were quantified using headspace solid-phase microextraction (HS-SPME) coupled with gas chromatography–mass spectrometry (GC-MS). The composition of VOCs changes according to the compression-molding temperature. It is, therefore, possible to optimize this temperature to obtain materials that are less odorous and more suitable for use in the automotive sector.
- *Neosinocalamus affinis* is a clumping bamboo. Cellulose is extractable from its powder through bio-enzymatic digestion [4]. Cellulose is the most abundant polysaccharide on Earth. Once extracted, it can be transformed into nanocellulose (CNM). The amount of extracted cellulose, purity, crystallinity, and thermal stability depend on the initial

- particle size of the powder and the bio-enzyme used. The result is cellulose nanofibrils (CNFs), which, as a hardening component, can be mixed with other materials to improve their toughness. The use of bamboo as an ingredient in composites could therefore expand in the years to come.
- Cellulose is also abundant in wood. Along with hemicelluloses and lignin, it can be used in tissue engineering and biomedicine because of their good properties. Wood fiber gel can be prepared as a 3D-printing material using a hybrid 3D printer with three nozzles and five degrees of freedom [5]. The printer is capable of multi-material and multi-degree-of-freedom printing, making the wood fiber gel suitable for multiple biomedical applications.
- Depending on the three-dimensional printing technology used, the structure of the composites produced can differ markedly. This technique can even be combined with another unconventional technology, electrospinning (ES), to produce composites with improved qualities. These could be used in future for the construction of filtering devices and in medical applications [6].
- In [1], one research article presented the use of sunflower proteins to produce controlled release fertilizers (CRFs) through injection-molding, with urea and/or new biopolymers (BPs) obtained via the hydrolysis of municipal biowastes acting as additional sources of nutrients for plants [7]. In the present Special Issue, these innovative CRFs were tested for spinach cultivation [8]. In the presence of BP, the work highlighted that composites yield the safest crop coupled with high biomass production, thus contributing to the development of a bio-based chemical industry exploiting biological wastes as a raw material for eco-friendly agriculture.
- Widely available, sugar cane bagasse fiber can be used to make polyvinyl alcohol (PVA)-based foamed composites [9]. Their static cushioning performance is comparable to that of expanded polystyrene (EPS), commonly used in the packaging industry. The foamed composites developed mainly have an open cell structure. Bagasse fiber is also compatible with PVA foam. Depending on their mechanical properties and static cushioning performance, some of these new materials could be suitable for the cushioned packaging of light, fragile products.

In the textile industry, agricultural by-products can also be used as dyes for environmentally friendly fabric finishing. In [10], for example, pomegranate peel extract was used to produce a natural dye. This new, bio-based dye is particularly recommended for dyeing polyamide fabrics. Despite their fair light fastness, finished samples are resistant to rubbing and washing, and have good antibacterial properties. Such an extract could soon be used in the textile industry.

At the frontier between the textile industry and the medical field, the development of antimicrobial textiles is playing an increasingly important, protective role. Schiff bases and nanometric complexes have been in situ synthesized on bio-based conventional cotton fabrics to give highly effective and durable antibacterial and UV protection properties [11].

Raw materials of non-plant origin can also be used to produce innovative composite materials. For example, in the medical sector, hydroxyapatite (HAP) has great potential for wound healing. Mineral in origin, it is a calcium phosphate belonging to the apatite family. This wound-healing property can be explained by its high biocompatibility and angiogenic capacity. While traditional HAP-based materials are often fragile and not very mechanically resistant, it is possible to obtain dressings based on HAP nanowires, in the form of a flexible and superhydrophilic biopaper, which, due to the continuous release of Ca^{2+} ions, will promote the healing of skin lesions [12]. Safe and effective, this biocompatible material opens new prospects for clinical applications.

The final research article in this Special Issue presents the preparation of graphene oxide composites and evaluates their adsorption properties for lanthanum (III), a metallic element contained in low proportions in the earth, but widely used in optical glasses, alloys, catalysts, and ceramics [13]. Graphene occurs naturally in graphite crystals (defined as a stack of graphene sheets). The adsorption capacity of lanthanum was studied under different conditions; the results show that the carboxylated graphene oxide/diatomite/magnetic chitosan (GOH/DMCS) composite could be used in future as an adsorbent that is effective and reusable.

This Special Issue presents a wide range of topics. It provides an update on the current research in the field of natural-fiber-based composite materials, and more. These contributions will be a source of inspiration for the development of new composites. These new materials are environmentally friendly and will undoubtedly find numerous applications in the years to come in many sectors.

Acknowledgments: I would like to thank all the authors for their valuable contributions to this Special Issue, the reviewers for their reviews and useful comments which allowed for the submitted papers to be improved, and the journal editors for their kind support throughout the production of this Special Issue.

Conflicts of Interest: The authors declare no conflict of interest.

References

1. Evon, P. Special Issue "Natural Fiber Based Composites". *Coatings* **2021**, *11*, 1031. [CrossRef]
2. Alias, A.H.; Norizan, M.N.; Sabaruddin, F.A.; Asyraf, M.R.M.; Norrrahim, M.N.F.; Ilyas, A.R.; Kuzmin, A.M.; Rayung, M.; Shazleen, S.S.; Nazrin, A.; et al. Hybridization of MMT/Lignocellulosic Fiber Reinforced Polymer Nanocomposites for Structural Applications: A Review. *Coatings* **2021**, *11*, 1355. [CrossRef]
3. Barthod-Malat, B.; Hauguel, M.; Behlouli, K.; Grisel, M.; Savary, G. Influence of the Compression Molding Temperature on VOCs and Odors Produced from Natural Fiber Composite Materials. *Coatings* **2023**, *13*, 371. [CrossRef]
4. Zhao, J.; Wu, X.; Yuan, X.; Yang, X.; Guo, H.; Yao, W.; Ji, D.; Li, X.; Zhang, L. Nanocellulose and Cellulose Making with Bio-Enzymes from Different Particle Sizes of Neosinocalamus Affinis. *Coatings* **2022**, *12*, 1734. [CrossRef]
5. Chen, J.; Zhao, Q.; Wu, G.; Su, X.; Chen, W.; Du, G. Design and Analysis of a 5-Degree of Freedom (DOF) Hybrid Three-Nozzle 3D Printer for Wood Fiber Gel Material. *Coatings* **2022**, *12*, 1061. [CrossRef]
6. Kozior, T.; Mamun, A.; Trabelsi, M.; Sabantina, L. Comparative Analysis of Polymer Composites Produced by FFF and PJM 3D Printing and Electrospinning Technologies for Possible Filter Applications. *Coatings* **2022**, *12*, 48. [CrossRef]
7. Evon, P.; Labonne, L.; Padoan, E.; Vaca-Garcia, C.; Montoneri, E.; Boero, V.; Negre, M. A New Composite Biomaterial Made from Sunflower Proteins, Urea, and Soluble Polymers Obtained from Industrial and Municipal Biowastes to Perform as Slow Release Fertiliser. *Coatings* **2021**, *11*, 43. [CrossRef]
8. Padoan, E.; Montoneri, E.; Bordiglia, G.; Boero, V.; Ginepro, M.; Evon, P.; Vaca-Garcia, C.; Fascella, G.; Negre, M. Waste Biopolymers for Eco-Friendly Agriculture and Safe Food Production. *Coatings* **2022**, *12*, 239. [CrossRef]
9. Liu, B.; Huang, X.; Wang, S.; Wang, D.; Guo, H. Performance of Polyvinyl Alcohol/Bagasse Fibre Foamed Composites as Cushion Packaging Materials. *Coatings* **2021**, *11*, 1094. [CrossRef]
10. Bouaziz, A.; Dridi, D.; Gargoubi, S.; Chelbi, S.; Boudokhane, C.; Kenani, A.; Aroui, S. Analysis of the Coloring and Antibacterial Effects of Natural Dye: Pomegranate Peel. *Coatings* **2021**, *11*, 1277. [CrossRef]
11. Saeed, S.E.-S.; Al-Harbi, T.M.; Alhakimi, A.N.; Abd El-Hady, M.M. Synthesis and Characterization of Metal Complexes Based on Aniline Derivative Schiff Base for Antimicrobial Applications and UV Protection of a Modified Cotton Fabric. *Coatings* **2022**, *12*, 1181. [CrossRef]
12. Gao, J.; Hao, L.-S.; Ning, B.-B.; Zhu, Y.-K.; Guan, J.-B.; Ren, H.-W.; Yu, H.-P.; Zhu, Y.-J.; Duan, J.-L. Biopaper Based on Ultralong Hydroxyapatite Nanowires and Cellulose Fibers Promotes Skin Wound Healing by Inducing Angiogenesis. *Coatings* **2022**, *12*, 479. [CrossRef]
13. Zhou, J.; Song, X.; Shui, B.; Wang, S. Preparation of Graphene Oxide Composites and Assessment of Their Adsorption Properties for Lanthanum (III). *Coatings* **2021**, *11*, 1040. [CrossRef]

Disclaimer/Publisher's Note: The statements, opinions and data contained in all publications are solely those of the individual author(s) and contributor(s) and not of MDPI and/or the editor(s). MDPI and/or the editor(s) disclaim responsibility for any injury to people or property resulting from any ideas, methods, instructions or products referred to in the content.

Review

Hybridization of MMT/Lignocellulosic Fiber Reinforced Polymer Nanocomposites for Structural Applications: A Review

Aisyah Humaira Alias [1,2], Mohd Nurazzi Norizan [3], Fatimah Athiyah Sabaruddin [4], Muhammad Rizal Muhammad Asyraf [5], Mohd Nor Faiz Norrrahim [6], Ahmad Rushdan Ilyas [7,8], Anton M. Kuzmin [9], Marwah Rayung [10], Siti Shazra Shazleen [1], Asmawi Nazrin [1], Shah Faisal Khan Sherwani [1], Muhammad Moklis Harussani [1,2], Mahamud Siti Nur Atikah [11], Mohamad Ridzwan Ishak [4], Salit Mohd Sapuan [1,2] and Abdan Khalina [1,*]

1. Institute of Tropical Forestry and Forest Products (INTROP), Universiti Putra Malaysia (UPM), Seri Kembangan 43400, Malaysia; a.humaira.aisyah@gmail.com (A.H.A.); shazra.shazleen@yahoo.com (S.S.S.); nazrinnurariefmardi@gmail.com (A.N.); faisalsherwani786@gmail.com (S.F.K.S.); mmharussani17@gmail.com (M.M.H.); sapuan@upm.edu.my (S.M.S.)
2. Department of Mechanical and Manufacturing Engineering, Faculty of Engineering, Universiti Putra Malaysia (UPM), Seri Kembangan 43400, Malaysia
3. Centre for Defence Foundation Studies, Universiti Pertahanan Nasional Malaysia (UPNM), Kem Perdana Sungai Besi, Kuala Lumpur 57000, Malaysia; mohd.nurazzi@gmail.com
4. Faculty of Biotechnology and Biomolecular Sciences, Universiti Putra Malaysia (UPM), Seri Kembangan 43400, Malaysia; atiyah88@gmail.com (F.A.S.); mohdridzwan@upm.edu.my (M.R.I.)
5. Department of Aerospace Engineering, Universiti Putra Malaysia (UPM), Seri Kembangan 43400, Malaysia; asyrafriz96@gmail.com
6. Research Centre for Chemical Defence, Universiti Pertahanan Nasional Malaysia (UPNM), Kem Perdana, Sungai Besi, Kuala Lumpur 57000, Malaysia; faiznorrrahim@gmail.com
7. School of Chemical and Energy Engineering, Faculty of Engineering, Universiti Teknologi Malaysia (UTM), Skudai 81310, Malaysia; ahmadilyas@utm.my
8. Centre for Advanced Composite Materials (CACM), Universiti Teknologi Malaysia, UTM Johor Bahru, Skudai 81310, Malaysia
9. Department of Mechanization and Agricultural Products Processing, Ogarev Mordovia State University, 430005 Saransk, Russia; kuzmin.a.m@yandex.ru
10. Faculty of Science, Universiti Putra Malaysia (UPM), Seri Kembangan 43400, Malaysia; marwahrayung@yahoo.com
11. Department of Chemical and Environmental Engineering, Universiti Putra Malaysia (UPM), Seri Kembangan 43400, Malaysia; sitinuratikah_asper7@yahoo.com
* Correspondence: khalina@upm.edu.my

Abstract: In the recent past, significant research effort has been dedicated to examining the usage of nanomaterials hybridized with lignocellulosic fibers as reinforcement in the fabrication of polymer nanocomposites. The introduction of nanoparticles like montmorillonite (MMT) nanoclay was found to increase the strength, modulus of elasticity and stiffness of composites and provide thermal stability. The resulting composite materials has figured prominently in research and development efforts devoted to nanocomposites and are often used as strengthening agents, especially for structural applications. The distinct properties of MMT, namely its hydrophilicity, as well as high strength, high aspect ratio and high modulus, aids in the dispersion of this inorganic crystalline layer in water-soluble polymers. The ability of MMT nanoclay to intercalate into the interlayer space of monomers and polymers is used, followed by the exfoliation of filler particles into monolayers of nanoscale particles. The present review article intends to provide a general overview of the features of the structure, chemical composition, and properties of MMT nanoclay and lignocellulosic fibers. Some of the techniques used for obtaining polymer nanocomposites based on lignocellulosic fibers and MMT nanoclay are described: (i) conventional, (ii) intercalation, (iii) melt intercalation, and (iv) in situ polymerization methods. This review also comprehensively discusses the mechanical, thermal, and flame retardancy properties of MMT-based polymer nanocomposites. The valuable properties of MMT nanoclay and lignocellulose fibers allow us to expand the possibilities of using polymer nanocomposites in various advanced industrial applications.

Keywords: bamboo; kenaf; MMT; natural fiber; polymer composites

1. Introduction

At present, researchers are aware of the necessity of developing new composite products with high mechanical and thermal stability for structural applications. In the service life of power transmission, marine, automobile, and aerospace structures, several factors can cause damage to the structural integrity of polymer composite and reduce their lifespan due to high impact, creep, and fatigue loads from the external environment [1–4]. Furthermore, the potential for penetration and other such impacts can lead to extensive delamination, which results in degradation of the structural performance of composites [5–7]. For high-performance applications such as fire extinguishers, armored vehicles, and aircraft, the main purpose of the product is to have a high ratio of strength-to-weight structure [8]. Generally, thermoplastic composite materials exhibit high strength properties, which commonly decrease to a low value in terms of damping capacity due to its viscoelastic properties [9]. This leads to valuable properties for structural purposes.

Currently one can observe growing popularity and market demand for natural fiber-reinforced polymer composites due to their advantages of low price, light weight and versatile applications [10,11]. However, lignocellulosic fibers such as sugar palm, kenaf, coir etc. exhibit low mechanical properties due to their hydrophilic properties and high moisture rate when used as reinforcement in polymer composites [12–14]. Moreover, the main drawback associated with the inclusion of lignocellulosic fiber in polymeric resin is the lack of surface adhesion between the two phases, which affects the flow properties of the final biocomposite laminate [15,16].

The previously mentioned drawbacks of lignocellulosic fiber composites can be addressed either by regulating the fiber loading and orientation, fabric stacking sequence, fiber treatment as well as the use of nanofiller additives [17–20]. Several fiber treatment methods either via physical approaches such as plasma or corona treatment and steam explosion as well as gamma irradiation [10,21] and chemical methods such as alkaline treatment, acetylation methods and coating techniques using hydrophobic polymers have been reported [22–26]. A significant decrease in water uptake of jute-reinforced polyester composites from the incorporation of synthetic fibers such as E-glass has also been discovered [27]. Research revealed that the inclusion of impermeable fillers along with lignocellulosic fiber inside polymer composites is one of the most significant approaches to lower the moisture uptake of composites. In this case, the filler factors such as filler concentration, distribution, shape and size would reflect on the water barrier, mechanical, and thermal properties of the final composites [28,29].

Heretofore, many studies and reviews have mentioned the superiority of reinforcement with smaller size particles which has great importance for the overall reinforcement effect in the matrix system [30,31]. An escalation of progress and interest can be seen in the application of plate-structured, nanofiller-filled materials represented by nanocomposites due to their high aspect ratio [32]. These nanofiller-filled materials, particularly when in particle sizes in the vicinity of 100 nm and below, have shown great effect in imparting a high degree of reinforcement. In fact, the reinforcement achieved by using 3–5% by weight of nano-scale reinforcement is similar to that achievable using 20–30% of a micro-scale filler. Therefore, the incorporation of nanofillers in composites systems is much more preferred due to their high specific interfacial area, which enables higher interfacial interactions, thus increasing the modulus properties of the final materials. In general, nanofillers are categorized based on their dimensions. Mrinal [31] has listed the types of nanofiller including one dimensional nanofillers such as nanotubes and nanowires, two dimensional ones like nanoclays and graphene, and three dimensional examples, for instance spherical and cubical nanoparticles. Carbonaceous nanofillers like nanotubes and graphene impart excellent properties that are attributed to their high mechanical strength and high aspect ratio, whilst, clays are naturally found as platelets, stacked from a few to as many as one thousand sheets. Montmorillonite (MMT), the most popular clay nanofiller, is comprised of two silicates layers of octahedral sheets of alumina that are sandwiched together. This aluminosilicate with low dimensions of 1–5 nm thickness and 100–500 nm in diameter will impart platelets with a high aspect ratio (>50), hence providing stiffness and strength to the composite [33].

Presently, global demand for nanofillers like MMT clay and organically modified montmorillonite (OMMT) is growing due to the potential of using nanoparticles to' extensively modify the overall properties of polymer composites [34,35]. Nanoclay particles are composed of layered silicates. Examples of nanoclays are compounds such as pyrophyllite, organoclay, hectorite, saponite and nontronite nanoclay and MMT clay. Among these, MMT clay, which is the main constituent of bentonite, is the most regularly used layered silicate in polymer matrices due to its properties. MMT clay is well known as having a good filler anisotropy, high aspect ratio, excellent barrier properties, great strength and stiffness, as well as good thermal stability [36,37]. Despite the many promising advantages of MMT fillers in composites, the hydrophilicity of the nanoclay leads to difficulties in homogenously distributing it in the polymer matrix. Subsequently, this would induce disparity in the distribution of electrical fields in the material and inferior electrical performance of the resulting nanocomposites. Nanoclays can be commonly modified via ion exchange reactions to form organophilic species which allow the polymer molecules to penetrate between the clay galleries [38]. Besides that, the use of organic moieties as modifiers of MMT silicates can lead to a decrease in the polymer crosslinking, consequently inhibiting the interfacial adhesion between the polymer matrix and the nanofiller [36].

The combination of bio-based fibers/matrices with synthetic fibers/matrices through hybridization provides a novel approach for overcoming the drawbacks of solely natural fibers or matrices. By combining MMT nanoclay as a reinforcement agent into a composite's composition, it is possible to address some of the drawbacks of one type of reinforcement compared to another. Hybridization of MMT clay with lignocellulosic fibers could improve the thermal, mechanical and physical characteristics of biocomposites, allowing them to be used in a variety of applications [39].

As mentioned in the previous paragraphs, MMT can be used as an additive for the cross-arm structure in transmission towers since it significantly promotes good electrical resistance performance for composites while elevating the composites' mechanical performance [38,40,41]. To increase the number of potential applications, lignocellulosic composites' properties need to be more easy to modulate and flexible in terms of improvement to match the required performance of various applications [42–46]. In this manuscript, we narrow our survey to the methods of extraction, treatment and modification, classification, and applications of MMT fillers. This work aims to review the recent progress on the impact of MMT fillers hybridised with lignocellulosic fibers on the mechanical and thermal properties of reinforced polymer composites.

2. MMT Nanoclay

MMT is a member of the natural smectite group and is categorized as a phyllosilicate mineral with a nanolayered structure that has high biocompatibility and biodegradability as well as high mechanical characteristics [47]. Table 1 lists the most essential physical and chemical characteristics of MMT [48]. The common substituents for MMT particles are iron (Fe), potassium (K), aluminium and other cations, where their ratios may vary greatly according to their source [49]. The precise theoretical formulation of MMT, according to Uddin, [49], had never been found in Nature due to the charge imbalance introduced by the cation substitution. The chemical formula for MMT is shown in Equation (1):

$$(Na, Ca)_{0.33}(Al, Mg)_2(Si_4O_{10})(OH)_2 \cdot nH_2O \tag{1}$$

The molecular formula of MMT is $Al_2H_2O_{12}Si_4$ [49]. and aluminium oxides are present in the oxide composition, while calcium is the man mineral in MMT. The basic molecular structure is made up of silica tetrahedra and aluminium octahedral units. The Si^{+4} cation has fourfold tetrahedral coordination with oxygen, whereas the cation Al^{+3} has six-fold or octahedral coordination with oxygen [49].

Table 1. Physical and chemical characteristics of MMT.

Properties	Description
Physical properties	
Color	White, grey, beige to buff
Surface area (m^2/g)	240
Bulk density (g/L)	370
Diameter of particles (μm)	1
Length of particles (nm)	100–150
Surface dimensions (nm)	300–600
Crystal system	Monoclinic
Transparency	Translucent
Fracture	Irregular
Aspect ratio	High
Nature	Hydrophilic
Chemical properties	
Chemical composition (%)	
SiO_2	73.0
Al_2O_3	14.0
Fe_2O_3	2.7
K_2O	1.9
MgO	1.1
Na_2O	0.6
CaO	0.2

MMT is comprised of two O-Si-O tetrahedral sheets of silica sandwiched by an alumina layer, O-Al(Mg)-O octahedral sheet, 2:1 clay [50]. Silica layers form a hexagonal network by sharing three corners with neighboring tetrahedra [50]. Each tetrahedron's remaining fourth corner connects to an adjacent octahedral sheet. Aluminium or magnesium, in six-fold coordination with oxygen from the tetrahedral layer and with hydroxyls, make up the octahedral sheets. To create the elementary particles of MMT, neighboring layers of approximately 10 μm-sized are stacked together assisted by van der Waals forces and electrostatic forces, or by hydrogen bonding [51]. As seen in Figure 1, the particles aggregate to produce micrometer-to millimeter-scale particles [52]. The layer structure of MMT may vary as a consequence of its changeable structure, allowing for the creation of a range of hybrids and composites [53]. Figure 1a shows a side view, where tetrahedral MMT units are assembled through weak van der Waals and electrostatic forces to form the primary particles, and Figure 1b shows a top view of the hexagonal structure of the oxygen and hydroxyl ligands of the octahedral layer.

Physical and chemical approaches can be used to overcome the van der Waals forces and the electrostatic forces that stack the layers together [49]. According to Nakato and Miyamoto, MMT particles have plate-like shapes with surface dimensions of around 300 to 600 nm and a mean diameter of around 1 μm [50]. Thus, MMT exhibits a large aspect ratio of around 50 to 1000. In addition, MMT shows great hydrophilic nature where its water content tendency increases when exposed to water [54]. However, because polymers are typically organophilic, researchers have developed modified MMT forms by altering the clay surface. MMT may very well be made compatible with common petroleum-based polymers and distributed easily within the polymers for improved characteristics in nanocomposites after surface compatibilization or intercalation [50]. Due to these characteristics, MMT is the most common filler utilized in biocomposites and nanocomposite applications, in order to improve a particular property of materials [55–62].

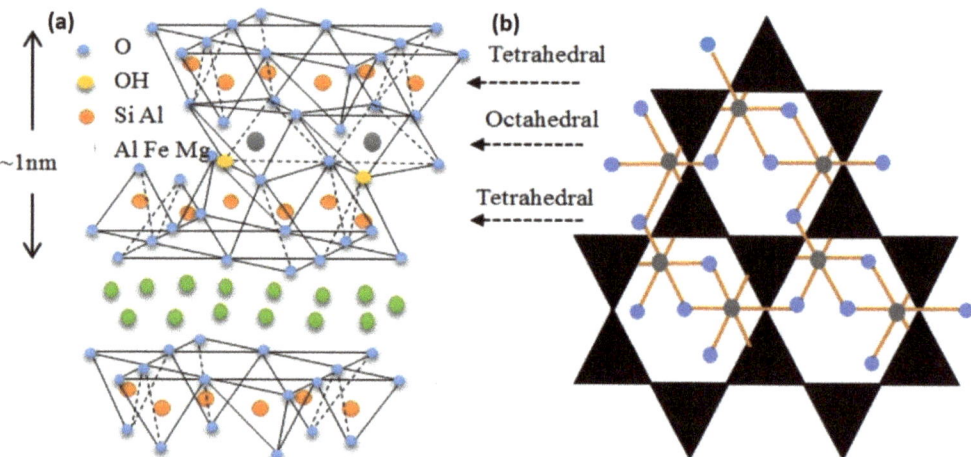

Figure 1. Schematic representation of the MMT structure. (a) Side view: tetrahedrons units of Mt assembled through weak van der Waals and electrostatic forces to form the primary particles, and (b) top view: hexagonal structure of oxygen and hydroxyl ligands of the octahedral layer. Reproduced from [63].

Due to the high aspect ratio and wide surface area of MMT nanolayers, it is extremely possible that the inclusion and dispersion of MMT nanolayers into the polymer matrix would result in MMT/polymer nanocomposites with significantly improved mechanical properties [52,64]. According to Zhu et al. [52], the addition and dispersion of MMT nanolayers into the polymer matrix results in MMT/polymer nanocomposites with significantly improved mechanical [65] and thermal properties [66,67], good oxygen and water barrier properties [68,69], as well as better flame retardancy [70]. As a result, intercalating MMT into nanolayers is critical for manufacturing effective MMT/polymer nanocomposites. Figure 2 shows the schematic diagram of in situ exfoliations of MMT to aid its intercalation into the polymer matrix. In situ exfoliation of MMT comprises the introduction of MMT to a liquid monomer or a monomer solution for it to expand and the monomer to reach the interlayer gap of MMT [52], called the polymerization process [71].

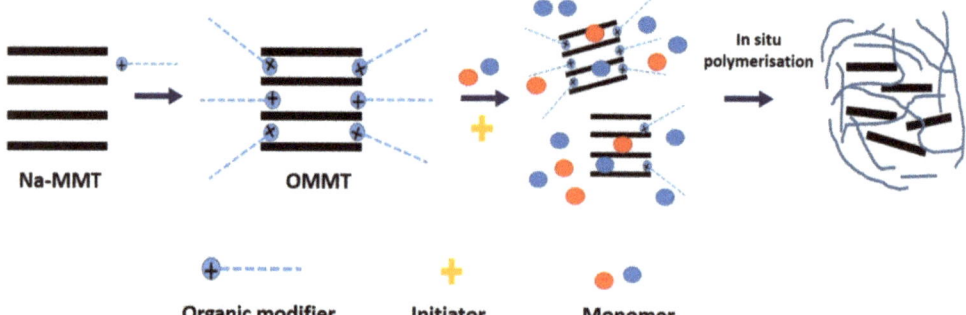

Figure 2. Schematic diagram of in situ exfoliation of MMT to aid its intercalation in polymer matrix. Reproduced from [72].

According to previous works, it was discovered that adding nanoclay and rice husk (RH) to high-density polyethylene (HDPE) improved its mechanical characteristics [73]. Nanoclay intercalation altered the crystallization behaviour of RH-filled HDPE by raising the crystallization temperature, enthalpy, and crystallinity. According to previous research, a combination of nanoclay and hemp has a synergistic impact on the flexibility and stiffness of PP [74]. With increasing nanoparticle concentration, there is a substantial reduction in water absorption and thickness swelling. The presence of nanoparticles improved dynamic mechanical behaviour, as well as fire retardancy and dimensional stability. Numerous studies have been done on the influence of nanoclays on the thermal stability of polymers such as polystyrene (PS) and ethylene vinyl acetate (EVA) [75]. The thermal stability of nanoclay-containing nanocomposites improved significantly as the nanoclay concentration increased. However, over a certain percentage of nanoclay (generally 5%), the thermal stability either plateaus or begins to abruptly decrease. The degree of dispersion, like that of most nanocomposites, has a direct impact on the thermal stability of PS/nanoclay composite systems. Polymer chains entrapped between clay platelets were linked to a significant increase in thermal stability [75].

The production of polymer clay nanocomposites is now one of the most important uses of nanoclays. These natural nanomaterials can also be used as rheological modifiers and additives in paints, inks, greases, and cosmetics, as well as pollution control bio-systems carriers and medication delivery systems [76]. When a small amount of MMT is incorporated into the polymer matrix, the polymer is modified, resulting typically in a significant improvement in the physical, mechanical, fracture, wear resistance, thermal stability, peak heat release rate, flame retardancy, biocompatibility, and chemical properties of the resulting composite compared to standard materials [77–79]. MMT nanoclays and acid-treated MMT are also often employed in catalytic cracking, acid-based catalysis, and materials applications. Organoclays might be utilized as gas absorbents, rheological modifiers, polymer nanocomposites, and drug delivery carriers, among other applications [80].

3. Lignocellulosic Fibers

Lignocellulosic fiber is the scientific name that refers to natural fibers. All natural fibers contain a few constituents, which are cellulose, hemicelluloses, and lignin. These constituents are mainly attached by hydrogen bonds [81]. The mechanical and physical properties of lignocellulosic fibers are influenced by their chemical composition. Moreover, lignocellulosic fibers are hydrophilic due to the presence of an abundance of hydroxyl groups, thus they can absorb water [82]. Figure 3 (reproduced from [79]) shows the schematic structure of lignocellulosic fibers. The chemical constituents of lignocellulosic fibers depend on the geographic location where the plants are grown [83]. Fibers that contain more cellulose might have higher mechanical and good thermal properties [83], while a high hemicellulose content will promote fiber degradation at low temperatures and absorb more moisture [84–86]. Nevertheless, lignocellulosic fiber-containing natural materials are commonly obstinately resistant to any external attack due to their high crystallinity and the high degree of polymerization of cellulose, which are protected by the lignin constituents that lower the accessible surface area and impart high fiber strength. In addition, lignin also linked to both hemicellulose and cellulose, has a function of providing structural support, impermeability and resistance against microbial attack (chemical and biological hydrolysis) and oxidative stress [81].

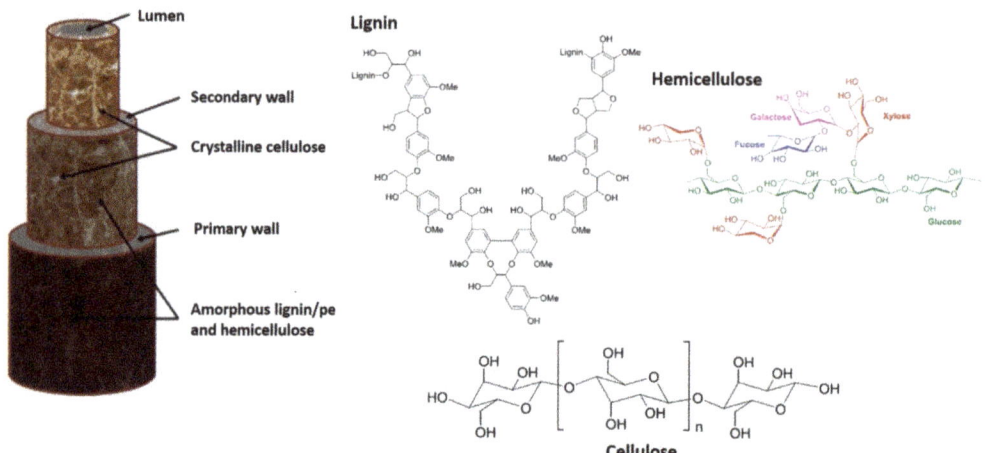

Figure 3. Schematic structure of lignocellulosic natural fiber. Reproduced from [84].

3.1. Cellulose

Cellulose is the primary component of lignocellulosic fibers [83]. Cellulose is made of long unbranched chains composed exclusively of glucose that are held together by hydrogen bonding. Cellulose fibrils have a particular cell geometry, which is a factor responsible for the properties of the fiber. The chemical formula of cellulose is $(C_6H_{10}O_5)_n$ (where n represents the number of glucose groups). The overall structure of cellulose consists of crystalline and amorphous regions. Cellulose is composed of carbon (44.44%), hydrogen (6.17%), and oxygen (49.39%) [85]. There are roughly 4000 to 8000 glucose molecules chained together. The polymer chain in cellulose is joined by glycosidic linkages at the C1 and C4 positions. Each repeating unit contains three hydroxyl groups. The presence of the hydroxyl group makes the cellulose hydrophilic. It is noted that these hydroxyl groups and their ability to form hydrogen bonds play a major role in directing the crystalline packing and also govern the physical properties of cellulose. The hydrogen bonding of many cellulose molecules to each other results in the formation of microfibers that can interact to form a fiber. Cellulose fibers usually have more than 500,000 cellulose molecules, therefore, cellulose fiber may contain 7000 to 15,000 glucose molecules per polymer [86]. The hydrogen bonding between cellulose fibrils is influenced by the mechanical properties of cellulose. Hydrogen bonding may not have the same strength as a covalent bond, however, the cumulative bonding energy of 2.5 billion hydrogen bonds is incredible. These properties, including its microcrystalline structure, make cellulose very difficult to dissolve or hydrolyze under natural conditions. In fact, the degree polymerization of cellulose also very high, in the range of 500 to 2500, indicating the high thermal stability and mechanical properties of cellulose [87]. The biodegradation of cellulose also requires more enzymes due to its greater complexity compared to other constituents.

3.2. Hemicellulose

Hemicelluloses are polysaccharides that differ from cellulose in that they consist of several sugar moieties. These sugars include glucose and other monomers such as hexoses (galactose, mannose) and pentoses (xylose and arabinose). The structure of hemicellulose is mostly a branched carbohydrate (hexoses and pentoses) and has a low molecular weight [86]. It is strongly bound to cellulose fibrils by hydrogen bonding. It is partly soluble in water and hydroscopic because of its open structure, which contains hydroxyl and acetyl groups [88]. This characteristic also allows natural fibers to absorb significant amounts of water, making it a weaker polymer compared to cellulose. Moreover, hemicellulose also has an amorphous structure and a low degree of polymerization of 100 to 200, which suggests low mechanical properties and a low degradation temperature [87,89].

3.3. Lignin

Lignin is a phenolic compound. Lignin is the second largest source of organic material in Nature [90]. It is a complicated amorphous polymer, hydrophobic in nature, formed of highly complex copolymers of aliphatic and aromatic constituents. Lignin contains hydroxyl, methoxyl and carbonyl groups [88,91]. The high carbon and low hydrogen content of lignin suggest that it is highly unsaturated or aromatic. The presence of hydroxyls and many polar groups in the lignin structure, resulting in strong intramolecular and intermolecular hydrogen bonds, making lignin essentially insoluble in any solvent. Phenolic hydroxyl and carboxyl groups allow lignin to be dissolved in alkaline solutions. Lignin binds the elementary fibers together with hemicellulose as cementing elements. Lignin imparts rigidity to the cell walls and fills the space between hemicellulose and cellulose [92]. Due to its hydrophobic character, lignin acts as a sealant to water, protection against biological attack and a stiffener to the fibers [93,94] Besides that, lignin can also be a compatibiliser between hydrophilic fibers and hydrophobic polymers, resulting in a stronger fiber-matrix interface [95]. In terms of degradation, lignin is degraded via oxidative process attributed to the secondary metabolism or to restricted availability of carbon, nitrogen, or sulphur which are commonly degraded as sole carbon, and energy sources [96], whilst in Nature, lignin is biodegraded by some white rot fungi, for instance from the basidiomycetes class, that degrade lignin more rapidly and extensively compared than other microorganisms. The white-rot fungi species, for example *Ceriporiopsis subvermispora*, *Phellinus pini*, *Phlebia sp.*, *Pleurotus sp.*, *Phanerochaete chrysosporium* etc. attack lignin more readily than cellulose and hemicellulose. In addition, these fungi also produce a set of ligninolytic enzymes that catalyze the oxidation of an array of aromatic substrates, producing aromatic radicals and changing the structure of the lignocellulose-containing raw materials and lignin [81].

Table 2 shows different lignocellulosic compositions for several natural fibers [97]. Generally, the cellulose, hemicellulose, and lignin in a typical lignocellulosic fiber fall within the range of 30% to 60%, 20% to 40%, and 15% to 25%, respectively. These lignocellulosic compositions greatly influence the mechanical properties of the fibers and result in significant improvements in the mechanical and thermal performance of polymer nanocomposites.

Table 2. Chemical composition of lignocellulosic fibers.

Natural Fiber	Lignocellulosic Components (%)			Ref.
	Cellulose	Hemicellulose	Lignin	
Sugar Palm	43.88	7.24	33.24	[98]
Bagasse	32 to 34	19 to 24	25 to 32	[99]
Bamboo	73.83	12.49	10.15	[100]
Flax	60 to 81	14 to 20.6	2.2 to 5	
Hemp	70 to 92	18 to 22	3 to 5	
Jute	51 to 84	12 to 20	5 to 13	[101]
Kenaf	44 to 87	22	15 to 19	
Ramie	68 to 76	13 to 15	0.6 to 1	
Sisal	65.8	12	9.9	[102]
Pineapple	66.2	19.5	4.2	[103]
Coir	32 to 43	0.15 to 0.25	40 to 45	[104]

4. Processing Techniques for MMT/Natural Fiber Reinforced Polymer Nanocomposites

Today, the advancements in nanotechnology have propelled the utilization of clay minerals as effective fillers and additives in polymers for desired applications. The reinforcement effect can produce significant improvements in polymers at very low filler contents of less than 5 wt.%, compared to those achieved using traditional micro-size fillers ($\geq$20 wt.%). However, natural sodium montmorillonite (NaMMT) must be surface modified to become OMMT for easier dispersion in the polymer matrix. The intercalation structure of poly(butylene) succinate (PBS)-reinforced OMMT and polylactic acid (PLA)-reinforced OMMT was observed and attributed to the strong interaction of hydroxyl groups in the clay structures with the carboxyl groups of the polyester. Meanwhile, the addition of NaMMT led to intercalation of TPS in the clay. MMT introduced in natural fiber reinforced polymer composites promotes resistance against chemical, heat, electricity, fire and UV light exposure [51].

There are some commonly utilized methods for incorporating nanoclays into polymer nanocomposites, such as intercalation of polymers, in-situ intercalative polymerization, and melt intercalation [105]. The mixing of clays, natural fibers and matrix generally applies three mixing techniques, namely mechanical mixing, magnetic stirring and sonication. Depending on the mixing technique used, the dispersion of nanoclays within the matrix can take the form of a phase-separated, intercalated or exfoliated nanocomposite structures. The most desired exfoliated structures present a complete separation of nanoclays due to the segregation of electrostatic forces between clay platelets by the polymer chain in the composites. Meanwhile, phase-separated structures indicate that the electrostatic forces between clay platelets cannot be overcome completely. This kind of structure is unfavorable to the properties of the nanocomposites as higher loading will be required for significant improvement, which otherwise can be achieved at lower loading for exfoliated and intercalated structures. An intercalated structure is formed when extended polymer chains are intercalated into the clay interlayers with the clay platelets still intact.

4.1. Conventional Composite Fabrication Techniques for Nanocomposites

Conventional techniques for fabricating composites, such as two-roll mill, twin-screw extruder, solution casting, compression and injection molding involve different clay incorporating techniques. Generally, twin-screw extruders and two-roll mills are the mixing machines used. The twin-screw extruder machine is a closed barrel containing two rotating screws mounted on splined shafts, which applies the melt intercalation technique utilizing the shear strength of mechanical mixing and high temperatures to mix clays and natural fibers into the matrix. The electrostatic binding force that clumped clay platelets together is broken down by the centrifugal force of rotating extruders in a high-pressure environment. Several studies [106,107] have shown that the equal channel angular extrusion (ECAE) method is able to re-engineer the nanofiller aspect ratio and orientation along with the crystalline lamellar structure. Polymer nanocomposite bars are extruded into a device made up of several pairs of channels of the same diameter, which intersect at the adjusted angles. Pairs of oblique channels within the plane can be rotated through the vertical axis to arrange different routes of deformation. Depending on the processing routes, nanoclays with shortened, closely packed, well-aligned and crystalline lamellae were compressed and diagonally well-oriented [108]. The two-roll mill consists of two rolls operating at different speeds that disentangle the clay particles and increase the clay galleries that enable the matrix to enter. Meanwhile, the solution casting method generally consists of three stages, including the dispersion of clay in a polymer solution, controlled removal of the solvent, and lastly composite film casting [109]. This solvent-based process involves mixing a polymer and prepolymer, which are soluble, and causing swelling of the clays layers [110]. The layered clays are then exfoliated using a solvent such as water, chloroform, or toluene. As the polymer/prepolymer are mixed with layered clays in a solution, the polymer chains intercalate and displace the solvent within the interlayers of the clay [109]. The solvent is then removed via vaporization or precipitation, and thus the intercalated sheets tend to reassemble and consequently form polymer/nanoclay composites. The dispersion of clay and natural fibers is usually done through sonication and magnetic stirring. Magnetic stirring uses magnetic force to break the electrostatic force between clay platelets. Sonication applies sound waves to agitate the particles in solution. In the case of high viscosity materials such as epoxy resins, high sonication amplitude is required to agitate the clays within the matrix before applying auxiliary natural fiber reinforcements. Meanwhile, compression and injection molding usually use premixed materials through intercalation of polymer and melt intercalation to optimize the dispersion of fillers within the matrices. As the name suggested, compression molding applies compressive strength to fabricate a specimen using male and female molds. First, materials are placed into the female mold to be melted down before applying compressive force joining both molds to distribute the desired amount of resin throughout the designed mold. The joined molds are then compressed again at lower temperature to solidify the resin. For injection molding, premixed materials are filled into a heated barrel, mixed and then injected into a designed mold. As the materials cool off, they will harden into the form of the mold cavity.

4.2. Intercalation of Polymer

Polymer intercalation techniques disperse MMT in a solution in which the polymer has been dissolved. In this form, the stacked clay layers can be easily dispersed in an adequate solvent. The polymer can then enter into the clay galleries and establish an ordered multilayer structure when the solvent is removed. Ayana et al. [111] added NaMMT during starch gelatinization with the addition of a glycerol later melt intercalated with polylactic acid (PLA), producing TPS/PLA/NaMMT (60/40, *w/w*) nanocomposites. The exfoliation of NaMMT in the thermoplastic starch (TPS) acts as a compatibiliser by reducing the size distribution of the PLA in the TPS matrix. The average diameter of the dispersed domain (Dn) value of the TPS/PLA blend was recorded at 21.3 µm, whereas the incorporation of 0.5 phr and 1.0 phr of NaMMT into TPS/PLA showed Dn values of 17.2 µm and 13.45 µm, respectively. The complete exfoliation of NaMMT was due to the initial preparation of clay dispersion in a starch suspension. The delayed insertion of plasticizer facilitates the penetration of starch molecules into the clay galleries, leading to complete exfoliation. Tian et al. [112] prepared MMT-reinforced PVA/starch (50:50) nanocomposites using an intercalation polymer method for better clay dispersion. TEM analysis showed that after incorporation of 5% MMT the clay nanolayers were well dispersed at random within the matrix, indicating a highly exfoliated structure. At 10% of MMT, poly(vinyl) alcohol (PVA) and starch macromolecules penetrated into the clay galleries and enlarged the spacing of clay nanolayers, forming intercalating structures. Through a similar method, Trifol et al. [113] reported that the reinforcement of 1% of OMMT and 1% of nanocellulose in PLA nanocomposites was sufficient to achieve a significant reduction of 57% in the water vapor permeability rate (WVTR) and 74% in the oxygen vapor permeability rate, whilst preserving good transparency, thermal stability and decent mechanical properties compared to neat PLA. In this work it seemed that the two types of nanofillers had a synergistic effect, whereby the fibrous nanocellulose and nanoclay platelets formed a strong percolated network maintaining the integrity of the nanocomposite films. TEM analysis indicated a highly exfoliated structure for low clay loading films (1%) while a higher clay loading (3%) showed an exfoliated and small proportion of intercalated nanoclay structures. Meanwhile, even higher nanocellulose loading films (5%) displayed a homogenous structure with no nanocellulose aggregates. Li et al. [114] firstly dispersed cellulose nanofibers (CNF) and NaMMT in an aqueous system, which was later mixed into corn starch CS) solution to obtain CS/NaMMT/CNF nanocomposite films through the solution casting method. CNF was observed to be distributed uniformly around or inserted into the lamella, indicating a good interaction between CNF and NaMMT. CNF bonded with starch by hydrogen bonds enters into the clay structures and enlarges the interlayer distance of NaMMT, thus forming ternary nanocomposites with a layered structure. At 3% of NaMMT, the addition of CNF up to 7% decreased the interlayer distance of NaMMT caused by the aggregation of CNF. Significant increments in mechanical properties and transparency values were recorded for the nanocomposite films with CNF loadings of up to 5% only. At a higher CNF content, the number of active CNF sites was unable to interact with CS and NaMMT, thus allowing a weak stress transfer between the reinforcements and matrix. The in-situ polymerization method synthesizes polymer nanocomposites by using monomers, where the low molecular weight monomer can easily diffuse into the clay galleries, forming either an intercalated or an exfoliated structure.

4.3. Melt Intercalation

The melt intercalation or melt mixing method provides better mixing of the nanoclay fillers and polymers that are typically used on a mass production scale through extrusion and injection molding. The processing parameters such as rotor speed, temperature profile, feed rate, mixing period, melting conditions, die pressure, materials grade and content as well as the chemical composition of the nanoclay filler and polymer are important to synthesize composites with desirable properties [109]. The melt intercalation approach involves melting the polymer at a high temperature and then the filler is blended with the polymer at a high temperature under shear. This method requires MMT to be surface-modified to weaken the electrostatic forces holding the platelets together. Boonprasith et al. [115] used NaMMT and OMMT to reinforce TPS/PBS blended nanocomposites through a melt intercalation technique. The polarity of NaMMT does not have good affinity with PBS/TPS as compared to OMMT. OMMT is compatible with PBS, forming a strong interaction between the hydroxyl groups in OMMT and the carboxylic groups of PBS. Owing to this, the higher content of PBS in TPS/PBS (25:75) blended nanocomposites ensured the majority of the clay was well dispersed within the PBS phase during the mixing process. Meanwhile, NaMMT was suspected to reside within the TPS dispersed phase of higher content of TPS in TPS/PBS (75:25) blended nanocomposites. NaMMT is known to be more hydrophilic than OMMT. The extra moisture within NaMMT promoted starch gelatinization, which led to better interfacial interaction and effective heat transfer between the clay and the TPS dispersed phase. Higher WVTR and OTR were shown by PBS/TPS/OMMT owing to the tortuous structure formed by the exfoliation and intercalation of clay.

Zahedi et al. [116] prepared polypropylene (PP)/walnut shell flour (WSF)/OMMT (50:43:3) nanocomposites directly mixed using a twin-screw extruder with maleated anhydride grafted polypropylene (MAPP) (4%) as a compatibiliser. SEM and TEM analysis for 3% of OMMT demonstrated a higher order of intercalation and good dispersion of clay layers in the nanocomposites as compared to 5% of OMMT. The well-dispersed clay layers within nanocomposites promoted the mechanical properties of the composite due to the high stiffness of clay platelets and lower percolation points. At 5%, OMMT was observed to agglomerate, which led to lower mechanical properties. Water absorption and thickness swelling rates decreased as the content of OMMT and MAPP increased. The dispersion of clay-filled up microvoids and fibers cavities creating a tortuous path preventing deeper water penetration through capillary action. Zhao et al. [117] investigated the relationship of untreated wood flour and silane-treated (WF) with OMMT in PVC-based composites through dry mixing using a high-speed mixer. TEM analysis showed that 0.5% of OMMT was homogenously dispersed as partially exfoliated structures in PVC/treated WF (100:70) composites. OMMT has better interfacial compatibility with treated WF resulting in a significant increase in impact strength and tensile strength (by 14.8% and 18.5%, respectively). The hydrophobic PVC and hydrophilic untreated WF displayed poor interfacial adhesion thus further incorporation of OMMT worsened the compatibility and increased the stress concentration points. At 3 wt.% of OMMT content, the size of dispersed clays became larger and in part aggregated, which led to a minor reduction in the impact strength.

4.4. In Situ Polymerization

The high surface energy of the clay platelets induces the penetration of monomer molecules into the clay interlayer spacing to establish an equilibrium state. As polymerization occurs, the interlayer spacing of clay platelets increases and separates the clay platelets into a disoriented state resulting in an exfoliated structure. The polymerization can be initiated by exposure to heat or radiation and incorporation of catalyst or initiator via ion-exchange reactions. It seems that the surface modification of MMT and natural fibers are essential to establish good compatibility between fillers and hydrophobic polymers.

Since MMT and natural fibers are rich in hydroxyl groups, they form strong hydrogen bonds with hydrophilic polymers such as TPS and PVA.

5. Mechanical Properties of MMT/Natural Fiber Reinforced Polymer Nanocomposites

Composites sample are subjected to mechanical evaluation to investigate the negative or positive hybrid effect of the composites. The common mechanical properties evaluated include Young's modulus, tensile strength, strain at yield, and shear stress, supported with morphological and rheological properties [118]. The hybridization of natural fibers and nanoclay fillers in the same type of matrix polymer has been the focus of researchers due to their easy availability, low density and low cost as well as their enhanced physical and mechanical properties [119]. The hybridizing technique between these two types of filler—natural fibers and nanoclay fillers—enables the reduction of some of the drawbacks of composite materials including low durability and low resistance to water absorption which make them incompatible with non-polar matrices [120]. It is expected that the modulus properties of hybrid composites will increase with the addition of a nanoclay filler loading due to its rigid properties, whilst natural fibers possess a high elastic modulus compared to that of the polymer matrix. Therefore, the addition of rigid filler like nanoclays to natural fiber-reinforced polymer composites enable improvement of the stiffness and strength which cannot be provided by the addition of natural fibers alone [35,120]. This occurrence is attributed to the action of the filler to fill the polymer voids and restrict the movement of polymer chains. This restriction then enhances the mechanical strength and rigidity, improving the load transfer at the matrix-fiber interface, reducing the gas and moisture penetration and improving the flammability properties [121,122].

Despite the enhanced properties provided by hybrid composites, the hydrophilic character of natural fibers and clay-based reinforcement fillers that causes poor interfacial adhesion creates difficulties in providing homogenous dispersions of filler in the composite system. Other than natural fibers, clay-based fillers are also known to have hydrophilic properties. Clays display interlayer formation where van der Waals gaps are created that normally present ions of alkali metals including Na+ and K+, or alkaline-earth ones like Ca^{2+}. These counterbalance the negatively charged platelets and causes more water particles to be chemically linked and attached within the clay structure. Therefore, to improve the affinity between filler and the matrix, the inorganic cations are generally exchanged by ammonium and phosphonium cations. The resulting clays are referred to as organo-modified layered silicates (OMLS) and, in the case of MMT) [122,123]. As reported by Rozman et al. [124], the introduction of nanoclay filler in natural fiber-based composites at levels as low as 1 wt.% is likely to offer better stress transfer throughout the matrix which leads to improved impact and flexural properties. Similar findings by Hetzer and Dee [125] showed that with the addition of 3 to 5 wt.%, nanoclay results in 30% better properties compared to conventional fillers.

However, the potential of agglomeration of filler can lead to deterioration of the composites, particularly at higher levels of nanoclay filler addition. Furthermore, the agglomeration of the filler also can lead to unstable stress concentrations and poor interfacial adhesion between natural fibers and polymers, which cause the brittleness of the composites [126]. Therefore, it is necessary to modify the surface of the fillers and utilize a compatibiliser to ensure their better dispersion and good adhesion with the polymer matrix [32,39,121]. For example, Majeed et al. [32] showed the potential of MMT in polypropylene PP/RH composites as a water-resistant strengthening filler. By integrating MMT into the PP/RH system, the mechanical characteristics were improved and the reinforcement effect due to the application of the compatibilizing agent was notable. With the addition of RH and MMT into a polypropylene matrix, the tensile and flexural modulus improved by 63% and 92%, which showed that higher stress was required for deformation as both RH and MMT are rigid and high-modulus materials. When MMT and PP-g-MAH were employed for the development of PP/RH-based nanocomposites, improved modulus values were achieved. The PP/RH composite tensile and flexural modulus rose by 5%

and 4%, by the addition of PP-g MAH and when MMT was used in the presence of the compatibilisers, it rose by 36% and 25%, respectively. The enhancement in the modulus by the addition of the MMT in the composite can be attributed to the presence of delaminated, stiffer platelets and the high aspect ratio of the MMT in the PP polymer matrix, leading to increased interactions between PP chains.

Zahedi et al. [127] addressed the effects of OMMT as a reinforcing agent on the mechanical properties of almond shell flour–polypropylene bio-nanocomposites. Maleic anhydride grafted polypropylene (MAPP) was applied as a coupling agent to improve the poor interface between the lignocellulosic material (hydrophilic) and the polypropylene matrix (hydrophobic). The tensile and flexural properties attained their maximum values when 3 wt.% OMMT was added. The increased population of OMMT molecules leads to agglomeration at a high degree of OMMT loading (5 wt.%). Flexural strength and modulus improved as the OMMT loading was increased by up to 3%. After 3% MMT, it started decreasing. A greater stiffness of clay platelets and lower percolation points can be attributed to this 3 wt.% OMMT load, which is the cause of the improved mechanical properties. This is due to the high aspect ratios of organo-clays and the high surface area of silicate polymer matrix layers, which lead to increased interactions between nano-scaled clay particles and the PP matrix, with good interfacial adhesion. The finding is evidenced in Figure 4 through SEM images of fracture surface of composites with and without addition of OMMT.

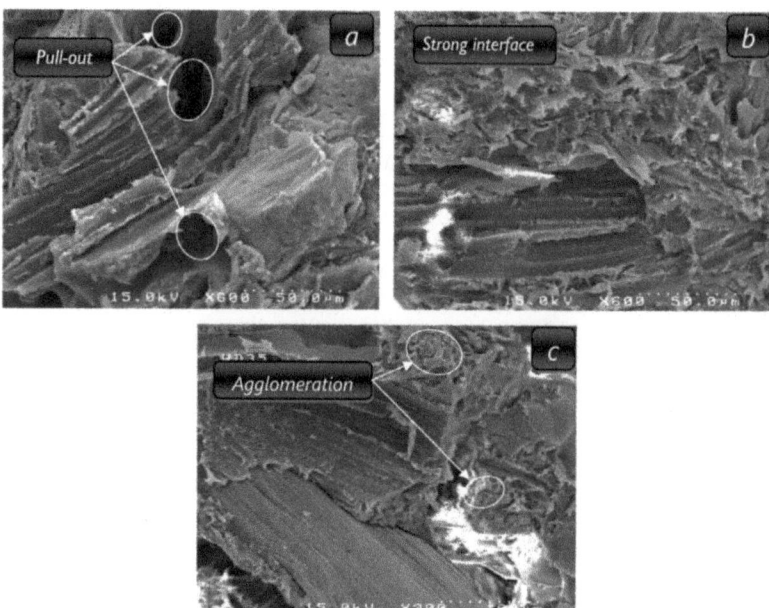

Figure 4. The SEM micrographs of fractured surfaces of composites types: (**a**) ASF/PP/MAPP composites, (**b**) ASF/PP/MAPP reinforced 3% OMMT and (**c**) ASF/PP/MAPP reinforced 5% OMMT. (ASF: Almond shell flour; PP: Polypropylene; MAPP: Maleic anhydride polypropylene; OMMT: organo-modified montmorillonite) [127].

Other than synthetic polymers, MMT also was utilized as a material for reinforcement in biopolymer-based composites. To date, to reduce the utilization of petroleum-based polymers, natural-based polymers known as biopolymers have become popular. Among all biopolymers, polylactic acid (PLA) has caught the attention of researchers due to its biodegradability, good processability and excellent biocompatibility properties. To enhance its properties, natural fibers and nanoclay fillers have been introduced into the system to develop cost-effective materials with superior properties compared to composites with carbon and glass fiber inclusion [34]. For example, Petersson and Oksman [128] have reported the effect of MMT and cellulose nanowhiskers (CNW) on PLA. Both tensile modulus and yield strength were increased with MMT, whereas the yield strength was increased by CNW. They had also found a higher ductility for PLA/CNW nanocomposites than for MMT/PLA nanocomposites.

The addition of MMT improved the tensile strength of PLA/MMT nanocomposite to reach a plateau of 5 phr filler content and thereafter it declined at a higher level [129]. Compared to PLA/MMT nanocomposites, neat PLA displayed 63% less tensile strength. This may be due to a significant reinforcing effect of the inorganic phase. The MMT filler is regularly distributed in the PLA and, therefore, sufficient percolation networks are established. An amount of 5 phr of filler was appropriate for the MMT filler to achieve a maximum tensile strength. In excess of 5 phr, the tensile strength value declined. This was because of filler aggregation that resulted in stress concentration spots conducive to fractures. However, with increased MMT filler content, the Young's modulus of the PLA/MMT nanocomposite increased steeply. It was worth noting that, compared to pure PLA film, the Young's modulus of PLA/MMT nanocomposites was increased by 18%. This increase was due to the rigid filling of MMT that restricts the movement of the segments in the polymer matrix [129].

It is observed that the tensile strength improves with the addition of CNW and reaches its maximum tensile strength at 3 phr, and then decreases with a higher filler content [129]. The tensile strength of an increase in the interaction between the filler and the matrix, thereby increasing the tensile strength. However, when the filler content is more than 3 phr, the tensile strength decreases. This reduction may be due to the accumulation of CNW in PLA, which then acts as a stress concentration point. However, the Young's modulus of PLA/CNW nanocomposites gradually increases with increasing CNW filler content. Compared to pure PLA, the Young's modulus of PLA/CNW nanocomposites is increased by 25%. This increase is attributed to the hardening effect of the high modulus CNW reinforcing material [129]. The tensile, flexural, and impact properties of PLA-hybrid biocomposites were improved by 5.72%, 6.08%, and 10.43% than PLA biocomposites, after adding 1 wt.% MMT [39].

Jalalvandi et al. [130], reported the effect of MMT nanoclay on mechanical properties of starch/PLA hybrid composites, indicated that the tensile strength, tensile modulus, and elongation at break increased after the addition of MMT clay. The tensile modulus values increase up to the MMT content of 4 phr. The MMT particles agglomerated over 4 phr and consequently reduced the tensile modulus. The sample contains 4 phr MMT of the highest Young's modulus observed. The Young's modulus of samples started to decrease with the further increase of MMT content in samples. The MMT particles had been unevenly spread across the PLA matrix when MMT content increased beyond 4 phr and there was a chance of agglomerating MMT particles. For the tensile strength graph, with 2 phr MMT loading, the tensile strength sample increased. Due to the exfoliated structure of the MMT-PLA/starch matrix, a significant rise in tensile values was considered. But after the MMT content was increased to 4 phr, there were no substantial changes. The tensile strength of 67 MPa was further increased by the MMT loads. Surprisingly, very slight changes in the tensile strength at 68 MPa were detected with the greatest MMT concentration, that is, 8 phr [130].

Saba et al. [131] worked on mechanical properties of oil palm empty fruit bunch (OPEFB) nanofiller/kenaf fiber/epoxy-hybrid nanocomposites and reported that the tensile strength of nano OPEFB filler/kenaf/epoxy hybrid nanocomposites improved by 24.9% with kenaf/epoxy composites, whereas 20.7% with MMT/kenaf/epoxy hybrid nanocomposites. By adding 3% OMMT to kenaf/epoxy its tensile strength improved by 29.3% with respect to MMT/kenaf/epoxy hybrid nanocomposites whereas a 56% increment was observed as compared to nano-OPEFB/kenaf/epoxy hybrid nanocomposites. This is because of the strong reinforcing effects, a high aspect ratio, good interlayer spacing, and the OMMT platelet structure. This allows the stress concentration to be reduced efficiently under the action of tensile loads on the nanocomposites. In contrast, MMT is a soft and hydrophilic substance that makes it incompatible with the hydrophobic epoxy polymer matrix, which results in a comparatively low tensile strength with the kenaf hybrid nanocomposites filled by OMMT. Consequently, there was no strong interphase between the epoxy matrix and the dispersion phase, which reduces the applied concentrations of tensile stress, in contrast to nano OPEFB filler/kenaf/epoxy hybrid nanocomposites. It was also noted, with regard to the MMT/kenaf/epoxy hybrid nanocomposites, that OMMT hybrid nanocomposites show the highest tensile modulus, which is raised by 14%. Good stiffening and clay layers' rigidity properties were the root cause of the improvement in tensile modulus of nanoclay (OMMT, MMT)-based kenaf hybrid nanocomposites than OPEFB-based kenaf hybrid nanocomposites. Similarly, the impact strength of OMMT/kenaf/epoxy hybrid nanocomposites improved by 25.9% as compared to MMT/kenaf/epoxy hybrid nanocomposites [131]. The combination of a nano filler and micro-size reinforcements in the polymer may be used to create stronger and lighter hybrid composites with improved mechanical characteristics for a range of housing and bridge applications.

In separate studies, instead of hybridizing the polymer composites using two types of fibers, the utilization of more types of fibers might impart better mechanical properties on the composite systems. In spite of being renewable, non-abrasive and low in cost, natural fiber hybrid composites also exhibit excellent mechanical properties and are environmentally friendly, which make them a material of choice for engineering markets, including the automotive and construction industries [132]. Besides, to complement current trends, the application of nanofillers like MMT nanoclay was also introduced into natural fibers hybrid composites to achieve advanced performance features suitable for future engineering applications. The hybridization of natural fiber reinforced polymer nanocomposites with nanoclay was studied by [133] using two types of natural fibers, viz. wood and coir fibers. The mechanical properties of the hybrid composites were found to be enhanced after the addition of MMT. The Fourier transform infrared spectra also proved the interaction between fiber, polymer and MMT by a new peak around 470 cm^{-1}. The tensile properties of the hybrid of coir fiber/wood polymer composites with the addition of MMT showed the highest value compared to the hybrid composites without MMT addition. This finding supports the surface morphology properties as the hybrid composite with MMT loading showed a smoother surface and indicated high interfacial adhesion attributed to enhance mechanical properties of the hybrid composites.

The PP/EVA/MMT nanocomposites were extensively investigated for the purposes of evaluating their mechanical properties, and it was predicted by different studies that the stiffness of the composites would increase [134]. The clay incorporation technique also had a substantial impact on the dynamic mechanical and rheological characteristics of PP-EVA/clay nanocomposites. These characteristics may be used in a variety of technical applications, such as the automotive sector, which often requires materials with high stiffness. The usage of these materials can also improve appearance, dimensional stability, and dimensional conformity [135].

Table 3 shows the Young's modulus, elongation at break and tensile strength of various polymeric blends and nanocomposites samples studied by Castro-Landinez et al. [136]. A 21% increase in stiffness was seen for blends with EVA28 while it was 15% with EVA12 as compared with EVOH systems. On comparing EVOH44 and EVOH38 blends, an increment of 0.8% and 10.2% in stiffness was observed. This may be because of addition of a stiffer copolymer (EVOH) than EVA. The results demonstrated that the MMT showed better interactions with PP/EVA blends because they operate as a reinforcing agent than the effect on the PP/EVOH blends. Interestingly, with the introduction of clay into polymeric blends for both PP/EVA and PP/EVOH, the elongation at break and the tensile strength showed no significant change [136].

Table 3. Mechanical properties (Young's modulus, elongation at break and tensile strength) of polymeric blends and nanocomposites samples.

Type of Material	Matrix	Young's Modulus (GPa)	Elongation at Break (%)	Tensile Strength (MPa)
Neat Polymers	PP	0.96 ± 0.04	690 ± 10	22.30 ± 4.00
	EVA12	0.10 ± 0.05	825 ± 60	14.88 ± 3.27
	EVA28	0.09 ± 0.04	900 ± 50	16.24 ± 4.21
	EVOH38	2.32 ± 0.21	43 ± 21	57.41 ± 3.12
	EVOH44	1.83 ± 0.23	48 ± 18	45.05 ± 7.34
Polymeric blends	PP/EVA12	0.65 ± 0.04	587 ± 45	17.62 ± 0.64
	PP/EVA28	0.60 ± 0.01	310 ± 42	16.85 ± 0.63
	PP/EVOH38	1.16 ± 0.05	6 ± 2	20.28 ± 2.51
	PP/EVOH44	1.13 ± 0.04	16 ± 7	19.01 ± 1.89
Nanocomposites materials	PP/EVA12/MMT	0.76 ± 0.02	574 ± 58	17.89 ± 0.61
	PP/EVA28/MMT	0.75 ± 0.01	353 ± 55	17.42 ± 0.45
	PP/EVOH38/MMT	1.29 ± 0.02	13 ± 8	19.95 ± 1.05
	PP/EVOH44/MMT	1.15 ± 0.04	19 ± 10	17.94 ± 0.83

PP—polypropylene. EVA12-Ethylene vinyl acetate (Elvax660) with 12 wt.% of vinyl acetate monomer content. EVA28-EVA28, with 28 wt.% of vinyl acetate monomer content. EVOH38-Ethylene vinyl alcohol with 38 mol.% of ethylene monomer content. EVOH44- 44 mol.% of ethylene monomer content.

Shikaleska et al. [137] reported that poly(ethylmethacrylate) (PEMA) and MMT clay particles have a considerable effect on the mechanical properties of poly(vinylchloride) (PVC)/PEMA 90/10 blend composite. A PVC/PEMA combination with a lower Young's modulus and higher tensile strength and elongation at break than clean PVC is observed. It is intended to achieve a better Young's modulus, less tensile strength and elongation at break by incorporating MMT clay particles into PVC/PEMA material. This indicated that the MMT clay addition had a greater effect on the tensile strength and elongation at break than on the elastic modulus. Sengwa et al. [138] revealed that the dispersion of MMT in poly(vinyl alcohol) (PVA), poly(vinyl pyrrolidone) (PVP), poly(ethylene oxide) (PEO), and poly(ethylene glycol) (PEG) matrices will improve the mechanical properties of the blended nanocomposites. These synthetic polymers, PVA, PVP, PEO, and PEG are hydrophilic in nature.

Recently, Ramesh et al. [139], worked on the combination of PLA-treated KF-MMT, proving that better properties were obtained in blend form than with the system alone. In order to obtain superior characteristics, it is therefore required to combine kenaf fiber (KF)—MMT and PLA. Currently, OMMT and MMT are of great importance. Both have potential tendency to improve the mechanical properties of any polymer [139]. In contrast to PLA/TKF biocomposites, the flexural, impact and tensile strengths were increased by 46.4%, 10.6% and 5.7%, respectively, when 1 wt.% MMT was added to PLA/TKF/MMT hybrid biocomposites [139]. Adding MMT above 1 wt.%, the above mechanical properties decreased. Ramesh et al. [34] indicated that tensile, flexural, and impact properties were increased by the addition of MMT clay with the treatment of (6%) NaOH-treated kenaf fiber/aloe vera fiber/PLA. The mechanical properties were improved by hybridised 15 wt.% kenaf, 15 wt.% aloe vera fibers and 1 wt.% MMT clay-incorporation. For hybrid nanocomposite, the tensile strength, flexural strength, and impact strength were improved by 5.24%, 2.46%, and 37.10% after adding 1 wt.% MMT clay compared to neat PLA, while the tensile and flexural modulus were increased by 24.61% and 108.09%. The tensile strength was decreased by 14.05% after the addition of 3 wt.% MMT clay as compared with neat PLA. The tensile strength was reduced because of the formation of agglomerations by 3 wt.% MMT clay and micro-evasion. For the same reason, the impact strength also decreased by 14.88% [34].

6. Thermal Properties of MMT/Natural Fiber Reinforced Polymer Nanocomposites

6.1. MMT-Reinforced Natural Fiber Polymer Nanocomposites

The biodegradability of polymeric materials can be accelerated by novel strategies developed as a result of consumer pressure as well as environmental legislation. In this scenario, the production of natural fiber-reinforced biocomposites is a step toward minimizing the environmental footprint of non-biodegradable polymeric materials [140,141]. A number of studies that have been conducted on the biodegradation natural fiber reinforced polymeric materials biodegradation have yielded promising results [142,143]. The advantages of biocomposites emerge from the properties of natural fibers, e.g., low cost, light weight, sustainability, tolerable specific mechanical properties, and environmental compatibility. Nevertheless, the lack of interfacial adhesion, lower processing temperature, as well as high water absorption potential have made natural fiber reinforced composites less appealing. Moisture penetration may negatively impact the mechanical properties and promote fungus growth, which can aid in composites' degradation [144]. As a result, characterization of the morphological, moisture exposure, and other performance characteristics of natural fiber-reinforced composites is critical for assessing their effectiveness when considering natural fibers as a feasible reinforcement and validating the longevity of such products. Nanoclays are regarded as among the most essential particles for the reinforcement of polymeric matrices and for promoting new and improved properties, e.g., increased stiffness, toughness, flame resistance, as well as gas barrier activity [145–148]. MMT is the most promising nanoclay of them all. The clay structure is composed of the stacking of hundreds of sheets or platelets in the manner of book pages [145]. The addition of nanoclays to natural fiber reinforced plastic materials as load-bearing components is expected to compensate for the deficiencies in natural fiber reinforced composite materials.

Another important piece of information needed to determine the thermomechanical performance of the hybrid composites is obtained by determining the elastic character of the material at a higher frequency and under the influence of heat. Dynamic mechanical analysis (DMA) is the most common analysis done to evaluate the rheological properties of hybrid composites represented by storage modulus (E'), loss modulus (E'') and tan δ. The E' focuses on the elastic behaviour of the material subjected to a sinusoidal stress which provides information about properties including the stiffness, load-bearing capacity, cross-link density and interfacial strength of between fiber and matrix [149–151], whilst the E'' indicates a higher ability to dissipate energy. The E'' also indicates better the damping properties to reduce damaging forces caused by mechanical energy. Meanwhile, tan δ provides information regarding the damping parameter of the composite material. A high tan δ value indicates a high non-elastic strain component, whilst its curve is attributed to the energy dissipation as heat during each deformation cycle [146].

A study conducted by Khaliq Majeed et al. [152] on the effect of maleic anhydride-grafted polyethylene (MaPE) compatibiliser on the thermal properties of rice husk (RH) and nanoclay-filled low-density polyethylene (LDPE) composite films demonstrated that adding RH and MMT nanofiller to the LDPE matrix increased the T_m by 1 °C. This could be because the dispersed nanoclay particles acted as a barrier to heat conduction to crystallites until the temperature rose to a point where the heat flow was sufficient to melt the crystallites. The addition of MaPE to RH and MMT-filled LDPE resulted in a clear shoulder near the main peak that became more noticeable with higher MaPE concentrations. No significant difference was observed in crystallization temperatures between both uncompatibilised and compatibilised composite films, indicating the absence of significant MMT nucleation activity in the LDPE-based system [152,153]. Seetharaman and co-workers [154] studied the influence of MMT nanoclay on jute fiber/unsaturated polyester nanocomposites in terms of dynamic mechanical analysis. The addition of nanoclay increased the E', as a result of the imparting effect of filler reinforcements that are more rigid compared to the polymer matrix [155]. The temperature of the glass transition was increased from 109 to 115 °C, but this resulted in a decrease in the damping factor. This might be because uniform molecular motion requires a large amount of thermal energy and the nanocomposites' energy accumulation capacity that can be improved by adding more fiber. The drop of the E' close to the glass transition temperature (T_g) was a result of the composite's softness. Incorporating nanoclay as the composites' filler improved the E' due to the nanoclay that functions as a stiffening agent, reducing the polymeric molecules' movement [156]. Besides, it was also observed that with a nanoclay content up to 5 wt.% in the polyester, along with 25 wt.% jute fiber content, significantly increased the E' more than with other MMT nanoclays. This finding resulted from cluster formation in the composite as the nanoclay content was increased. As a result, it developed heterogeneity and, as a consequence, a weak bond was formed between the matrix and the fiber. It was discovered that the 5 wt.% nanoclay and 25 wt.% jute content composites exhibited a lower tan δ peak, which was attributed to the enhanced fiber/matrix adhesion, decreased molecular mobility, and improved load carrying capacity. These observations were obtained from the additions of jute fiber and nanoclay as primary and secondary reinforcement, respectively, producing a positive composite hybridization effect. This phenomenon led to more loads and energy being borne by the hybrid composite, which increased the composite material's performance. By incorporating nanoclay, energy was dissipated more efficiently at the interface, thereby increasing the material's stiffness and T_g value [154].

In the work performed by Ramesh et al. [39], the preparation of PLA and PLA hybrid biocomposites was carried out using a twin-screw extruder, two-roll mills, and a compression molding method. 30% treated aloe vera fiber and 0 wt.%, 1 wt.%, 2 wt.%, and 3 wt.% MMT nanoclay filler were used in the PLA-based biocomposites fabrication to study the effect of MMT clay on the thermal properties of the produced PLA and PLA-hybrid biocomposites. The PLA-hybrid biocomposites' mechanical, thermal, and water resistance characteristics were enhanced with MMT clay addition. The findings of thermogravimetric analysis (TGA) analysis revealed improved decomposition temperature of the PLA-biocomposites with MMT clay inclusion. Adding 3 wt.% MMT clay improved the decomposition temperature of PLA biocomposites from 295 °C to 299 °C in T 10%, similarly, from 338 °C to 350 °C in T 75%.

TGA was employed to study the thermal behaviour and stability of PLA and PLA-hybrid biocomposites. The 10% and 75% weight loss temperatures [157] were set as the baselines for analyzing the thermal stability of both PLA and PLA-hybrid biocomposites. Improvements in the T10% and T75% of PLA-hybrid biocomposites were observed from 295 °C to 299 °C and 338 °C to 350 °C, respectively (Table 4). The thermal decomposition took place in three stages: (1) moisture evaporation at temperatures of up to 150 °C, (2) lignin, cellulose, and hemicellulose degradations, and (3) depolymerization of PLA and MMT clay [158–163]. The thermal stability of neat PLA is reduced with natural fiber addition that could be generally be explained by the decrease in polymer thermal stability with natural fiber addition. For PLA (P) and PLA-hybrid biocomposites (A1-containing 1 wt.% MMT, A2-containing 2 wt.% MMT, and A3-containing 3 wt.% MMT) 10% mass losses were degraded at 295 °C, 296 °C, 299 °C, and 299 °C, respectively. Approximately 75% of the biocomposites' weights were loss at 338 °C, 342 °C, 344 °C, and 350 °C, respectively. For 10% and 75% weight losses, the neat PLA degraded at 327 °C and 358 °C, correspondingly. The PLA-hybrid biocomposites (A3) containing 3 wt.% MMT exhibited the highest decomposition temperatures (299 °C in 10% and 350 °C in 75%) than other biocomposites (A, A1, and A2). These enhancements occurred due to the MMT clay that acted as a barrier, restricted chain mobility, and hindered the decomposition progress. A similar finding was found in a previous study, where hybridization resulted in enhanced thermal stability of the hybrid biocomposites [164].

Table 4. TGA analysis sample of MMT clay reinforced treated aloe vera fiber/PLA hybrid biocomposites.

Sample	Denotation	Weight Loss, Decomposition Temperature (°C)	
		10%	75%
PLA	P	327	358
PLA-30TAF	A	295	338
PLA-30TAF-1MMT	A1	296	342
PLA-30TAF-2MMT	A2	299	344
PLA-30TAF-3MMT	A3	299	350

PLA = polylactic acid, TAF = treated aloe vera fiber, MMT = montmorillonite.

Sajna et al. [165] successfully prepared green nanocomposites of poly(lactic acid) (PLA)/banana fiber/MMT nanoclay using melt-blending followed by injection molding techniques. Untreated and chemically modified banana fibers as well as organically modified MMT nanoclay with methyltallow bis(2-hydroxyethyl) ammonium (Cloisite 30B) were added as reinforcement into the PLA matrix. Numerous chemical modifications were performed on the banana fibers, including mercerization, silane, sodium lauryl sulphate and permanganate treatments, and a combination of mercerization and silane treatments. To study the effects of banana fiber chemical modification and nanoclay incorporation, the bionanocomposites and biocomposites were characterized via dynamic mechanical analysis (DMA), differential scanning calorimetry (DSC), TGA, and heat deflection temperature (HDT) studies. DMA results showed that the bionanocomposites' E' increased with respect to the neat PLA biocomposites. In general, higher dissipation energy is exhibited by composites having poor interfacial bonding, thus, demonstrating a high damping peak magnitude than composites possessing better matrix-filler interfacial bonding that bears higher stress transfer and less dissipation energy [166]. The damping peak of the PLA/SiB/C30B bionanocomposites showed decreasing tan δ compared to neat PLA, PLA/UTB, and PLA/SiB biocomposites that was probably due to the excellent filler-matrix interfacial bonding [165]. The DSC findings demonstrated insignificant changes in glass transition temperature. However, both silane-treated fiber and C30B nanoclay additions improved the PLA crystallites nucleation and raised the melting temperatures. There is a relationship between the degree of crystallinity of semicrystalline polymer PLA and its mechanical properties as well as thermal degradation temperature [167]. A sample with a low crystallinity degree degraded rapidly and possessed a lower strength [168]. TGA and derivative thermogravimetry (DTG) analyses concluded that the fillers addition increased the initial and maximum degradation temperatures of biocomposites and bionanocomposites. Banana fiber and C30B both played a significant role in preventing gas release during the thermal degradation process, which resulted in a delay in weight loss [169].

Kumar and Singh [170] prepared hybrid composite materials by incorporating MMT modified with dimethyl dehydrogenated tallow quaternary ammonium cation (Cloisite), layered silicates, and microcrystalline cellulose (MCC) into the thermoplastic polymer. Three types of composites were prepared using an ethylene–propylene (EP) copolymer as the thermoplastic polymer matrix and a maleated EP (MEP) copolymer as the compatibiliser: (i) cellulose melt mixing with thermoplastics [I], (ii) clay melt mixing with thermoplastics [II], and (iii) cellulose melt mixing with thermoplastic clay nanocomposites [III] (Table 5). Characterisation of the composites was conducted via DSC and TGA. The marginal increase in melting point of composites containing clay (TC05–TC15) might be ascribed to the heat deflection behaviour of layered silicates and the constraint of thermal motion of macromolecular chains within the silicate layers [171,172]. According to Tyan et al. [173], thermal expansion in nanocomposites was also reduced during melting because macromolecular chains sandwiched between silicate layers and oriented in a plane (platelet) direction tended to relax in the opposite direction of their original orientation. It might be due to silicate layers that are significantly more rigid than polymer molecules and do not deform, making it difficult for polymer molecules to relax.

Transitions beyond the matrix melting point in ternary composites can be described by the presumption that the layered silicates and rising cellulose fiber concentration increased phase separation and appeared to accelerate heterogeneous nucleation to crystallize at higher temperatures via their role as nucleation centers [174]. Table 6 presents the maximum decomposition of (host) matrix T_d (°C) for all samples. T_d was found to be approximately 5 °C higher in cellulose composites (CC05) and nanocomposites NC05 than in the neat polymer matrix. A similar effect was observed with the ternary composite (TC05), with an increase of approximately 13 °C as compared to the polymer matrix. The rise in T_d for composites containing clay was well explained by some authors who showed that the degraded products diffusion could be limited/slowed down in between the silicate layers [175,176]. The effect of silicate layers on the thermal mobility of molecules is dependent on the orientation of the polymer molecules, rigidity, and stability of the silicate layers. The increment in decomposition temperature, on the other hand, might be insignificant in terms of thermal stability improvement for natural fiber reinforced composites.

Table 5. Composition of composites prepared [a].

Sample No.	Neat EP Copolymer (wt.%)	Maleated EP Copolymer (wt.%)	Cellulose (wt.%)	Cloisite® 20A (wt.%)	Code
1	100	-	-	-	Neat EP
2	-	100	-	-	MEP
3	57	38	5	-	CC05
4	54	36	10	-	CC10
5	51	34	15	-	CC15
6	57	38	-	5	NC05
7	54	36	5	5	TC05
8	51	34	10	5	TC10
9	48	32	15	5	TC15

[a] Neat EP/MEP ratio is 60/40. Ethylene–propylene (EP); Maleated ethylene–propylene (MEP); CC = cellulose composite; TC = ternary composite; NC = nanocomposites (i.e., without cellulose).

Table 6. Melting and decomposition temperatures.

Samples	T_m (°C)	T_d (°C) EP
Neat EP	165.8	453.3
MEP	166.5	443.2
CC05	166.7	458.1
CC10	168.2	461.1
CC15	168.7	463.1
NC05	165.7	458.9
TC05	167.6	466.4
TC10	169.3	483.3
TC15	170.0	482.1

Ethylene–propylene (EP); Maleated ethylene–propylene (MEP); CC = cellulose composite; TC = ternary composite; NC = nanocomposites (i.e., without cellulose).

6.2. MMT-Reinforced Hybrid Natural Fiber Polymer Nanocomposite

Compression molding was used to develop hybrid fiber reinforced polymer composites from kenaf, and aloe vera fibers, PLA, and MMT clay (Table 7). The fiber and MMT clay hybridization effects on their water absorption, mechanical, biodegradability, and thermal properties were investigated [34]. To enhance the bonding and compatibility of kenaf and aloe vera fibers with the PLA matrix, they were treated with a 6% sodium hydroxide solution. The results revealed that the addition of MMT clay improved the thermal properties of the biocomposites. Adding 3 wt.% MMT clay to the biocomposite increased its decomposition temperature from 280 to 307 °C at T_{10} and 337 °C to 361 °C at T_{75}. SEM analysis revealed that MMT clay significantly enhanced the compatibility and bonding of fibers and PLA. The MMT clay at 3 wt.% is evenly distributed throughout the PLA matrix and the TEM result indicated that as the MMT content increased up to 6 wt.%, the quality of the MMT dispersion declined due to higher number of agglomerations. The key objective of the research was to study the effects of fiber hybridization and MMT clay content on the thermal characteristics and internal bonding behaviour of PLA/TKF/TAF/MMT hybrid bionanocomposite that was crucial to synchronizing the properties.

Table 7. TGA Characterisation of P, S, A, H, H1, and H3 composites.

Code	Sample	Weight Loss, Decomposition Temperature (°C) 10 (%)	75 (%)
P	Polylactic acid (PLA)	327	358
S	Polylactic acid/treated kenaf fiber (PLA/TKF)	280	337
A	Polylactic acid/treated aloe vera fiber PLA/TAF	295	338
H	Polylactic acid/ treated kenaf fiber/ treated aloe vera fiber (PLA/TKF/TAF)	291	343
H1	Polylactic acid/treated kenaf fiber/treated aloe vera fiber/1 wt.% MMT (PLA/TKF/TAF/1MMT)	301	346
H3	Polylactic acid/ treated kenaf fiber/treated aloe vera fiber/3 wt.% (MMT.PLA/TKF/TAF/3MMT)	307	361

From the findings, hybridization improved the biocomposites' thermal stability, as evidenced by the corresponding TGA curves. The temperature deviations of 10% and 75% weight losses from the baselines, T_{10} and T_{75}, respectively, were the indicators used for assessing the manufactured biocomposites' thermal stability that was adapted on a previous work [157]. Improvements in the T_{10} and T_{75} of PLA/TKF biocomposite from 280 °C to 291 °C and 337 °C to 343 °C after hybridization, respectively, were observed. The TAF's high thermal stability was found to balance the TKF's poor thermal stability. The thermal decomposition took place in three stages: (1) moisture evaporation at temperatures up to 150 °C, (2) lignin, cellulose, and hemicelluloses degradation, and (3) PLA depolymerization. Typically, the polymer's thermal stability declines with natural fiber addition [177–179]. A similar phenomenon was observed with TAF and TKF additions to the neat PLA. However, adding MMT clay resulted in a thermal stability improvement. The 3 wt.% MMT clay reinforced PLA/TKF/TAF composite showed better thermal stability than other bionanocomposites with 280 °C to 307 °C at T10 and 337 °C to 361 °C at T75 improvements than the neat PLA that degraded at 327 °C and 358 °C, respectively. Similarly, the 3 wt.% MMT clay reinforced PLA/TKF/TAF hybrid bionanocomposite was higher than neat PLA in terms of thermal stability, with an improvement of from 358 °C to 361 °C at T75. This finding was due to MMT clay that functioned as a barrier, constrained chain mobility, and hindered the degradation process. This was aligned with previous studies [39,146,171]. It was concluded that the MMT clay incorporation improved the PLA-based biocomposites' thermal stability.

A thermal properties investigation of MMT-incorporated polypropylene (PP)/RH hybrid nanocomposites was performed by Majeed et al. [32]. TGA was employed to study the thermal behaviour of neat PP composites, and PP nanocomposites with and without the presence of compatibiliser. The temperatures of thermal degradation at 10% and 50% weight losses (T10 and T50) attained from the thermogravimetry (TG) and DTG curves (Figure 5) are presented in Table 8. A one-step degradation process was exhibited by the PP, from 425 °C to 550 °C. On the other hand, RH followed a three-step thermal decomposition (water evaporation, cellulosic substances degradation, and noncellulosic materials degradation), as presented in TGA/DTG curves. From the observation, RH demonstrated similar thermal behaviour with other lignocellulosic fibers [180–182]. PP degradation was initiated at 299 °C and was increased with RH incorporation into the matrix that could be ascribed with the lower RH thermal degradation than PP. 10% and 50% weight losses of RH-reinforced PP (PR) occurred at 311 °C and 437 °C, correspondingly, and adding more PP-g-MAH into the PP/RH (PRC) composites slightly altered the PP/RH decomposition temperature. This observation might be due to the improved interaction between RH fibers and the PP bridged by PP-g-MAH. Nevertheless, the MMT/PP-g-MAH addition resulted in a more significant thermal stability improvement in the PP/RH system owing to the existence of the uniformly dispersed delaminated MMT platelets. Namely, thermal degradation improvement is highly influenced by MMT delamination and exfoliation in nanocomposites [183], as reported in [184]. MMT addition also limits the polymer chains movement owing to the presence of dispersed MMT platelets that are rigid and impermeable and assumed to decrease heat conduction [185].

Table 8 presents the peaks of melting and crystallization for neat PP observed at approximately 163 °C and 117 °C, respectively. The melting temperature, T_m, remained unchanged with further PP-g-MAH and/or MMT incorporations. In contrast, a slight decrement in the crystallization peak, T_c was observed with MMT and/or PP-g-MAH incorporations. The MMT addition to the PP/RH composite system yielded partial increment that later became more significant with compatibiliser addition, similar to the PRMC nanocomposites. Changes were found in the characteristics and the nucleation effect that might be associated with the incorporations of MMT and PP-g-MAH into PP. These findings, however, were inconsistent with the other studies [186]. The crystallinity increase from MMT/PP-g-MAH incorporation could be caused by the higher crystallinity of PP-g-MAH and/or the improved interfacial adhesion between MMT and PP [152,187]. Therefore, the increased crystallinity can be ascribed to the intercalation of PP chains between MMT platelets and their possible interaction, with MMT platelets acting as nucleation sites. Impermeable crystalline regions were thought to enhance stress transfer, which affected the overall mechanical properties of the composite. Consequently, the increase in crystallinity enhanced the composite system's mechanical properties in both dry and wet conditions.

Table 8. Thermal properties of neat PP and its composites.

Sample Designation	T_m (°C)	T_c (°C)	X_c (%)	T_{10} (°C)	T_{50} (°C)
PP	163.2	117.3	27.7	474	510
PR	162.9	120.1	29.6	311	437
PRC	163.1	119.2	30.1	318	441
PRM	163.0	116.1	29.9	315	440
PRMC	162.8	116.5	31.5	327	451

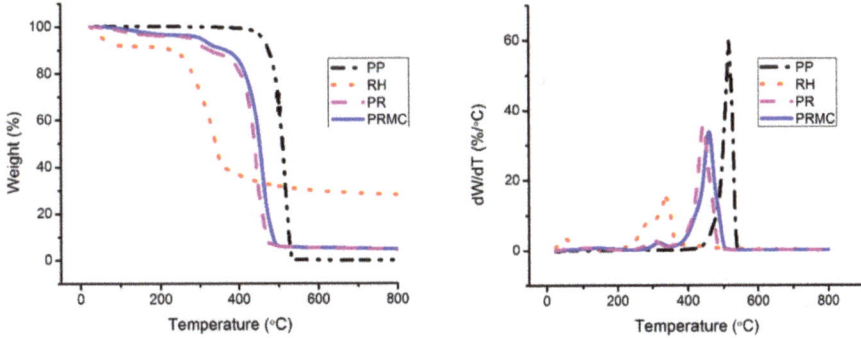

Figure 5. TG/DTG curves for neat PP, RH, and the representative composites. Reproduced from [32].

The study on the dynamic mechanical properties of the hybrid composites was done by Rajini et al. [188] by utilizing the coconut sheath/polyester (CS/PS) composites with different additions of MMT (0 to 5 wt.%) to analyze their rheological properties. The authors discussed the effect of nanoclay dispersion on the composite system in the exfoliated and intercalated states, which contribute to the efficient stress transition resulting in secondary reinforcement in CS/PS composites. The authors also found improvement in the modulus of the composites, indicating the effect of the high surface-to-contact ratio of nanoclay on the polymer matrix. The strengthening effect of the nanoscale second phase also led to an improvement in the thermal stability of the composites. At 100 °C, the E' of CS/PS composites was found to be 812 MPa and increased to 1580 MPa after the addition of 2% and 3% of MMT. However, the value of E' decreases after the addition of 5% of MMT due to the agglomeration of the MMT filler.

Another thorough study on the dynamic mechanical properties of hybrid natural composites with nanoclay addition was done by Chee et al. [146] with the combination of non-woven bamboo and woven kenaf reinforcement in epoxy composites. The dynamic mechanical properties of these hybrid composites were analyzed based on different types of nanoclays addition viz. organically modified OMMT, MMT and halloysite nanotube (HNT). The E' of the composites in the glassy region were found to increase by 98.4%, 41.5% and 21.7% with the addition of OMMT, MMT and HNT, respectively. This improvement can be attributed to the toughening effect by the nanoclay, which limits the movement of the polymer chain. A similar trend was also observed for E'' value. The E'' value increase by 159%, 65.2% and 53.6% for OMMT, MMT and HNT, respectively, which indicates higher internal friction has been induced by the addition of nanoclay that leads to higher energy dissipation. Meanwhile, the tan δ peak was found to decrease after the addition of nanoclay, which was attributed to the higher damping properties of non-elastic deformation. The reduction in the magnitude of the peak can also be referred to as the interlocking mechanism between nanoclay, fibers and epoxy polymer which restricts the polymer chain movement. Among all samples, the BK/E-OMMT composite represents the lowest tan δ peak, indicating strong interfacial adhesion interaction that reduces energy dissipation at the interface.

7. Flame Retardancy Properties of MMT/Natural Fiber Reinforced Polymer Nanocomposites

Flame retardancy of composites is another important characteristic that can determine their applicability. This information can be obtained by several characterization methods such as by using vertical burning tester, horizontal burning tester, oxygen index detector, cone calorimeter, etc. From vertical burning, the burning rating of composites can be determined based on burning time, afterglow time and dripping behaviors and reported as V-0, V-1, and V-2. Meanwhile, burning rate can be obtained from horizontal burning. Additionally, an oxygen index detector can be used to check the limiting oxygen index (LOI) of the composites. Some combustion properties like time to ignition (TTI), heat release rate (HRR), peak of release heat (PHR), average heat release rate (HRR), total heat release (THR), mass loss rate (MLR) and char yield can be acquired by using cone calorimeter tests. In general, the flammability of the composites is determined by several factors including chemical, physical, microscopic, and phase structure. There are various ways that can be used to reduce the flammability of composites. One of the most popular techniques is to introduce flame retardant materials. Another way is to use a non-flammable filler that will reduce combustible materials in the systems [189].

Suwanniroj and Suppakarn conducted a study on flame retardancy enhancement of polybutylene succinate (PBS) by addition of hybrid filler. In this study water hyacinth fiber (WHF) and MMT was used as the fillers and ammonium polyphosphate (APP) was added as the flame retardant. APP is a phosphorus-nitrogen containing flame retardant and was chosen due to its low toxicity, high efficiency and effective in reducing smoke. Phosphoric acid, water, and ammonia gas are formed as APP decomposes. Ammonia gas dilutes the concentration of oxygen gas in the gas phase, whereas phosphoric acid reacts with the carbonaceous component to create carbon char on the composites surface. This char functions as a protective layer, preventing heat and mass transfer between the gas and condensed phases. PBS and four other compositions of composites were prepared and their flame retardancy properties were characterized by vertical and horizontal burning test and the details were presented in Table 9. The results show that the neat PBS is a flammable material. PBS has 16.39 mm/min horizontal burning rate and unclassified (NC) rating for vertical burning test with severe melt-dripping and flaming material. The dripping material led to the ignition of cotton placed below the sample. Upon the addition of fillers, the horizontal fire-spread rate of the composites improved significantly. The APP/WHF/PBS and MMT/APP/WHF/PBS did not give horizontal burning rate as the composites extinguished themselves on removal from the ignition source. In addition, they obtained V-2 rating for vertical burning test with self-extinguish was less than 30 s. Moreover, melt dripping and ignition of absorbent cotton was still observed for these composites despite the addition of APP and MMT [190].

Table 9. Flammability properties of PBS and PBS composites.

Sample	Vertical Burning Test				Horizontal Burning Rate (mm/min)
	t_1 (s)	Dripping	Cotton Ignition	Rating	
PBS	>30	Yes	Yes	NC	16.39 ± 0.34
APP/WHF/PBS	<15	Yes	Yes	V-2	No burning
1MMT/APP/WHF/PBS	<30	Yes	Yes	V-2	No burning
3MMT/APP/WHF/PBS	<30	Yes	Yes	V-2	No burning
5MMT/APP/WHF/PBS	<30	Yes	Yes	V-2	No burning

Subasinghe et al., prepared PP/kenaf composites reinforced with an intumescent flame retardant (IFR), and two different types of nanofillers which were HNT and MMT to investigate their flammability properties. The neat PP, PP/kenaf composites (KeC), PP/Kenaf/IFR (KeC-IFR), PP/Kenaf/IFR/HNT (KeC-IFR-HNT) and PP/Kenaf/IFR/MMT (KeC-IFR-MMT) was studied using a vertical burning tester. From the analysis, they found out that the neat PP showed an intense dripping starting within 15 s of first flame application and the time substantially increase with the addition of kenaf (52 s) and IFR (85 s). However, addition of 3% HNT reduced the dripping time to 82 s and lowered the flammability performance. On the other hand, addition of 3% MMT had increased the flame retardancy properties at which the flame extinguished during the first 3 s of the first flame and the dripping delayed until 65 s of the second flame application. They deduced that in the presence of MMT, the sample's sustained combustion improved significantly. Further, the composites were analyzed by cone calorimeter and the details are presented in Figure 6. Based on the HRR data on Figure 6a, it can be seen that neat PP completely burned within 4 min with a value of 1145 kW/m^2. Upon the addition of kenaf fiber (30 wt.%) the HRR value reduced by 36%. Further reduction up to 67% was observed after the incorporation of IFR. This is due to the presence of a more stable intumescent char stabilized the unstable lignocellulose ash layer, which is form during combustion of kenaf. The presence of HNT and MMT nanoparticles enhanced the HRR value due to the formation of phosphocarbonaceous compound-containing aluminosilicate structures. As a comparison, KeC-IFR-MMT shows better performance with lower HRR value compare to the system with HNT. This can be linked to the viscosity and effective particle dispersion of the nanoparticles. From Figure 6b, the THR value is decreased after the addition of kenaf, IFR, HNT and MMT. Other important factors in the fire atmosphere for the composites are the amount of smoke release (SPR) and carbon monoxide production (COP) during combustion process. Figure 6c shows KeC-IFR-MMT produce the least amount of smoke compare to other system. Meanwhile, the KeC system produced the least CO during combustion and the addition of IFR significantly increases CO production as depicted in Figure 6d [191].

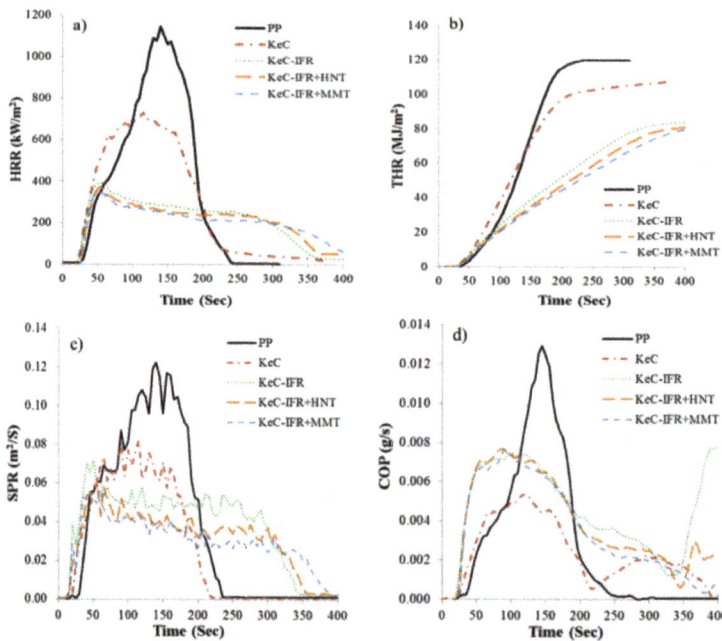

Figure 6. Flammability analysis by cone calorimeter: (**a**) HRR, (**b**) THR, (**c**) SPR, and (**d**) (COP). Reproduced from [191].

A study on the thermo-oxidative stability and flammability properties of bamboo/kenaf/nanoclay/epoxy hybrid nanocomposites was performed by Chee et al. Four systems were prepared in this study with three types of nanoclay which are HNT, MMT and organically modified MMT (OMMT). The ratio of bamboo (B) to kenaf (K) was fixed at 50:50 and the amount of nanoclay added is 1 wt.%. Initially, the epoxy/nanoclay was mixed by in situ polymerization method. Then, the fiber reinforced epoxy/nanoclay was obtained through a hand lay-up technique. The samples were denoted as B/K/HNT, B/K/MMT and B/K/OMMT. The control sample (B/K/epoxy) was prepared omitting the nanoclay. The first analysis for flame retardancy was conducted by using horizontal burning test. All samples did not extinguish and were able to pass first and second gauge length. There was also no flame dripping observed for all samples. They suggest that the mat-form reinforcing fibers provide better structural integrity to the composites. Besides, addition of nanoclay delays the flame spread. The flammability rating for all samples were HB 40 indicates that their burning rate is less than 40 mm/min and they follow the sequence of B/K/epoxy > B/K/HNT > B/K/MMT > B/K/OMMT. The flame delay was caused by the formation of a protective carbonaceous layer. This layer acts as a heat and flame insulator that limits the outgoing volatile gases and oxygen diffusion into the material. The B/K/OMMT composites shows better performance because of the presence of organic modifier enhances the catalytic effect during charring process; thereby produced a denser and more cohesive char. The second characterization was conducted to study the LOI of the nanocomposites. LOI can be defined as the lowest oxygen concentration needed to support flaming combustion of a material. From this analysis the B/K/epoxy, B/K/HNT, B/K/MMT and B/K/OMMT recorded LOI value of 19.8%, 22.9%, 22.9% and 27.7%, respectively. The B/K/epoxy is classified as combustible material as its LOI value is less than air oxygen content (21%) whereas all the nanoclay hybrid composites are self-extinguished as they have LOI value greater than 21%. This proves that addition of HNT and MMT nanoclays improve the flame retardant properties of the composites. This study found out that the LOI value increase with their ability to yield char. Higher amount of char will reduce the emission of flammable volatile and thus the oxygen level required to maintain flaming combustion will increase [192]. In addition, from the SEM analysis, the presence of filler, agglomerates and large tactoid particles on the surfaces of MMT/epoxy and HNT/epoxy nanocomposites fractioned items can be observed, while OMMT/epoxy nanocomposites showed more uniform dispersion.

Dutta and Maji prepared composites from PVC with microcrystalline cellulose (MCC) from RH and MMT by melt blending process at different MCC and MMT loadings. From the study, the LOI values for PVC/MCC and PVC/MMT are 44.4% and 46.7%, respectively. Interestingly, hybrid PVC/MCC/MMT has a higher LOI value (50.0%) compared to the individual systems. This is due to the formation of carbonaceous-silicate charred layer on the surface of the composites during combustion that retarded the heat transfer and thus delayed the decomposition rate. They suggested that the lower LOI value for single systems could be attributed to the agglomeration and poor dispersion of the components in the composites [193].

Composites from PLA/recycled bamboo chopstick fiber with multifunctional additive systems were produced by Wang and Shih. In this study, MMT was added as nanofiller, while APP and expandable graphene (EG) was used as flame retardants. Prior to mixing, the chopstick fiber (CF) underwent an alkaline treatment and then was chopped to obtain an average length of 2–4 mm. Both CF and MMT were treated with a silane coupling agent to obtain modified fibers and MMT (MCF and MOMMT). The composites were prepared by varying the amounts of APP and EG. The flammability characteristics were analyzed by a vertical burner tester. Details of the analysis are presented in Table 10. Based on the result, PLA can be classified as a flammable material with no rating in the UL94 standard. Similar pattern was found for the PLA containing 13 phr of EG and 4 phr of APP. A similar composition added with MCF and MOMMT produced a composite that passed the vertical burning test V-1 rating. Though dripping still occurred, the sample did not ignite the cotton. This proved that the addition of MCF and MOMMT enhanced the flammability of the composites. By modifying the amount EG and APP component, both PLA/23EG-4APP/MCF/MOMMT and PLA/13EG-8APP/MCF/MOMMT passed the V-0 rating, no dripping occurred and the cotton did not ignite.

Table 10. Flame retardance properties of PLA and PLA composites.

Sample	Ignite the Cotton	Dripping	UL-94 Rating
Polylactic acid (PLA)	Yes	Yes	NR
PLA + 13 phr expandable graphite + 4 phr ammonium polyphosphate (PLA/13EG-4APP)	Yes	Yes	NR
PLA + 13 phr expandable graphite + 4 phr ammonium polyphosphate + modified chopstick fiber + modified montmorillonite (PLA/13EG-4APP/MCF/MOMMT)	No	Yes	V-1
PLA + 23 phr expandable graphite + 4 phr ammonium polyphosphate + modified chopstick fiber + modified montmorillonite (PLA/23EG-4APP/MCF/MOMMT)	No	No	V-0
PLA + 13 phr expandable graphite + 8 phr ammonium polyphosphate + modified chopstick fiber + modified montmorillonite (PLA/13EG-8APP/MCF/MOMMT)	No	No	V-0

Further, photographic images of the composites after combustion process are shown in Figure 7. It is evident from the image, that the PLA had severe dripping problems. Char residues formation can be observed for the composites, especially for the PLA/23EG-4APP/MCF/MOMMT. They deduced that the addition of EG promoted the char yield in the presence of APP. During burning, EG was the first to degrade and expanded into a worm-like and loose porous char. This will inhibit oxygen and heat from permeating the substrate materials. After that, APP decomposed and produced polyphosphoric and ultraphosphoric acids, which catalyzed PLA to form a dense, continuous, and sealed char layer. This char layer protected the inner materials and limited the degradation of PLA [194].

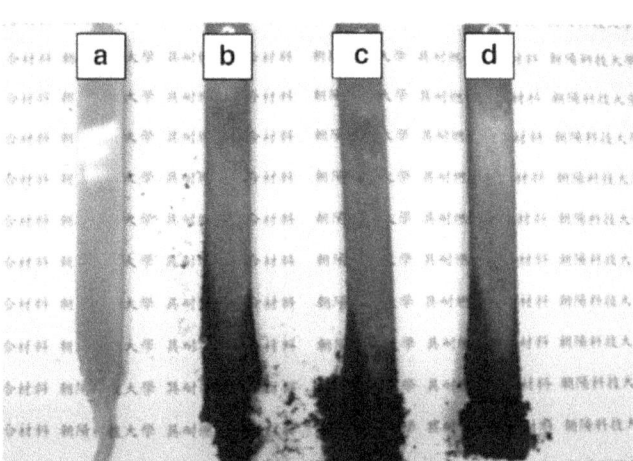

Figure 7. Photographic images of char residues after combustion of (**a**) neat PLA (**b**) PLA/13EG-4APP/MCF/MOMMT, (**c**) PLA/23EG-4APP/MCF/MOMMT, and (**d**) PLA/13EG 8APP/MCF/MOMMT composites. Reproduced from ref. [194].

The effect of MMT loading on the properties of methacrylic anhydride modified epoxidized soybean oil (MAESO), wood flour (WF) and divinyl acrylicpimaric acid (DAPA) was investigated by Mandal et al. The properties of composites with different percentages of nano-clay were investigated and reported. The composites were prepared by compression molding at loadings of 0, 1, 3 and 5 wt.% of MMT. The flame retardance characteristic was by determined by the LOI value. Based on the result, the LOI value of the MAESO/DAPA/WF composites increased with the increase in the MMT loading. The recorded LOI value for MAESO/DAPA/WF/MMT0, MAESO/DAPA/WF/MMT1, and MAESO/DAPA/WF/MMT3 were 55%, 57% and 60%, respectively. Beyond the 3 wt.% level, the LOI value decreased as observed for the MAESO/DAPA/WF/MMT5 with 58%. Similar like the previously reported study, the MMT produced silicate char on the surface of the composites and provided a thermal barrier to the oxygen and heat. This resulted to the improvement of LOI value and the flame retardancy of the composites. However, at high MMT loading, there is a tendency of the nanoclay to agglomerate and thus decrease the interactions and barrier properties [195].

8. Biodegradation Properties of MMT/Natural Fiber Reinforced Polymer Nanocomposites

The incorporation of reinforcement in composite systems could enormously change the overall degradation characteristics of the resulting polymer composites systems. This is attributed to the superior properties of natural fiber reinforced polymer composites with nanoparticles due to their nano-level characteristics which maximize the interfacial adhesion. The addition of nanoclays not only can significantly improve the mechanical properties of the biocomposites, but also enhance the barrier properties and chemical hindrance, which leads to the reduction of the biodegradation time for the composites [195,196]. In addition, the incorporation of nanoparticles with antibacterial properties like those that MMTs can possibly affect the antibacterial properties of biocomposite systems by diminishing the quantity and propagation of microbes (bacteria, fungi).

The study on the effect of MMT loading in wood/plant oil composites was studied by Moon et al. [195] with a focus on the biodegradation properties of the composites.

Composite samples were loaded with 0, 1, 3 and 5 wt.% of MMT and treated with a cellulolytic bacterial strain in liquid broth culture medium for the biodegradation study. In general, the addition of MMT resulted in declines in the collapse of wood composites caused by its antimicrobial activity. The weight loss of the composites was observed after 3 month of incubation. It is started with the bacterial growth of the bacterium *Pseudomonas* sp. that initially degraded the lignin part of the wood and continued to degrade cellulose and pectin after the production of cellulose and pectinase enzymes. Through the evaluation on the SEM results, the physical breakdown of composite on exposure to microbes were reported, attributed to the bacterial degradation activity. The degradation also shown to be prominent for a control sample without addition of MMT. The biodegradation activity was shown to decrease as the MMT loading increased due to the strong interconnected network of MMT in the composite which slowed down the accessibility and reactivity of the microorganisms. The addition of nanoclay also hinders the water absorption process thus further inhibiting the penetration of microorganisms within the composites.

Another study has been done by Islam et al. [197] to evaluate the effect of MMT addition on the biodegradability properties of kenaf/coir/PP composites. Similar findings were also reported as the biodegradability properties decrease as the MMT loading increased in the composite systems. The enhancement of the mechanical properties of the composites after the addition of MMT filler was seen to reduce the possibility of the weight loss of the composites thus diminishing the biodegradation process. These properties are very important to enhance the quality of the natural fiber-reinforced polymer composites to make them worthy materials for various applications, especially those that require strength and high durability.

9. Applications and Potential Use of MMT/Natural Fiber Reinforced Polymer Nanocomposites

To date, extensive studies has been conducted on lignocellulosic fiber-reinforced polymer composites hybridised with MMT nanoclays and they are growing in popularity in the development of advanced constructional and structural materials with superior performance at a low cost [34]. For instance, Islam et al. [133] revealed that after incorporating MMT, hybrid coir/wood/PP composites had higher tensile strength and modulus than wood/PP and coir/PP composites. The nanoclay improved the mechanical properties of the composites by enhancing interfacial interactions and the adhesion between the fiber and the polymer matrix. Majeed et al. [32] have studied the hybridization effect of MMT and RH on the mechanical and thermal properties of PP nanocomposites and discovered that the tensile and flexural modulus of PP increased by 63% and 92%, respectively, when RH and MMT were incorporated simultaneously into the PP matrix. The authors also found that adding solely RH to the PP matrix reduced the thermal stability of the composites. However, introducing MMT to PP/RH composites enhanced the thermal stability. DSC studies demonstrated that adding RH and MMT to the PP matrix increased crystallinity while maintaining melting and crystallization temperatures. Hybrid nanotechnology brings in a new era in material science, contributing to the development of ultra-high-tech advanced composites for future engineering applications. In regards to this matter, material scientists aim to diversify the applicability of MMT/lignocellulosic fiber nanocomposites in various areas, including electronics, automotive, outdoors, and adsorbents, as discussed in the subsections that follow.

9.1. Automotive Applications

With significant advances in science and technology, the manufacturing industries are shifting to more sustainable and environmentally friendly economic production. In relation to this matter, researchers are focusing their attention on developing novel materials for the benefit of society. Hybridizing lignocellulosic fiber with MMT presently offers tremendous potential in terms of features that can transcend conventional composites with the inclusion of only a small amount (5%) of nanoparticles, including superior mechanical strength, higher elastic modulus, high thermal stability, and fire resistance [198].

Hybridised nanocomposites have opened up new possibilities in the automotive industry due to their hassle-free fabrication and overall property improvements.

According to Galimberti et al. [199], the main driving forces for using clay polymer nanocomposites in the automotive field can be summarized as follows: (i) lighter weight of the vehicle, with fuel savings and reductions in CO_2 emissions: (ii) greater safety, (iii) better drivability, and (iv) increased comfort. These aims can be achieved by introducing MMT as a nanofiller in the polymer nanocomposites of a car. The MMT nanoclay hybridised with lignocellulosic fiber-reinforced polymer composites can be applied to several parts of a car such as suspension and braking systems, engines and power train, exhaust systems and catalytic converters, frames and body parts, paints and coatings, lubrication, tires, electrical and electronic equipment. By referring to Table 11, several thermoplastic or elastomeric polymers have been used in hybridised nanocomposites containing about 80% of clay minerals [199].

Table 11. Parts of a car and polymer matrices for clay polymer nanocomposite application.

Part of Car	Polymer
Timing belt/engine cover	Nylon-6
Step-assist, doors, center bridge, sail panel, seat backs and box-rail protector	PP, thermoplastic olefin
Rear floor	Thermoset polymer matrix (with glass fiber)
Tire	SBR
Thread	SBR, NBR, BR
Inner liner	Isoprene isobutylene copolymer, NBR
Internal compounds	NBR

SBR-Styrene butadiene rubber. NBR-Nitrile butadiene rubber. BR- Butadiene rubber.

Another important feature that needs to be highlighted is the flammability properties of automotive parts. Nowadays, most parts in cars have been replaced with plastic-based materials for lighter and more fuel-efficient cars. However, for this reason, the severity of collisions and non-crash auto fires is likely to worsen. Despite the implementation of rules imposed by the Federal Motor Vehicle Safety Standard and Flammability of Interior Materials issued by the National Highway Traffic Safety Administration (NHTSA) in 1968 and 1972, respectively, by the US government, fires have continued to be a serious threat. The combustible plastics in the form of electrical wiring, upholstery, and miscellaneous components are the most likely parts to ignite in automobile fires (47%) which is worsened by 27% by the presence of fuel for ignition. Therefore, due to the fire hazard caused by the plastic materials in motor vehicles and the speed at which plastics spread flames, smoke and combustion into the passenger compartment, automakers have continued to seek solutions to reduce the risk of fires occurring rather than relying on rescue efforts once the fire is initiated [200]. The effectiveness of nanoclay to improve the flammability properties of the lignocellulosic reinforced polymer composites was studied by Chee et al. [192] using bamboo/kenaf/epoxy nanocomposites reinforced with different types of nanoclay viz. MMT and O-MMT. Through the study, all hybrid nanocomposites achieved an HB40 rating in the UL94 horizontal burning test. However, the LOI value of the nanocomposites was observed to increase from 20% to 28% after the addition of nanoclays. The smoke generation was also remarkably reduced after the addition of nanoclays. The authors stated the mechanism of improvement imparted by the addition of nanoclays relies on the formation of an insulator barrier that provides a better charring effect which limits the migration of mass and heat between the gas and solid phase during the combustion process. The authors also found the effectiveness of organically modified MMT (O-MMT) compared to unmodified nanoclay due to better dispersion level with less agglomeration.

9.2. Outdoor Applications

Composite materials are generally designed and developed for use outdoors. Various types of composite technologies have been accessible in the market over the last few decades, but not all of them are suitable for integration or incorporation in building envelopes. These materials must be able to withstand temperatures of up to 60 °C or higher in particular to be suited for outdoor applications [201]. Moreover, composites should be able to meet certain key standards in terms of mechanical stability, fire resistance, sound insulation, thermal insulation, and other features as building components. Nevertheless, most polymers used for outdoor applications are light sensitive and frequently subjected to weathering, making them susceptible to degradation. Weathering aging, or more precisely photo-oxidation of polymer structures, is another aging mechanism that negatively impacts the service life of polymer materials under high-temperature conditions or in outdoor applications [202]. Weathering aging occurs when a polymer absorbs ultraviolet solar photons and atmospheric oxygen, resulting in discoloration and undesirable loss of mechanical properties caused by chemical bond scission and chain crosslinking of the polymer structure with exposure time [203]. Generally, weathering aging begins on the surface and eventually penetrates all the polymer material, depending on their chemical composition, environmental conditions, and exposure time [204]. However, it should be emphasized that degradation can be slowed but not totally avoided. This exposure causes catastrophic failure or severely jeopardizes the structural integrity of composite materials, diminishing their overall thermomechanical characteristics [205].

Kord et al. [206] found that exposing weathered samples (virgin polymers) for a longer time in an accelerated weathering environment elevated color change, lightness and water absorption. Hybrid nanocomposites, on the other hand, showed minor alterations in lightness and water absorption. Water cannot penetrate further into the composite because nanosized MMT could fill micro voids and fiber lumens, providing indirect weathering protection by reducing water absorption and oxidative activity, which results in less lightening [207]. Therefore, they can withstand the impacts of the sun, rain, and other weather conditions. Zahedi and his co-workers [116] observed a similar finding. The hybridization of MMT with lignocellulosic fiber also improves the nanocomposites resistance to electricity, chemicals, heat and flame, allowing these materials to be used as potential components for building enclosures including roofs, walls, windows, and shadings [51,208].

Most natural fiber reinforced composites (NFRPCs) find usage in outdoor applications, such as for construction products and building elements to comply with the fire safety codes such as EN 13501-1-2009 that emphasize fire resistance properties as necessary. Therefore, it is very important to develop and optimize fire-retardant composite materials at laboratory scale for evaluation through standard techniques such as TGA, Cone calorimetry based on ISO 5660-1 standard, limited oxygen index (LOI), burning test (UL 94 vertical or horizontal burning test) etc. [209,210]. The studies on NFRPC for outdoor applications as an alternative for conventional materials such as for components like roofs and door panels have been widely reported. Despite evaluating the mechanical and physical performance of the NFRPC, it is necessary to evaluate the flame and thermal resistance of the fabricated composite in case of any fire accidents. Through a series of laboratory analyses, Rajini et al. [211] proved the improvement in the thermal and flammability properties of the coconut sheath (CS) reinforced PE composite. The hybridization of the CS/PE composite with 5 wt.% of MMT nanoclay was observed to remarkably decrease the heat release rate and the mass loss rate of the composites by the char formation mechanism. The char formation acted as a heat barrier, thus lowered the flammability properties of the composites. Moreover, treated CS fiber also influenced the thermal properties of the composites by lowering the degradation temperature due to the presence of more hydroxyl group to attract water. This finding proved the ability of hybrid lignocellulosic/MMT polymer composites to be applied for outdoor applications.

9.3. Coating Applications

The integration of filler such as nanoclays, cellulose micro- or nanofibers, and MMT into coating products is generating a lot of attention in terms of improving general physical characteristics in plastics and bioplastics. The use of cellulose biofibers in melt blending and solution casting processing methods to enhance the barrier characteristics to gases and vapors and to impart new functions to biopackaging plastics has resulted in the creation of nanocomposites for food packaging applications. It has also been found that reinforcing biopolymers with organically modified MMT improves their characteristics significantly. MMT have been used as possible packaging materials in many studies. Chowdhury [212] investigated the effect of adding mica nanoclay to a PLA matrix vs. using MMT nanoclay and found that the oxygen barrier was improved. Mohan and Kanny [213] in their study was found that the MMT-treated fibers have a higher surface area, resulting in less moisture absorption. They also found that untreated and MMT-treated fibers have sigmoidal isotherm sorption behaviour, and sorption hysteresis is dependent on crystallinity and MMT infusion.

10. Environmental Concerns, Health, and Safety Issues of MMT

The inclusion of MMT into polymeric systems improves their barrier, thermal, and mechanical resistance, making them ideal for a multitude of industrial applications, including environmental remediation, engineering and construction, and food industry. Moreover, MMT has been widely utilized for biological applications since it has demonstrated very low cytotoxicity both in vitro and in vivo, and owing to its chemical properties, MMT can intercalate biologically active molecules, resulting in efficient systems for oral or topical drug delivery [214]. Nevertheless, MMT extraction, development, and application, like most other materials, are prone to environmental and safety concerns.

There have been relatively few experimental investigations into the exposure and possible toxicity of MMT to living organisms up to this moment. Even so, there are possibly two key aspects that should be considered when utilizing MMT: particle size and chemical composition. Epidemiological research has shown that morbidity and fatality have been related to an elevation in airborne particles, primarily those within the ultrafine size range (<0.1 µm) [51]. These nanoparticles have the tendency to penetrate the human body and circulate via the bloodstream to vital organs, causing toxicity and health issues such as tissue damage. Based on the findings of Mainasaba et al. [215], it can be inferred that MMT is potentially genotoxic since it induces the formation of micronuclei at non-cytotoxic concentrations, and thus poses a threat to human health, particularly when long-term exposure is considered. Nonetheless, MMT incorporated into bulk materials or polymer products will not cause these problems.

Till now, there seem to be no experimental investigations into MMT exposure in natural environments and its potential toxicity. There is virtually no known information regarding the exposure pathways, the limits of MMT exposure, and their toxicity towards the environment in occupation-related scenarios all over the world since this data has not been reported elsewhere. To monitor nanosafety for the environment in the real world, environmentally aware product design frameworks and life cycle assessment (LCA) approaches are recommended. The most essential steps in characterizing and evaluating MMT in the environment are the development of MMT databases to analyze the LCA of products: gathering real MMT data for the development of exposure limits for MMT and formulation of advanced models, as well as the implementation of standardized protocols [216]. These procedures will aid in determining the risks associated with the application of MMT in the real world, therefore improving worker safety.

11. Conclusions and Future Outlook

The implementation of MMT to modify composites for structural applications is gradually expanding. The inclusion of the nanofillers not only takes place in pure polymer and synthetic fiber reinforced polymer composites, but moreover in biocomposites which are mainly made up of lignocellulosic fibers. Most research studies found that the introduction of MMT enhanced the mechanical and water barrier properties of biocomposites as well as their thermal capability. These behaviour alterations depend mainly on the exfoliated/intercalated balance of the nanoclay filler introduced. The review article also specifies that flexural, tensile, compression, and impact strength were investigated for a wide range of solutions and uses.

According to the literature, MMT can be a well dispersed in polymer resins with the aid of mechanical stirrers to improve its homogeneity. Compared with the pure polymer matrix, the addition of the MMT in composites results in low dielectric permittivity, low dielectric loss, and enhanced dielectric strength. Moreover, the MMT nanofillers allow the enhancement in terms of thermal conductivity and stability. These improvement properties are highly influenced by the compatibility of interfaces between MMT and polymer resin. Other than that, the MMT modified with chemical treatments such as GPTMS and g-APTES promote greater influence in improving the interface between the nanoparticles and polymer matrices, thus causing lower dielectric loss, electric conductivity, lower thermal conductivity, as well as higher thermal stability and breakdown strength. From this point of view, it can be concluded that the inclusion of MMT could potentially benefit the structural industries, especially in the electrical transmission sector, since it promotes good electrical insulation properties in composites. In conclusion, this review can benefit many composite structural sectors in order to stimulate lower raw material and manufacturing costs of composite products with great strength and stiffness.

Author Contributions: Conceptualization, M.N.N. and A.R.I.; validation, M.N.N., A.R.I. and A.K.; writing—original draft preparation, A.H.A., M.N.N., F.A.S., M.R.M.A., M.N.F.N., A.R.I., A.M.K., M.R., S.S.S., A.N., S.F.K.S., M.M.H. and M.S.N.A.; supervision, M.R.I., S.M.S. and A.K.; project administration, A.H.A.; funding acquisition, A.K. All authors have read and agreed to the published version of the manuscript.

Funding: This research was funded by Higher Education Center of Excellence (HICoE), Ministry of Higher Education, Malaysia (Grant number 6369109).

Institutional Review Board Statement: Not applicable.

Informed Consent Statement: Not applicable.

Data Availability Statement: Not applicable.

Acknowledgments: The authors gratefully acknowledge the technical and financial support from the Universiti Putra Malaysia (UPM).

Conflicts of Interest: The authors declare no conflict of interest.

Abbreviations

The following abbreviations are used in this manuscript:

APTES	Aminopropyltriethoxysilane
APP	Ammonium polyphosphate (APP)
B	Bamboo
BR	Butadiene rubber
CF	Chopstick fiber
CNF	Cellulose nanofibers
CNW	Cellulose nanowhiskers
COP	Carbon monoxide production
CS	Coconut sheath
CS	Corn starch

DAPA	Divinyl acrylicpimaric acid
DMA	Dynamic mechanical analysis
DSC	Differential scanning calorimetry
DTG	Derivative thermogravimetry
E″	Loss modulus
E′	Storage modulus
ECAE	Equal channel angular extrusion
EP	Ethylene–propylene
EVA	Ethylene-vinyl acetate
EVOH	Ethylene vinyl alcohol
EG	Expandable graphene
GPTMS	Glicydoxy propyl trimethoxysilane
HDPE	High density polyethylene
HDT	Heat deflection temperature
HNT	Halloysite nanotube
HRR	Heat release rate
iPP	isotactic PP
IFR	Intumescent flame retardant
K	Kenaf
KeC	Kenaf composites
KF	Kenaf fiber
LCA	Life cycle assessment
LDPE	Low-density polyethylene
LOI	Limiting oxygen index
MAESO	Methacrylic anhydride modified epoxidized soybean oil
MAH	Maleic anhydride
MaPE	Maleic anhydride-grafted polyethylene
MAPP	Maleated anhydride grafted polypropylene
MCC	Microcrystalline cellulose
MCF	Modified chopstick fiber
MEP	Maleated ethylene–propylene
MLR	Mass loss rate
MMT	Montmorillonite
MOMMT	Modified montmorillonite
NaMMT	natural sodium montmorillonite
NBR	Nitrile butadiene rubber
NFRPC	Natural fiber reinforced composites
NHTSA	National highway traffic safety administration
OMLS	Organo-modified layered silicates
OMMT	Organically modified montmorillonite
OPEFB	Oil palm empty fruit bunch
OTR	Oxygen transmission rate
PBS	Poly(butylene succinate)
PEG	Poly(ethylene glycol)
PEMA	Poly(ethylmethacrylate)
PEO	Poly(ethylene oxide)
PHR	Peak of release heat
PLA	Polylactic acid
PP-g-MAH	Maleic anhydride grafted polypropylene
PP	Polypropylene
PS	Polystyrene
PVA	Polyvinyl alcohol
PVC	Poly(vinylchloride)
PVP	Poly(vinyl pyrrolidone)
RH	Rice husk
SBR	Styrene butadiene rubber
SiB	Silane treated banana fiber

SPR	Smoke release
SEM	Scanning electron microscope
TAF	Treated aloe vera fiber
TEM	Transmission electron microscopy
TKF	Treated kenaf fiber
TG	Thermogravimetry
TGA	Thermogravimetric analysis
THR	Total heat release
TPS	Tapioca starch
TTI	Time to ignition
WF	Wood flour
WHF	Water hyacinth fiber
WSF	Walnut shell flour
WVTR	Water vapour permeability rate

References

1. Asyraf, M.R.M.; Ishak, M.R.; Sapuan, S.M.; Yidris, N. Utilization of Bracing Arms as Additional Reinforcement in Pultruded Glass Fiber-Reinforced Polymer Composite Cross-Arms: Creep Experimental and Numerical Analyses. *Polymers* **2021**, *13*, 620. [CrossRef] [PubMed]
2. Ogin, S.L.; Brøndsted, P.; Zangenberg, J. *Composite Materials: Constituents, Architecture, and Generic Damage*; Elsevier: Amsterdam, The Netherlands, 2016.
3. Nurazzi, N.M.; Asyraf, M.R.M.; Khalina, A.; Abdullah, N.; Aisyah, H.A.; Rafiqah, S.A.; Sabaruddin, F.A.; Kamarudin, S.H.; Norrrahim, M.N.F.; Ilyas, R.A.; et al. A Review on Natural Fiber Reinforced Polymer Composite for Bullet Proof and Ballistic Applications. *Polymers* **2021**, *13*, 646. [CrossRef]
4. Asyraf, M.R.M.; Ishak, M.R.; Sapuan, S.M.; Yidris, N. Influence of Additional Bracing Arms as Reinforcement Members in Wooden Timber Cross-Arms on Their Long-Term Creep Responses and Properties. *Appl. Sci.* **2021**, *11*, 2061. [CrossRef]
5. Amir, A.L.; Ishak, M.R.; Yidris, N.; Zuhri, M.Y.M.; Asyraf, M.R.M. Advances of composite cross arms with incorporation of material core structures: Manufacturability, recent progress and views. *J. Mater. Res. Technol.* **2021**, *13*, 1115–1131. [CrossRef]
6. Asyraf, M.R.M.; Ishak, M.R.; Sapuan, S.M.; Yidris, N.; Ilyas, R.A. Woods and composites cantilever beam: A comprehensive review of experimental and numerical creep methodologies. *J. Mater. Res. Technol.* **2020**, *9*, 6759–6776. [CrossRef]
7. Asyraf, M.R.M.; Ishak, M.R.; Sapuan, S.M.; Yidris, N. Comparison of Static and Long-term Creep Behaviors between Balau Wood and Glass Fiber Reinforced Polymer Composite for Cross-arm Application. *Fibers Polym.* **2021**, *22*, 793–803. [CrossRef]
8. Ilyas, R.A.; Sapuan, S.M.; Harussani, M.M.; Hakimi, M.Y.A.Y.; Haziq, M.Z.M.; Atikah, M.S.N.; Asyraf, M.R.M.; Ishak, M.R.; Razman, M.R.; Nurazzi, N.M.; et al. Polylactic Acid (PLA) Biocomposite: Processing, Additive Manufacturing and Advanced Applications. *Polymers* **2021**, *13*, 1326. [CrossRef]
9. Kyriazoglou, C.; Guild, F.J. Quantifying the effect of homogeneous and localized damage mechanisms on the damping properties of damaged GFRP and CFRP continuous and woven composite laminates-an FEA approach. *Compos. Part A Appl. Sci. Manuf.* **2005**, *36*, 367–379. [CrossRef]
10. Hamidon, M.H.; Sultan, M.T.H.; Ariffin, A.H.; Shah, A.U.M. Effects of fibre treatment on mechanical properties of kenaf fibre reinforced composites: A review. *J. Mater. Res. Technol.* **2019**, *8*, 3327–3337. [CrossRef]
11. Ali, S.S.S.; Razman, M.R.; Awang, A.; Asyraf, M.R.M.; Ishak, M.R.; Ilyas, R.A.; Lawrence, R.J. Critical Determinants of Household Electricity Consumption in a Rapidly Growing City. *Sustainability* **2021**, *13*, 4441. [CrossRef]
12. Ilyas, R.A.; Sapuan, S.M.; Atiqah, A.; Ibrahim, R.; Abral, H.; Ishak, M.R.; Zainudin, E.S.; Nurazzi, N.M.; Atikah, M.S.N.; Ansari, M.N.M.; et al. Sugar palm (*Arenga pinnata* [Wurmb.] Merr) starch films containing sugar palm nanofibrillated cellulose as reinforcement: Water barrier properties. *Polym. Compos.* **2020**, *41*, 459–467. [CrossRef]
13. Ilyas, R.; Sapuan, S.; Atikah, M.; Asyraf, M.; Rafiqah, S.A.; Aisyah, H.; Nurazzi, N.M.; Norrrahim, M. Effect of hydrolysis time on the morphological, physical, chemical, and thermal behavior of sugar palm nanocrystalline cellulose (*Arenga pinnata* (Wurmb.) Merr). *Text. Res. J.* **2021**, *91*, 152–167. [CrossRef]
14. Omran, A.A.B.; Mohammed, A.A.B.A.; Sapuan, S.M.; Ilyas, R.A.; Asyraf, M.R.M.; Koloor, S.S.R.; Petrů, M. Micro- and Nanocellulose in Polymer Composite Materials: A Review. *Polymers* **2021**, *13*, 231. [CrossRef]
15. Alsubari, S.; Zuhri, M.Y.M.; Sapuan, S.M.; Ishak, M.R.; Ilyas, R.A.; Asyraf, M.R.M. Potential of Natural Fiber Reinforced Polymer Composites in Sandwich Structures: A Review on Its Mechanical Properties. *Polymers* **2021**, *13*, 423. [CrossRef]
16. Ilyas, R.A.; Sapuan, S.M.; Asyraf, M.R.M.; Atikah, M.S.N.; Ibrahim, R.; Dele-Afolabia, T.T. Introduction to biofiller reinforced degradable polymer composites. In *Biofiller Reinforced Biodegradable Polymer Composites*; Sapuan, S.M., Jumaidin, R., Hanafi, I., Eds.; CRC Press: Boca Raton, FL, USA, 2020; pp. 1–23.
17. Nurazzi, N.M.; Asyraf, M.R.M.; Khalina, A.; Abdullah, N.; Sabaruddin, F.A.; Kamarudin, S.H.; Ahmad, S.; Mahat, A.M.; Lee, C.L.; Aisyah, H.A.; et al. Fabrication, functionalization, and application of carbon nanotube-reinforced polymer composite: An overview. *Polymers* **2021**, *13*, 1047. [CrossRef]

18. Asyraf, M.R.M.; Rafidah, M.; Azrina, A.; Razman, M.R. Dynamic mechanical behaviour of kenaf cellulosic fibre biocomposites: A comprehensive review on chemical treatments. *Cellulose* **2021**, *28*, 2675–2695. [CrossRef]
19. Amir, A.L.; Ishak, M.R.; Yidris, N.; Zuhri, M.Y.M.; Asyraf, M.R.M. Potential of Honeycomb-Filled Composite Structure in Composite Cross-Arm Component: A Review on Recent Progress and Its Mechanical Properties. *Polymers* **2021**, *13*, 1341. [CrossRef]
20. Johari, A.N.; Ishak, M.R.; Leman, Z.; Yusoff, M.Z.M.; Asyraf, M.R.M. Influence of CaCO3 in pultruded glass fibre/unsaturated polyester composite on flexural creep behaviour using conventional and TTSP methods. *Polimery* **2020**, *65*, 46–54. [CrossRef]
21. El-zayat, M.M.; Mohamed, M.A. Effect of gamma radiation on the physico mechanical properties of recycled HDPE/modified sugarcane bagasse composite. *J. Macromol. Sci. Part A* **2019**, *56*, 1–9. [CrossRef]
22. Nurazzi, N.M.; Harussani, M.M.; Aisyah, H.A.; Ilyas, R.A.; Norrrahim, M.N.F.; Khalina, A.; Abdullah, N. Treatments of natural fiber as reinforcement in polymer composites—a short review. *Funct. Compos. Struct.* **2021**, *3*, 024002. [CrossRef]
23. Zin, M.H.; Abdan, K.; Norizan, M.N.; Mazlan, N. The effects of alkali treatment on the mechanical and chemical properties of banana fibre and adhesion to epoxy resin. *Pertanika J. Sci. Technol.* **2018**, *26*, 161–176.
24. Norizan, M.N.; Abdan, K.; Salit, M.S.; Mohamed, R. The effect of alkaline treatment on the mechanical properties of treated sugar palm yarn fibre reinforced unsaturated polyester composites reinforced with different fibre loadings of sugar palm fibre. *Sains Malaysiana* **2018**, *47*, 699–705. [CrossRef]
25. Abdel-Hakim, A.; El-Wakil, A.E.A.A.; El-Mogy, S.; Halim, S. Effect of fiber coating on the mechanical performance, water absorption and biodegradability of sisal fiber/natural rubber composite. *Polym. Int.* **2021**, *70*, 1356–1366. [CrossRef]
26. Abdel-Hakim, A.; Mourad, R.M. Mechanical, water uptake properties, and biodegradability of polystyrene-coated sisal fiber-reinforced high-density polyethylene. *Polym. Compos.* **2020**, *41*, 1435–1446. [CrossRef]
27. Akil, H.M.; Santulli, C.; Sarasini, F.; Tirillò, J.; Valente, T. Environmental effects on the mechanical behaviour of pultruded jute/glass fibre-reinforced polyester hybrid composites. *Compos. Sci. Technol.* **2014**, *94*, 62–70. [CrossRef]
28. Ilyas, R.A.; Sapuan, S.M.; Asyraf, M.R.M.; Dayana, D.A.Z.N.; Amelia, J.J.N.; Rani, M.S.A.; Norrrahim, M.N.F.; Nurazzi, N.M.; Aisyah, H.A.; Sharma, S.; et al. Polymer composites filled with metal derivatives: A review of flame retardants. *Polymers* **2021**, *13*, 1701. [CrossRef]
29. Azman, M.A.; Asyraf, M.R.M.; Khalina, A.; Petrů, M.; Ruzaidi, C.M.; Sapuan, S.M.; Wan Nik, W.B.; Ishak, M.R.; Ilyas, R.A.; Suriani, M.J. Natural Fiber Reinforced Composite Material for Product Design: A Short Review. *Polymers* **2021**, *13*, 1917. [CrossRef]
30. Edwards, D.C. Polymer-filler interactions in rubber reinforcement. *J. Mater. Sci.* **1990**, *25*, 4175–4185. [CrossRef]
31. Bhattacharya, M. Polymer Nanocomposites—A Comparison between Carbon Nanotubes, Graphene, and Clay as Nanofillers. *Materials* **2016**, *9*, 262. [CrossRef]
32. Majeed, K.; Ahmed, A.; Abu Bakar, M.S.; Mahlia, T.M.I.; Saba, N.; Hassan, A.; Jawaid, M.; Hussain, M.; Iqbal, J.; Ali, Z. Mechanical and thermal properties of montmorillonite-reinforced polypropylene/rice husk hybrid nanocomposites. *Polymers* **2019**, *11*, 1557. [CrossRef]
33. Okada, A.; Usuki, A. Twenty years of polymer-clay nanocomposites. *Macromol. Mater. Eng.* **2006**, *291*, 1449–1476. [CrossRef]
34. Ramesh, P.; Prasad, B.D.; Narayana, K.L. Effect of fiber hybridization and montmorillonite clay on properties of treated kenaf/aloe vera fiber reinforced PLA hybrid nanobiocomposite. *Cellulose* **2020**, *27*, 6977–6993. [CrossRef]
35. Khan, A.; Asiri, A.M.; Jawaid, M.; Saba, N. Inamuddin Effect of cellulose nano fibers and nano clays on the mechanical, morphological, thermal and dynamic mechanical performance of kenaf/epoxy composites. *Carbohydr. Polym.* **2020**, *239*, 116248. [CrossRef] [PubMed]
36. Shahroze, R.M.; Ishak, M.R.; Salit, M.S.; Leman, Z.; Asim, M.; Chandrasekar, M. Effect of organo-modified nanoclay on the mechanical properties of sugar palm fiber-reinforced polyester composites. *BioResources* **2018**, *13*, 7430–7444. [CrossRef]
37. Al-Samhan, M.; Samuel, J.; Al-Attar, F.; Abraham, G. Comparative Effects of MMT Clay Modified with Two Different Cationic Surfactants on the Thermal and Rheological Properties of Polypropylene Nanocomposites. *Int. J. Polym. Sci.* **2017**, *2017*, 5717968. [CrossRef]
38. Jia, Z.R.; Gao, Z.G.; Lan, D.; Cheng, Y.H.; Wu, G.L.; Wu, H.J. Effects of filler loading and surface modification on electrical and thermal properties of epoxy/montmorillonite composite. *Chinese Phys. B* **2018**, *27*, 117806. [CrossRef]
39. Ramesh, P.; Prasad, B.D.; Narayana, K.L. Effect of MMT Clay on Mechanical, Thermal and Barrier Properties of Treated Aloevera Fiber/ PLA-Hybrid Biocomposites. *Silicon* **2020**, *12*, 1751–1760. [CrossRef]
40. Rozman, H.D.; Musa, L.; Azniwati, A.A.; Rozyanty, A.R. Tensile properties of kenaf/unsaturated polyester composites filled with a montmorillonite filler. *J. Appl. Polym. Sci.* **2011**, *119*, 2549–2553. [CrossRef]
41. Asyraf, M.R.M.; Ishak, M.R.; Sapuan, S.M.; Yidris, N.; Rafidah, M.; Ilyas, R.A.; Razman, M.R. Potential application of green composites for cross arm component in transmission tower: A brief review. *Int. J. Polym. Sci.* **2020**, *2020*, 8878300. [CrossRef]
42. Nik Baihaqi, N.M.Z.; Khalina, A.; Mohd Nurazzi, N.; Aisyah, H.A.; Sapuan, S.M.; Ilyas, R.A. Effect of fiber content and their hybridization on bending and torsional strength of hybrid epoxy composites reinforced with carbon and sugar palm fibers. *Polimery/Polymers* **2021**, *66*, 36–43. [CrossRef]
43. Norrrahim, M.N.F.; Nurazzi, N.M.; Jenol, M.A.; Farid, M.A.A.; Janudin, N.; Ujang, F.A.; Yasim-Anuar, T.A.T.; Syed Najmuddin, S.U.F.; Ilyas, R.A. Emerging development of nanocellulose as an antimicrobial material: An overview. *Mater. Adv.* **2021**, *2*, 3538–3551. [CrossRef]

44. Sabaruddin, F.A.; Paridah, M.T.; Sapuan, S.M.; Ilyas, R.A.; Lee, S.H.; Abdan, K.; Mazlan, N.; Roseley, A.S.M.; Abdul Khalil, H.P.S. The Effects of Unbleached and Bleached Nanocellulose on the Thermal and Flammability of Polypropylene-Reinforced Kenaf Core Hybrid Polymer Bionanocomposites. *Polymers* **2020**, *13*, 116. [CrossRef]
45. Ilyas, R.A.; Sapuan, M.S.; Norizan, M.N.; Norrrahim, M.N.F.; Ibrahim, R.; Atikah, M.S.N.; Huzaifah, M.R.M.; Radzi, A.M.; Izwan, S.; Azammi, A.M.N.; et al. Macro to nanoscale natural fiber composites for automotive components: Research, development, and application. In *Biocomposite and Synthetic Composites for Automotive Applications*; Sapuan, M.S., Ilyas, R.A., Eds.; Woodhead Publishing Series: Amsterdam, The Netherlands, 2020.
46. Zin, M.H.; Abdan, K.; Mazlan, N.; Zainudin, E.S.; Liew, K.E.; Norizan, M.N. Automated spray up process for Pineapple Leaf Fibre hybrid biocomposites. *Compos. Part B Eng.* **2019**, *177*, 107306. [CrossRef]
47. Bergaya, F.; Detellier, C.; Lambert, J.-F.; Lagaly, G. Introduction to clay–polymer nanocomposites (CPN). In *Developments in Clay Science*; Elsevier: Amsterdam, The Netherlands, 2013; pp. 655–677.
48. Said, M.; Abu Hasan, H.; Mohd Nor, M.T.; Mohammad, A.W. Removal of COD, TSS and colour from palm oil mill effluent (POME) using montmorillonite. *Desalin. Water Treat.* **2016**, *57*, 10490–10497. [CrossRef]
49. Uddin, F. Clays, nanoclays, and montmorillonite minerals. *Metall. Mater. Trans. A* **2008**, *39*, 2804–2814. [CrossRef]
50. Nakato, T.; Miyamoto, N. Liquid crystalline behavior and related properties of colloidal systems of inorganic oxide nanosheets. *Materials* **2009**, *2*, 1734–1761. [CrossRef]
51. Uddin, F. *Montmorillonite: An introduction to properties and utilization*; IntechOpen: London, UK, 2018.
52. Zhu, T.T.; Zhou, C.H.; Kabwe, F.B.; Wu, Q.Q.; Li, C.S.; Zhang, J.R. Exfoliation of montmorillonite and related properties of clay/polymer nanocomposites. *Appl. Clay Sci.* **2019**, *169*, 48–66. [CrossRef]
53. Zhou, C.H.; Keeling, J. Fundamental and applied research on clay minerals: From climate and environment to nanotechnology. *Appl. Clay Sci.* **2013**, *74*, 3–9. [CrossRef]
54. Chan, M.; Lau, K.; Wong, T.; Ho, M.; Hui, D. Mechanism of reinforcement in a nanoclay/polymer composite. *Compos. Part B Eng.* **2011**, *42*, 1708–1712. [CrossRef]
55. Zhu, B.; Wang, Y.; Liu, H.; Ying, J.; Liu, C.; Shen, C. Effects of interface interaction and microphase dispersion on the mechanical properties of PCL/PLA/MMT nanocomposites visualized by nanomechanical mapping. *Compos. Sci. Technol.* **2020**, *190*, 108048. [CrossRef]
56. Shehap, A.M.; Nasr, R.A.; Mahfouz, M.A.; Ismail, A.M. Preparation and characterizations of high doping chitosan/MMT nanocomposites films for removing iron from ground water. *J. Environ. Chem. Eng.* **2021**, *9*, 104700. [CrossRef]
57. Zare, Y.; Rhee, K.Y.; Park, S.-J. A modeling methodology to investigate the effect of interfacial adhesion on the yield strength of MMT reinforced nanocomposites. *J. Ind. Eng. Chem.* **2019**, *69*, 331–337. [CrossRef]
58. Harussani, M.M.; Sapuan, S.M.; Rashid, U.; Khalina, A. Development and Characterization of Polypropylene Waste from Personal Protective Equipment (PPE)-Derived Char-Filled Sugar Palm Starch Biocomposite Briquettes. *Polymers* **2021**, *13*, 1707. [CrossRef] [PubMed]
59. Saba, N.; Paridah, T.M.; Abdan, K.; Ibrahim, N.A. Preparation and characterization of fire retardant nano-filler from oil palm empty fruit bunch fibers. *BioResources* **2015**, *10*, 4530–4543. [CrossRef]
60. Kumar, P.; Sandeep, K.P.; Alavi, S.; Truong, V.D.; Gorga, R.E. Effect of type and content of modified montmorillonite on the structure and properties of bio-nanocomposite films based on soy protein isolate and montmorillonite. *J. Food Sci.* **2010**, *75*, N46–N56. [CrossRef]
61. Fan, Q.; Han, G.; Cheng, W.; Tian, H.; Wang, D.; Xuan, L. Effect of intercalation structure of organo-modified montmorillonite/polylactic acid on wheat straw fiber/polylactic acid composites. *Polymers* **2018**, *10*, 896. [CrossRef]
62. Xue, M.-L.; Yu, Y.-L.; Li, P. Preparation, dispersion, and crystallization of the poly (trimethylene terephthalate)/organically modified montmorillonite (PTT/MMT) nanocomposites. *J. Macromol. Sci. Part B Phys.* **2010**, *49*, 1105–1116. [CrossRef]
63. Swearingen, C.; Macha, S.; Fitch, A. Leashed ferrocenes at clay surfaces: Potential applications for environmental catalysis. *J. Mol. Catal. A Chem.* **2003**, *199*, 149–160. [CrossRef]
64. Jia, F.; Song, S. Exfoliation and characterization of layered silicate minerals: A review. *Surf. Rev. Lett.* **2014**, *21*, 1430001. [CrossRef]
65. Asgari, M.; Abouelmagd, A.; Sundararaj, U. Silane functionalization of sodium montmorillonite nanoclay and its effect on rheological and mechanical properties of HDPE/clay nanocomposites. *Appl. Clay Sci.* **2017**, *146*, 439–448. [CrossRef]
66. Nistor, M.-T.; Vasile, C. Influence of the nanoparticle type on the thermal decomposition of the green starch/poly (vinyl alcohol)/montmorillonite nanocomposites. *J. Therm. Anal. Calorim.* **2013**, *111*, 1903–1919. [CrossRef]
67. Ejder-Korucu, M.; Gürses, A.; Karaca, S. Poly (ethylene oxide)/clay nanaocomposites: Thermal and mechanical properties. *Appl. Surf. Sci.* **2016**, *378*, 1–7. [CrossRef]
68. Yang, F.; Mubarak, C.; Keiegel, R.; Kannan, R.M. Supercritical carbon dioxide (scCO2) dispersion of poly (ethylene terephthalate)/clay nanocomposites: Structural, mechanical, thermal, and barrier properties. *J. Appl. Polym. Sci.* **2017**, *134*. [CrossRef]
69. Ammar, A.; Elzatahry, A.; Al-Maadeed, M.; Alenizi, A.M.; Huq, A.F.; Karim, A. Nanoclay compatibilization of phase separated polysulfone/polyimide films for oxygen barrier. *Appl. Clay Sci.* **2017**, *137*, 123–134. [CrossRef]
70. Wu, H.; Krifa, M.; Koo, J.H. Flame retardant polyamide 6/nanoclay/intumescent nanocomposite fibers through electrospinning. *Text. Res. J.* **2014**, *84*, 1106–1118. [CrossRef]
71. Roghani-Mamaqani, H.; Haddadi-Asl, V.; Najafi, M.; Salami-Kalajahi, M. Preparation of nanoclay-dispersed polystyrene nanofibers via atom transfer radical polymerization and electrospinning. *J. Appl. Polym. Sci.* **2011**, *120*, 1431–1438. [CrossRef]

72. Zhang, G.; Ke, Y.; He, J.; Qin, M.; Shen, H.; Lu, S.; Xu, J. Effects of organo-modified montmorillonite on the tribology performance of bismaleimide-based nanocomposites. *Mater. Des.* **2015**, *86*, 138–145. [CrossRef]
73. Kord, B. Nanofiller reinforcement effects on the thermal, dynamic mechanical, and morphological behavior of HDPE/rice husk flour composites. *BioResources* **2011**, *6*, 1351–1358.
74. Kord, B. Effect of nanoparticles loading on properties of polymeric composite based on hemp fiber/polypropylene. *J. Thermoplast. Compos. Mater.* **2012**, *25*, 793–806. [CrossRef]
75. Sinha Ray, S. *Clay-Containing Polymer Nanocomposites*; Elsevier: Amsterdam, The Netherlands, 2013.
76. Kenned, J.J.; Sankaranarayanasamy, K.; Kumar, C.S. Chemical, biological, and nanoclay treatments for natural plant fiber-reinforced polymer composites: A review. *Polym. Polym. Compos.* **2020**, *29*. [CrossRef]
77. Norrrahim, M.N.F.; Ariffin, H.; Yasim-Anuar, T.A.T.; Hassan, M.A.; Ibrahim, N.A.; Yunus, W.M.Z.W.; Nishida, H. Performance Evaluation of Cellulose Nanofiber with Residual Hemicellulose as a Nanofiller in Polypropylene-Based Nanocomposite. *Polymers* **2021**, *13*, 1064. [CrossRef]
78. Norrrahim, M.N.F.; Ariffin, H.; Yasim-Anuar, T.A.T.; Ghaemi, F.; Hassan, M.A.; Ibrahim, N.A.; Ngee, J.L.H.; Yunus, W.M.Z.W. Superheated steam pretreatment of cellulose affects its electrospinnability for microfibrillated cellulose production. *Cellulose* **2018**, *25*, 3853–3859. [CrossRef]
79. Norrrahim, M.N.F.; Ariffin, H.; Hassan, M.A.; Ibrahim, N.A.; Yunus, W.M.Z.W.; Nishida, H. Utilisation of superheated steam in oil palm biomass pretreatment process for reduced chemical use and enhanced cellulose nanofibre production. *Int. J. Nanotechnol.* **2019**, *16*, 668–679. [CrossRef]
80. Matsuda, D.K.M.; Verceheze, A.E.S.; Carvalho, G.M.; Yamashita, F.; Mali, S. Baked foams of cassava starch and organically modified nanoclays. *Ind. Crops Prod.* **2013**, *44*, 705–711. [CrossRef]
81. Andlar, M.; Rezić, T.; Marđetko, N.; Kracher, D.; Ludwig, R.; Šantek, B. Lignocellulose degradation: An overview of fungi and fungal enzymes involved in lignocellulose degradation. *Eng. Life Sci.* **2018**, *18*, 768–778. [CrossRef]
82. Mohammed, L.; Ansari, M.N.; Pua, G.; Jawaid, M.; Islam, M.S. A review on natural fiber reinforced polymer composite and its applications. *Int. J. Polym. Sci.* **2015**, *2015*, 243947. [CrossRef]
83. Komuraiah, A.; Kumar, N.S.; Prasad, B.D. Chemical composition of natural fibers and its influence on their mechanical properties. *Mech. Compos. Mater.* **2014**, *50*, 359–376. [CrossRef]
84. Nurazzi, N.M.; Khalina, A.; Sapuan, S.M.; Ilyas, R.A.; Rafiqah, S.A.; Hanafee, Z.M. Thermal properties of treated sugar palm yarn/glass fiber reinforced unsaturated polyester hybrid composites. *J. Mater. Res. Technol.* **2020**, *9*, 1606–1618. [CrossRef]
85. Rahul, K.C. Chemical Analysis of Spruce Needles. Available online: https://www.theseus.fi/handle/10024/280086 (accessed on 20 May 2021).
86. Brunner, G. Processing of biomass with hydrothermal and supercritical water. In *Supercritical Fluid Science and Technology*; Elsevier: Amsterdam, The Netherlands, 2014; pp. 395–509.
87. Malherbe, S.; Cloete, T.E. Lignocellulose biodegradation: Fundamentals and applications. *Rev. Environ. Sci. Biotechnol.* **2002**, *1*, 105–114. [CrossRef]
88. Mohanty, A.K.; Misra, M.; Hinrichsen, G. Biofibres, biodegradable polymers and biocomposites: An overview. *Macromol. Mater. Eng.* **2000**, *24*, 1–24. [CrossRef]
89. Patel, J.P. and Parsania, P.H. Characterization, testing, and reinforcing materials of biodegradable composites. In *Biodegradable and Biocompatible Polymer Composites*; Elsevier: Amsterdam, The Netherlands, 2018; pp. 55–79.
90. Calvo-Flores, F.G.; Dobado, J.A. Lignin as renewable raw material. *ChemSusChem* **2010**, *3*, 1227–1235. [CrossRef]
91. John, M.J.; Thomas, S. Biofibres and biocomposites. *Carbohydr. Polym.* **2008**, *71*, 343–364. [CrossRef]
92. Liu, Q.; Luo, L.; Zheng, L. Lignins: Biosynthesis and biological functions in plants. *Int. J. Mol. Sci.* **2018**, *19*, 335. [CrossRef]
93. Brebu, M.; Vasile, C. Thermal degradation of lignin - A review. *Cellul. Chem. Technol.* **2010**, *44*, 353–363.
94. Ramamoorthy, S.K.; Skrifvars, M.; Persson, A. A review of natural fibers used in biocomposites: Plant, animal and regenerated cellulose fibers. *Polym. Rev.* **2015**, *55*, 107–162. [CrossRef]
95. Rozman, H.D.; Tan, K.W.; Kumar, R.N.; Abubakar, A.; Mohd. Ishak, Z.A.; Ismail, H. The effect of lignin as a compatibilizer on the physical properties of coconut fiber–polypropylene composites. *Eur. Polym. J.* **2000**, *36*, 1483–1494. [CrossRef]
96. Silva, I.S.; de Menezes, C.R.; Franciscon, E.; dos Santos, E.d.C.; Durrant, L.R. Degradation of lignosulfonic and tannic acids by ligninolytic soil fungi cultivated under icroaerobic conditions. *Brazilian Arch. Biol. Technol.* **2010**, *53*, 693–699. [CrossRef]
97. Nurazzi, N.; Khalina, A.; Sapuan, S.; Laila, A.H.D.; Mohamed, R. Curing behaviour of unsaturated polyester resin and interfacial shear stress of sugar palm fibre. *J. Mech. Eng. Sci.* **2017**, *11*, 2650–2664. [CrossRef]
98. Norizan, M.N.; Abdan, K.; Salit, M.S.; Mohamed, R. Physical, mechanical and thermal properties of sugar palm yarn fibre loading on reinforced unsaturated polyester composites. *J. Phys. Sci.* **2017**, *28*, 115–136. [CrossRef]
99. Haghdan, S.; Renneckar, S.; Smith, G.D. Sources of Lignin. In *Lignin in Polymer Composites*; William Andrew: Norwich, NY, USA, 2016; pp. 1–11. ISBN 9780323355667.
100. Azeez, M.A.; Orege, J.I. Bamboo, Its Chemical Modification and Products. In *Bamboo—Current and Future Prospects*; Abdul Khalil, H.P.S., Ed.; InTech: London, UK, 2018.
101. Ansell, M.P.; Mwaikambo, L.Y. The structure of cotton and other plant fibres. In *Handbook of Textile Fibre Structure*; Elsevier: Amsterdam, The Netherlands, 2009; pp. 62–94.

102. Barreto, A.C.H.; Rosa, D.S.; Fechine, P.B.A.; Mazzetto, S.E. Properties of sisal fibers treated by alkali solution and their application into cardanol-based biocomposites. *Compos. Part A Appl. Sci. Manuf.* **2011**, *42*, 492–500. [CrossRef]
103. Daud, Z.; Mohd Hatta, M.Z.; Mohd Kassi, A.S.; Mohd Aripi, A. Analysis of the Chemical Compositions and Fiber Morphology of Pineapple (Ananas comosus) Leaves in Malaysia. *J. Appl. Sci.* **2014**, *14*, 1355–1358. [CrossRef]
104. Zainudin, E.S.; Yan, L.H.; Haniffah, W.H.; Jawaid, M.; Alothman, O.Y. Effect of coir fiber loading on mechanical and morphological properties of oil palm fibers reinforced polypropylene composites. *Polym. Compos.* **2014**, *35*, 1418–1425. [CrossRef]
105. Agubra, V.; Owuor, P.; Hosur, M. Influence of Nanoclay Dispersion Methods on the Mechanical Behavior of E-Glass/Epoxy Nanocomposites. *Nanomaterials* **2013**, *3*, 550–563. [CrossRef] [PubMed]
106. Li, H.; Huang, X.; Huang, C.; Zhao, Y. An investigation about solid equal channel angular extrusion on polypropylene/organic montmorillonite composite. *J. Appl. Polym. Sci.* **2012**, *123*, 2222–2227. [CrossRef]
107. Creasy, T.S.; Kang, Y.S. Fibre fracture during equal-channel angular extrusion of short fibre-reinforced thermoplastics. *J. Mater. Process. Technol.* **2005**, *160*, 90–98. [CrossRef]
108. Beloshenko, V.A.; Voznyak, A.V.; Voznyak, Y.V.; Novokshonova, L.A.; Grinyov, V.G.; Krasheninnikov, V.G. Processing of Polypropylene-Organic Montmorillonite Nanocomposite by Equal Channel Multiangular Extrusion. *Int. J. Polym. Sci.* **2016**, *2016*, 8564245. [CrossRef]
109. Guo, F.; Aryana, S.; Han, Y.; Jiao, Y. A review of the synthesis and applications of polymer-nanoclay composites. *Appl. Sci.* **2018**, *8*, 1696. [CrossRef]
110. Alexandre, M.; Dubois, P. Polymer-layered silicate nanocomposites: Preparation, properties and uses of a new class of materials. *Mater. Sci. Eng. R Rep.* **2000**, *28*, 1–63. [CrossRef]
111. Ayana, B.; Suin, S.; Khatua, B.B. Highly exfoliated eco-friendly thermoplastic starch (TPS)/poly (lactic acid)(PLA)/clay nanocomposites using unmodified nanoclay. *Carbohydr. Polym.* **2014**, *110*, 430–439. [CrossRef]
112. Tian, H.; Wang, K.; Liu, D.; Yan, J.; Xiang, A.; Rajulu, A.V. Enhanced mechanical and thermal properties of poly (vinyl alcohol)/corn starch blends by nanoclay intercalation. *Int. J. Biol. Macromol.* **2017**, *101*, 314–320. [CrossRef]
113. Trifol, J.; Plackett, D.; Sillard, C.; Szabo, P.; Daugaard, A.E. Hybrid poly (lactic acid)/nanocellulose/nanoclay composites with synergistically enhanced barrier properties and improved thermomechanical resistance. *Polym. Int.* **2016**, *65*, 988–995. [CrossRef]
114. Li, J.; Zhou, M.; Cheng, G.; Cheng, F.; Lin, Y.; Zhu, P.X. Fabrication and characterization of starch-based nanocomposites reinforced with montmorillonite and cellulose nanofibers. *Carbohydr. Polym.* **2019**, *210*, 429–436. [CrossRef]
115. Boonprasith, P.; Wootthikanokkhan, J.; Nimitsiriwat, N. Mechanical, thermal, and barrier properties of nanocomposites based on poly(butylene succinate)/thermoplastic starch blends containing different types of clay. *J. Appl. Polym. Sci.* **2013**, *130*, 1114–1123. [CrossRef]
116. Zahedi, M.; Pirayesh, H.; Khanjanzadeh, H.; Tabar, M.M. Organo-modified montmorillonite reinforced walnut shell/polypropylene composites. *Mater. Des.* **2013**, *51*, 803–809. [CrossRef]
117. Zhao, Y.; Wang, K.; Zhu, F.; Xue, P.; Jia, M. Properties of poly(vinyl chloride)/wood flour/montmorillonite composites: Effects of coupling agents and layered silicate. *Polym. Degrad. Stab.* **2006**, *91*, 2874–2883. [CrossRef]
118. Essabir, H.; Raji, M.; Bouhfid, R.; Qaiss, A.E.K. Nanoclay and Natural Fibers Based Hybrid Composites: Mechanical, Morphological, Thermal and Rheological Properties. In *Nanoclay Reinforced Polymer Composites*; Mohammad, J., Abouel, K.Q., Rachid, B., Eds.; Springer: Berlin/Heidelberg, Germany, 2016; pp. 175–207.
119. Ku, H.; Wang, H.; Pattarachaiyakoop, N.; Trada, M. A review on the tensile properties of natural fiber reinforced polymer composites. *Compos. Part B Eng.* **2011**, *42*, 856–873. [CrossRef]
120. Saba, N.; Tahir, P.M.; Jawaid, M. A review on potentiality of nano filler/natural fiber filled polymer hybrid composites. *Polymers* **2014**, *6*, 2247–2273. [CrossRef]
121. Yousefian, H.; Rodrigue, D. Hybrid Composite Foams Based on Nanoclays and Natural Fibres. In *Nanoclay Reinforced Polymer Composites*; Mohammad, J., Abouel, K.Q., Rachid, B., Eds.; Springer: Berlin/Heidelberg, Germany, 2016; pp. 175–207.
122. Santulli, C. Nanoclay Based Natural Fibre Reinforced Polymer Composites: Mechanical and Thermal Properties. In *Nanoclay Reinforced Polymer Composites*; Mohammad, J., Abouel, K.Q., Rachid, B., Eds.; Springer: Berlin/Heidelberg, Germany, 2016; pp. 175–207.
123. Nisini, E.; Santulli, C.; Liverani, A. Mechanical and impact characterization of hybrid composite laminates with carbon, basalt and flax fibres. *Compos. Part B Eng.* **2017**, *127*, 92–99. [CrossRef]
124. Rozman, H.D.; Rozyanty, A.R.; Musa, L.; Tay, G.S. Ultra-violet radiation-cured biofiber composites from kenaf: The effect of montmorillonite on the flexural and impact properties. *J. Wood Chem. Technol.* **2010**, *30*, 152–163. [CrossRef]
125. Hetzer, M.; De Kee, D. Wood/polymer/nanoclay composites, environmentally friendly sustainable technology: A review. *Chem. Eng. Res. Des.* **2008**, *86*, 1083–1093. [CrossRef]
126. Chen, Z.; Kuang, T.; Yang, Z.; Ren, X. Effect of Nanoclay on Natural Fiber/Polymer Composites. In *Nanoclay Reinforced Polymer Composites*; Mohammad, J., Abouel, K.Q., Rachid, B., Eds.; Springer: Berlin/Heidelberg, Germany, 2016; pp. 175–207.
127. Zahedi, M.; Khanjanzadeh, H.; Pirayesh, H.; Saadatnia, M.A. Utilization of natural montmorillonite modified with dimethyl, dehydrogenated tallow quaternary ammonium salt as reinforcement in almond shell flour-polypropylene bio-nanocomposites. *Compos. Part B Eng.* **2015**, *71*, 143–151. [CrossRef]
128. Petersson L, O.K. Biopolymer based nanocomposites: Comparing layered silicates and microcrystalline cellulose as nanoreinforcement. *Compos. Sci. Technol.* **2006**, *66*, 2187–2196. [CrossRef]

129. Arjmandi, R.; Hassan, A.; Haafiz, M.K.M.; Zakaria, Z. Effect of microcrystalline cellulose on biodegradability, tensile and morphological properties of montmorillonite reinforced polylactic acid nanocomposites. *Fibers Polym.* **2015**, *16*, 2284–2293. [CrossRef]
130. Jalalvandi, E.; Majid, R.A.; Ghanbari, T.; Ilbeygi, H. Effects of montmorillonite (MMT) on morphological, tensile, physical barrier properties and biodegradability of polylactic acid/starch/MMT nanocomposites. *J. Thermoplast. Compos. Mater.* **2015**, *28*, 496–509. [CrossRef]
131. Saba, N.; Paridah, M.T.; Abdan, K.; Ibrahim, N.A. Effect of oil palm nano filler on mechanical and morphological properties of kenaf reinforced epoxy composites. *Constr. Build. Mater.* **2016**, *123*, 15–26. [CrossRef]
132. Thakur, V.K.; Thakur, M.K.; Gupta, R.K. Review: Raw Natural Fiber-Based Polymer Composites. *Int. J. Polym. Anal. Charact.* **2014**, *19*, 256–271. [CrossRef]
133. Islam, M.S.; Ahmad, M.B.; Hasan, M.; Aziz, S.A.; Jawaid, M.; Haafiz, M.K.M.; Zakaria, S.A.H. Natural fiber-reinforced hybrid polymer nanocomposites: Effect of fiber mixing and nanoclay on physical, mechanical, and biodegradable properties. *BioResources* **2015**, *10*, 1394–1407. [CrossRef]
134. Goodarzi, V.; Jafari, S.H.; Khonakdar, H.A.; Seyfi, J. Morphology, rheology and dynamic mechanical properties of PP/EVA/clay nanocomposites. *J. Polym. Res.* **2011**, *18*, 1829–1839. [CrossRef]
135. Mata-Padilla, J.M.; Medellín-Rodríguez, F.J.; Ávila-Orta, C.A.; Ramírez-Vargas, E.; Cadenas-Pliego, G.; Valera-Zaragoza, M.; Vega-Díaz, S.M. Morphology and chain mobility of reactive blend nanocomposites of PP-EVA/Clay. *J. Appl. Polym. Sci.* **2014**, *131*, 1–14. [CrossRef]
136. Castro-Landinez, J.F.; Salcedo-Galan, F.; Medina-Perilla, J.A. Polypropylene/ethylene—and polar—monomer-based copolymers/montmorillonite nanocomposites: Morphology, mechanical properties, and oxygen permeability. *Polymers* **2021**, *13*, 705. [CrossRef]
137. Shikaleska, A.V.; Pavlovska, F.P. Advanced materials based on polymer blends/polymer blend nanocomposites. *IOP Conf. Ser. Mater. Sci. Eng.* **2012**, *40*, 012009. [CrossRef]
138. Sengwa, R.J.; Choudhary, S. Structural characterization of hydrophilic polymer blends/montmorillonite clay nanocomposites. *J. Appl. Polym. Sci.* **2014**, *131*, 1–11. [CrossRef]
139. Ramesh, P.; Prasad, B.D.; Narayana, K.L. Influence of Montmorillonite Clay Content on Thermal, Mechanical, Water Absorption and Biodegradability Properties of Treated Kenaf Fiber/ PLA-Hybrid Biocomposites. *Silicon* **2021**, *13*, 109–118. [CrossRef]
140. Gwon, J.G.; Lee, S.Y.; Chun, S.J.; Doh, G.H.; Kim, J.H. Physical and mechanical properties of wood–plastic composites hybridized with inorganic fillers. *J. Compos. Mater.* **2012**, *46*, 301–309. [CrossRef]
141. Khare, A.; Deshmukh, S. Studies Toward Producing Eco-Friendly Plastics. *J. Plast. Film Sheeting* **2006**, *22*, 193–211. [CrossRef]
142. Tajeddin, B.; Rahman, R.A.; Abdulah, L.C. The effect of polyethylene glycol on the characteristics of kenaf cellulose/low-density polyethylene biocomposites. *Int. J. Biol. Macromol.* **2010**, *47*, 292–297. [CrossRef]
143. George, J.; Kumar, R.; Jayaprakash, C.; Ramakrishna, A.; Sabapathy, S.N.; Bawa, A.S. Rice bran-filled biodegradable low-density polyethylene films: Development and characterization for packaging applications. *J. Appl. Polym. Sci.* **2006**, *102*, 4514–4522. [CrossRef]
144. Kalia, S.; Kaith, B.S.; Kaur, I. Pretreatments of Natural Fibers and their Application as Reinforcing Material in Polymer Composites — A Review. *Polym. Eng. Sci.* **2009**, *49*, 1253–1272. [CrossRef]
145. Majeed, K.; Jawaid, M.; Hassan, A.; Abu Bakar, A.; Abdul Khalil, H.P.S.; Salema, A.A.; Inuwa, I. Potential materials for food packaging from nanoclay/natural fibres filled hybrid composites. *Mater. Des.* **2013**, *46*, 391–410. [CrossRef]
146. Chee, S.S.; Jawaid, M.; Alothman, O.Y.; Fouad, H. Effects of nanoclay on mechanical and dynamic mechanical properties of bamboo/kenaf reinforced epoxy hybrid composites. *Polymers* **2021**, *13*, 395. [CrossRef]
147. Wei, J.; Wang, B.; Li, Z.; Wu, Z.; Zhang, M.; Sheng, N.; Liang, Q.; Wang, H.; Chen, S. A 3D-printable TEMPO-oxidized bacterial cellulose/alginate hydrogel with enhanced stability via nanoclay incorporation. *Carbohydr. Polym.* **2020**, *238*, 116207. [CrossRef]
148. Mohan, T.P.; Kanny, K. Compressive characteristics of unmodified and nanoclay treated banana fiber reinforced epoxy composite cylinders. *Compos. Part B Eng.* **2019**, *169*, 118–125. [CrossRef]
149. Chee, S.S.; Jawaid, M.; Sultan, M.T.H.; Alothman, O.Y.; Abdullah, L.C. Thermomechanical and dynamic mechanical properties of bamboo/woven kenaf mat reinforced epoxy hybrid composites. *Compos. Part B Eng.* **2019**, *163*, 165–174. [CrossRef]
150. Anand, G.; Alagumurthi, N.; Elansezhian, R.; Venkateshwaran, N. Dynamic mechanical, thermal and wear analysis of Ni-P coated glass fiber/Al_2O_3 nanowire reinforced vinyl ester composite. *Alexandria Eng. J.* **2018**, *57*, 621–631. [CrossRef]
151. Joseph, S.; Appukuttan, S.P.; Kenny, J.M.; Puglia, D.; Thomas, S.; Joseph, K. Dynamic mechanical properties of oil palm microfibril-reinforced natural rubber composites. *J. Appl. Polym. Sci.* **2010**, *117*, 1298–1308. [CrossRef]
152. Majeed, K.; Hassan, A.; Bakar, A.A. Influence of maleic anhydride-grafted polyethylene compatibiliser on the tensile, oxygen barrier and thermal properties of rice husk and nanoclay-filled low-density polyethylene composite films. *J. Plast. Film Sheeting* **2014**, *30*, 120–140. [CrossRef]
153. Ali Dadfar, S.M.; Alemzadeh, I.; Reza Dadfar, S.M.; Vosoughi, M. Studies on the oxygen barrier and mechanical properties of low density polyethylene/organoclay nanocomposite films in the presence of ethylene vinyl acetate copolymer as a new type of compatibilizer. *Mater. Des.* **2011**, *32*, 1806–1813. [CrossRef]
154. Arulmurugan, S.; Venkateshwaran, N. Effect of nanoclay addition and chemical treatment on static and dynamic mechanical analysis of jute fibre composites. *Polímeros* **2019**, *29*, 4–11. [CrossRef]

155. Palanivel, A.; Veerabathiran, A.; Duruvasalu, R.; Iyyanar, S.; Velumayil, R. Dynamic mechanical analysis and crystalline analysis of hemp fiber reinforced cellulose filled epoxy composite. *Polímeros* **2017**, *27*, 309–319. [CrossRef]
156. Jesuarockiam, N.; Jawaid, M.; Zainudin, E.S.; Hameed Sultan, M.T.; Yahaya, R. Enhanced thermal and dynamic mechanical properties of synthetic/natural hybrid composites with graphene nanoplateletes. *Polymers* **2019**, *11*, 1085. [CrossRef] [PubMed]
157. Yussuf, A.A.; Massoumi, I.; Hassan, A. Comparison of Polylactic Acid/Kenaf and Polylactic Acid/Rise Husk Composites: The Influence of the Natural Fibers on the Mechanical, Thermal and Biodegradability Properties. *J. Polym. Environ.* **2010**, *18*, 422–429. [CrossRef]
158. Ilyas, R.A.; Sapuan, S.M.; Ibrahim, R.; Abral, H.; Ishak, M.R.; Zainudin, E.S.; Asrofi, M.; Atikah, M.S.N.; Huzaifah, M.R.M.; Radzi, A.M.; et al. Sugar palm (*Arenga pinnata* (Wurmb.) Merr) cellulosic fibre hierarchy: A comprehensive approach from macro to nano scale. *J. Mater. Res. Technol.* **2019**, *8*, 2753–2766. [CrossRef]
159. Ilyas, R.A.; Sapuan, S.M.; Ishak, M.R. Isolation and characterization of nanocrystalline cellulose from sugar palm fibres (*Arenga pinnata*). *Carbohydr. Polym.* **2018**, *181*, 1038–1051. [CrossRef]
160. Ilyas, R.A.; Sapuan, S.M.; Ishak, M.R.; Zainudin, E.S. Sugar palm nanofibrillated cellulose (*Arenga pinnata* (Wurmb.) Merr): Effect of cycles on their yield, physic-chemical, morphological and thermal behavior. *Int. J. Biol. Macromol.* **2019**, *123*, 379–388. [CrossRef]
161. Ilyas, R.A.; Sapuan, S.M.; Ishak, M.R.; Zainudin, E.S. Effect of delignification on the physical, thermal, chemical, and structural properties of sugar palm fibre. *BioResources* **2017**, *12*, 8734–8754. [CrossRef]
162. Ayu, R.S.; Khalina, A.; Harmaen, A.S.; Zaman, K.; Isma, T.; Liu, Q.; Ilyas, R.A.; Lee, C.H. Characterization Study of Empty Fruit Bunch (EFB) Fibers Reinforcement in Poly(Butylene) Succinate (PBS)/Starch/Glycerol Composite Sheet. *Polymers* **2020**, *12*, 1571. [CrossRef]
163. Syafri, E.; Sudirman; Mashadi; Yulianti, E.; Deswita; Asrofi, M.; Abral, H.; Sapuan, S.M.; Ilyas, R.A.; Fudholi, A. Effect of sonication time on the thermal stability, moisture absorption, and biodegradation of water hyacinth (Eichhornia crassipes) nanocellulose-filled bengkuang (Pachyrhizus erosus) starch biocomposites. *J. Mater. Res. Technol.* **2019**, *8*, 6223–6231. [CrossRef]
164. Ismail, H.; Pasbakhsh, P.; Fauzi, M.N.A.; Abu Bakar, A. Morphological, thermal and tensile properties of halloysite nanotubes filled ethylene propylene diene monomer (EPDM) nanocomposites. *Polym. Test.* **2008**, *27*, 841–850. [CrossRef]
165. Sajna, V.; Mohanty, S.; Nayak, S.K. Hybrid green nanocomposites of poly(lactic acid) reinforced with banana fibre and nanoclay. *J. Reinf. Plast. Compos.* **2014**, *33*, 1717–1732. [CrossRef]
166. Huda, M.S.; Drzal, L.T.; Mohanty, A.K.; Misra, M. Effect of fiber surface-treatments on the properties of laminated biocomposites from poly(lactic acid) (PLA) and kenaf fibers. *Compos. Sci. Technol.* **2008**, *68*, 424–432. [CrossRef]
167. Du, Y.; Wu, T.; Yan, N.; Kortschot, M.T.; Farnood, R. Fabrication and characterization of fully biodegradable natural fiber-reinforced poly(lactic acid) composites. *Compos. Part B Eng.* **2014**, *56*, 717–723. [CrossRef]
168. Yuzay, I.E.; Auras, R.; Soto-Valdez, H.; Selke, S. Effects of synthetic and natural zeolites on morphology and thermal degradation of poly(lactic acid) composites. *Polym. Degrad. Stab.* **2010**, *95*, 1769–1777. [CrossRef]
169. Mróz, P.; Białas, S.; Mucha, M.; Kaczmarek, H. Thermogravimetric and DSC testing of poly(lactic acid) nanocomposites. *Thermochim. Acta* **2013**, *573*, 186–192. [CrossRef]
170. Pratheep Kumar, A.; Pal Singh, R. Novel hybrid of clay, cellulose, and thermoplastics. I. Preparation and characterization of composites of ethylene–propylene copolymer. *J. Appl. Polym. Sci.* **2007**, *104*, 2672–2682. [CrossRef]
171. Blumstein, A. Polymerization of adsorbed monolayers. II. Thermal degradation of the inserted polymer. *J. Polym. Sci. Part A Gen. Pap.* **1965**, *3*, 2665–2672. [CrossRef]
172. Burnside, S.D.; Giannelis, E.P. Synthesis and properties of new poly(dimethylsiloxane) nanocomposites. *Chem. Mater.* **1995**, *7*, 1597–1600. [CrossRef]
173. Tyan, H.-L.; Liu, Y.-C.; Wei, K.-H. Thermally and Mechanically Enhanced Clay/Polyimide Nanocomposite via Reactive Organoclay. *Chem. Mater.* **1999**, *11*, 1942–1947. [CrossRef]
174. Amash, A. Morphology and properties of isotropic and oriented samples of cellulose fibre-polypropylene composites. *Polymer* **2000**, *41*, 1589–1596. [CrossRef]
175. Bharadwaj, R.K.; Mehrabi, A.R.; Hamilton, C.; Trujillo, C.; Murga, M.; Fan, R.; Chavira, A.; Thompson, A.K. Structure–property relationships in cross-linked polyester–clay nanocomposites. *Polymer* **2002**, *43*, 3699–3705. [CrossRef]
176. Tidjani, A.; Wald, O.; Pohl, M.-M.; Hentschel, M.P.; Schartel, B. Polypropylene–graft–maleic anhydride-nanocomposites: I—Characterization and thermal stability of nanocomposites produced under nitrogen and in air. *Polym. Degrad. Stab.* **2003**, *82*, 133–140. [CrossRef]
177. Ohkita, T.; Lee, S.-H. Thermal degradation and biodegradability of poly (lactic acid)/corn starch biocomposites. *J. Appl. Polym. Sci.* **2006**, *100*, 3009–3017. [CrossRef]
178. Jumaidin, R.; Diah, N.A.; Ilyas, R.A.; Alamjuri, R.H.; Yusof, F.A.M. Processing and Characterisation of Banana Leaf Fibre Reinforced Thermoplastic Cassava Starch Composites. *Polymers* **2021**, *13*, 1420. [CrossRef]
179. Jumaidin, R.; Khiruddin, M.A.A.; Asyul Sutan Saidi, Z.; Salit, M.S.; Ilyas, R.A. Effect of cogon grass fibre on the thermal, mechanical and biodegradation properties of thermoplastic cassava starch biocomposite. *Int. J. Biol. Macromol.* **2020**, *146*, 746–755. [CrossRef]
180. Threepopnatkul, P.; Kaerkitcha, N.; Athipongarporn, N. Effect of surface treatment on performance of pineapple leaf fiber–polycarbonate composites. *Compos. Part B Eng.* **2009**, *40*, 628–632. [CrossRef]

181. Mofokeng, J.P.; Luyt, A.S.; Tábi, T.; Kovács, J. Comparison of injection moulded, natural fibre-reinforced composites with PP and PLA as matrices. *J. Thermoplast. Compos. Mater.* **2012**, *25*, 927–948. [CrossRef]
182. Ahmed, A.; Hidayat, S.; Abu Bakar, M.S.; Azad, A.K.; Sukri, R.S.; Phusunti, N. Thermochemical characterisation of Acacia auriculiformis tree parts via proximate, ultimate, TGA, DTG, calorific value and FTIR spectroscopy analyses to evaluate their potential as a biofuel resource. *Biofuels* **2021**, *12*, 9–20. [CrossRef]
183. Majeed, K.; Hassan, A.; Bakar, A.A.; Jawaid, M. Effect of montmorillonite (MMT) content on the mechanical, oxygen barrier, and thermal properties of rice husk/MMT hybrid filler-filled low-density polyethylene nanocomposite blown films. *J. Thermoplast. Compos. Mater.* **2016**, *29*, 1003–1019. [CrossRef]
184. Bertini, F.; Canetti, M.; Audisio, G.; Costa, G.; Falqui, L. Characterization and thermal degradation of polypropylene–montmorillonite nanocomposites. *Polym. Degrad. Stab.* **2006**, *91*, 600–605. [CrossRef]
185. Leszczyńska, A.; Njuguna, J.; Pielichowski, K.; Banerjee, J.R. Polymer/montmorillonite nanocomposites with improved thermal properties. Part, I. Factors influencing thermal stability and mechanisms of thermal stability improvement. *Thermochim. Acta* **2007**, *453*, 75–96. [CrossRef]
186. Chow, W.S.; Mohd Ishak, Z.A.; Karger-Kocsis, J.; Apostolov, A.A.; Ishiaku, U.S. Compatibilizing effect of maleated polypropylene on the mechanical properties and morphology of injection molded polyamide 6/polypropylene/organoclay nanocomposites. *Polymer* **2003**, *44*, 7427–7440. [CrossRef]
187. Zaman, H.U.; Hun, P.D.; Khan, R.A.; Yoon, K.-B. Polypropylene/clay nanocomposites. *J. Thermoplast. Compos. Mater.* **2014**, *27*, 338–349. [CrossRef]
188. Rajini, N.; Winowlin Jappes, J.T.; Rajakarunakaran, C.; Siva, I. Tensile and Flexural Properties of MMT-Clay/ Unsaturated Polyester Using Robust Design Concept. *Nano Hybrids* **2012**, *2*, 87–101. [CrossRef]
189. Sienkiewicz, A.; Czub, P. Flame retardancy of biobased composites—research development. *Materials* **2020**, *13*, 5253. [CrossRef]
190. Suwanniroj, A.; Suppakarn, N. Enhancement of flame retardancy and mechanical properties of poly(butylene succinate) composites by adding hybrid fillers. *AIP Conf. Proc.* **2020**, *2279*, 070004. [CrossRef]
191. Subasinghe, A.; Das, R.; Bhattacharyya, D. Study of thermal, flammability and mechanical properties of intumescent flame retardant PP/kenaf nanocomposites. *Int. J. Smart Nano Mater.* **2016**, *7*, 202–220. [CrossRef]
192. Chee, S.S.; Jawaid, M.; Alothman, O.Y.; Yahaya, R. Thermo-oxidative stability and flammability properties of bamboo/kenaf/nanoclay/epoxy hybrid nanocomposites. *RSC Adv.* **2020**, *10*, 21686–21697. [CrossRef]
193. Dutta, N.; Maji, T.K. Synergic effect of montmorillonite and microcrystalline cellulose on the physicochemical properties of rice husk/PVC composite. *SN Appl. Sci.* **2020**, *2*, 439. [CrossRef]
194. Wang, Y.Y.; Shih, Y.F. Flame-retardant recycled bamboo chopstick fiber-reinforced poly(lactic acid) green composites via multifunctional additive system. *J. Taiwan Inst. Chem. Eng.* **2016**, *65*, 452–458. [CrossRef]
195. Mandal, M.; Halim, Z.; Maji, T.K. Mechanical, moisture absorption, biodegradation and physical properties of nanoclay-reinforced wood/plant oil composites. *SN Appl. Sci.* **2020**, *2*, 250. [CrossRef]
196. Glaskova-Kuzmina, T.; Starkova, O.; Gaidukovs, S.; Platnieks, O.; Gaidukova, G. Durability of Biodegradable Polymer Nanocomposites. *Polymers* **2021**, *13*, 3375. [CrossRef]
197. Islam, M.S.; Hasbullah, N.A.B.; Hasan, M.; Talib, Z.A.; Jawaid, M.; Haafiz, M.K.M. Physical, mechanical and biodegradable properties of kenaf/coir hybrid fiber reinforced polymer nanocomposites. *Mater. Today Commun.* **2015**, *4*, 69–76. [CrossRef]
198. Kumar, S.; Krishnan, S.; Samal, S.K. Recent Developments of Epoxy Nanocomposites Used for Aerospace and Automotive Application. In *Diverse Applications of Organic-Inorganic Nanocomposites: Emerging Research and Opportunities*; Clarizia, G., Bernardo, P., Eds.; IGI Global: Pittsburgh, PA, USA, 2020; pp. 162–190.
199. Galimberti, M.; Cipolletti, V.R.; Coombs, M. Applications of clay–polymer nanocomposites. In *Developments in Clay Science*; Elsevier: Amsterdam, The Netherlands, 2013; pp. 539–586.
200. Lyon, R.E.; Walters, R.N. Flammability of automotive plastics. *SAE Tech. Pap.* **2006**. [CrossRef]
201. Jiang, J.; Michaels, H.; Vlachopoulos, N.; Freitag, M. Beyond the Limitations of Dye-Sensitized Solar Cells. In *Dye-Sensitized Solar Cells: Mathematical Modeling, and Materials Design and Optimization*; Soroush, M., Lau, K.K.S., Eds.; Academic Press: Cambridge, UK, 2019; pp. 285–323.
202. Beg, M.D.H.; Pickering, K.L. Accelerated weathering of unbleached and bleached Kraft wood fibre reinforced polypropylene composites. *Polym. Degrad. Stab.* **2008**, *93*, 1939–1946. [CrossRef]
203. Searle, N.D.; McGreer, M.; Zielnik, A. Weathering of Polymeric Materials. In *Encyclopedia of Polymer Science and Technology*; Matyjaszewski, K., Ed.; John Wiley & Sons, Inc.: Hoboken, NJ, USA, 2010.
204. Brown, R.P.; Kockott, D.; Trubiroha, P.; Ketola, W.; Shorthouse, J. *A Review of Accelerated Durability Tests*; National Physical Laboratory Teddington: Middlesex, UK, 1995.
205. Raji, M.; Zari, N.; Bouhfid, R.; el kacem Qaiss, A. Durability of composite materials during hydrothermal and environmental aging. In *Durability and Life Prediction in Biocomposites, Fibre-Reinforced Composites and Hybrid Composites*; Jawaid, M., Thariq, M., Saba, N., Eds.; Woodhead Publishing Series: Cambridge, UK, 2019; pp. 83–119.
206. Kord, B.; Malekian, B.; Ayrilmis, N. Weathering Performance of Montmorillonite/Wood Flour-Based Polypropylene Nanocomposites. *Mech. Compos. Mater.* **2017**, *53*, 271–278. [CrossRef]
207. Eshraghi, A.; Khademieslam, H.; Ghasemi, I.; Talaiepoor, M. Effect of weathering on the properties of hybrid composite based on polyethylene, woodflour, and nanoclay. *BioResources* **2013**, *8*, 201–210. [CrossRef]

208. High Efficiency Plants and Building Integrated Renewable Energy Systems. In *Handbook of Energy Efficiency in Buildings: A Life Cycle Approach*; Asdrubali, F.; Desideri, U. (Eds.) Butterworth-Heinemann: Oxford, UK, 2019; pp. 441–595.
209. Nguyen, Q.; Ngo, T.; Mendis, P.; Tran, P. Composite Materials for Next Generation Building Façade Systems. *Civ. Eng. Archit.* **2013**, *1*, 88–95. [CrossRef]
210. Nguyen, Q.T.; Tran, P.; Ngo, T.D.; Tran, P.A.; Mendis, P. Experimental and computational investigations on fire resistance of GFRP composite for building façade. *Compos. Part B Eng.* **2014**, *62*, 218–229. [CrossRef]
211. Rajini, N.; Winowlin Jappes, J.T.; Siva, I.; Varada Rajulu, A.; Rajakarunakaran, S. Fire and thermal resistance properties of chemically treated ligno-cellulosic coconut fabric–reinforced polymer eco-nanocomposites. *J. Ind. Text.* **2017**, *47*, 104–124. [CrossRef]
212. Chowdhury, S.R. Some important aspects in designing high molecular weight poly(L-lactic acid)-clay nanocomposites with desired properties. *Polym. Int.* **2008**, *57*, 1326–1332. [CrossRef]
213. Mohan, T.P.; Kanny, K. Sorption studies of few selected raw and nanoclay infused lignocellulosic fibres. *Indian J. Fibre Text. Res.* **2019**, *44*, 263–270.
214. Massaro, M.; Cavallaro, G.; Lazzara, G.; Riela, S. Covalently modified nanoclays: Synthesis, properties and applications. In *Clay Nanoparticles: Properties and Applications*; Cavallaro, G., Fakhrullin, R., Pasbakhsh, P., Eds.; Elsevier: Amsterdam, The Netherlands, 2020; pp. 305–333.
215. Maisanaba, S.; Hercog, K.; Filipic, M.; Jos, Á.; Zegura, B. Genotoxic potential of montmorillonite clay mineral and alteration in the expression of genes involved in toxicity mechanisms in the human hepatoma cell line HepG2. *J. Hazard. Mater.* **2016**, *304*, 425–433. [CrossRef]
216. Shafique, M.; Luo, X. Nanotechnology in Transportation Vehicles: An Overview of Its Applications, Environmental, Health and Safety Concerns. *Materials* **2019**, *12*, 2493. [CrossRef]

Article

Influence of the Compression Molding Temperature on VOCs and Odors Produced from Natural Fiber Composite Materials

Benjamin Barthod-Malat [1,2], Maxime Hauguel [2], Karim Behlouli [2], Michel Grisel [1] and Géraldine Savary [1,*]

1. URCOM, Université Le Havre Normandie, 76600 Le Havre, France
2. Ecotechnilin SAS-ZA Caux Multipôles, RD 6015, 76190 Valliquerville, France
* Correspondence: geraldine.savary@univ-lehavre.fr

Abstract: In the automotive sector, the use of nonwoven preforms consisting of natural and thermoplastic fibers processed by compression molding is well known to manufacture vehicle interior parts. Although these natural fiber composites (NFCs) have undeniable advantages (lightweight, good life cycle assessment, recyclability, etc.), the latter release volatile organic compounds (VOCs) and odors inside the vehicle interior, which remain obstacles to their wide deployment. In this study, the effect of the compressing molding temperature on the VOCs and odors released by the flax/PP nonwoven composites was examined by heating nonwoven preforms in a temperature range up to 240 °C. During the hot-pressing process, real-time and in situ monitoring of the composite materials' core temperature has been carried out using a thermocouples sensor. A chemical approach based on headspace solid-phase microextraction (HS-SPME) coupled with gas chromatography—mass spectrometry (GC-MS) was used for the VOCs analysis. The olfactory approach is based on the odor intensity scale rated by expert panelists trained in olfaction. The results demonstrate marked changes in the VOCs composition with temperature, thus making it possible to understand the changes in the NFCs odor intensity. The results allow for optimizing the molding temperature to obtain less odorous NFC materials.

Keywords: Natural Fiber Composite; flax; polypropylene; nonwoven; HS-SPME-GC-MS method; VOCs; odors; intensity; thermocouple; in situ monitoring

Citation: Barthod-Malat, B.; Hauguel, M.; Behlouli, K.; Grisel, M.; Savary, G. Influence of the Compression Molding Temperature on VOCs and Odors Produced from Natural Fiber Composite Materials. *Coatings* **2023**, *13*, 371. https://doi.org/10.3390/coatings13020371

Academic Editors: Yong Gan and Jinyang Xu

Received: 30 December 2022
Revised: 23 January 2023
Accepted: 30 January 2023
Published: 6 February 2023

Copyright: © 2023 by the authors. Licensee MDPI, Basel, Switzerland. This article is an open access article distributed under the terms and conditions of the Creative Commons Attribution (CC BY) license (https://creativecommons.org/licenses/by/4.0/).

1. Introduction

Facing the challenges of climate change [1], the car manufacturers are forced by the European Union (EU) to reduce CO_2 emissions by 55% among new cars placed on the market by 2030 and by 100% by 2035 [2]. In the EU, cars represent 12% of total CO_2 emissions [3]. Consequently, manufacturers will have to drastically reduce the CO_2 emissions of their new thermal vehicles and quickly switch to electric vehicles. In order to reduce the carbon footprint both in manufacturing and in use, switching from thermal to electric motorization is not enough, so it is also necessary to lighten the vehicles [4–6]. Beyond making less powerful vehicles [7], car manufacturers actually have two action levers for reducing vehicle weight. The first is by compacting the vehicle as much as possible in order to reduce the consumption of materials. The second is by manufacturing lighter parts by optimizing and reducing the thickness, surface and density of its components notably by using lighter materials. Usually, vehicle interior parts are made from polymer or glass fibers and polypropylene nonwoven composites. Many original equipment manufacturers (OEMs) and researchers seek to substitute glass fibers with natural fibers [8,9]. Such a substitution is initially motivated by the low density (around 40% lower than that of glass fibers) and the interesting specific mechanical properties for a relatively competitive price [10] compared to glass or synthetic fibers. Then, the use of natural fibers is advantageous because of their availability, their accessibility and their stable supply chain thanks to a fully structured sector, particularly for flax and hemp [11]. Finally, their renewability and low environmental impact are considered overall. In a life cycle assessment, the climate change indicator of hackled flax fibers production is lower than that of the glass fibers, and it is also established to be beneficial to acidification, abiotic depletion, terrestrial ecotoxicity, ozone layer depletion, human toxicity as well as non-renewable energy consumption indicators [12]. With their low density, good specific strength and stiffness, low carbon footprint [13,14], good acoustic properties [15–17], proven recyclability [18,19] and reasonable cost of raw materials [20], the use of natural fiber (NF) and polypropylene (PP) nonwoven composites as lightweight materials for vehicle interior parts is increasingly widespread in the automotive sector [21]. There are two types of polymers: thermo-plastic and thermosetting polymers. Thermo-plastic polymers are largely used for car interior parts due to their processability and their recyclability. In the car interiors, the main parts made from NF-PP nonwoven composites are: headliners, parcel shelfs, door panels, dashboards and trunk floors. For a door panel made from NFPP nonwoven composites, a weight gain of 20% can be achieved compared to a petro-sourced door panel with iso-mechanical performance [14].

Although these NFPP nonwoven composites have undeniable advantages in being used in vehicle interiors, the latter release volatile organic compounds (VOCs) that are potentially odorous and can affect the vehicle interior air quality (VIAQ). The World Health Organization (WHO) has indicated that vehicle interiors are a potential threat to people's health [22], as some vehicle interiors may contain up to 250 different VOCs. Among these VOCs, some can be harmful, causing serious pathologies during long and/or regular exposure in confined spaces and/or at high concentrations [23,24]. A variety of symptoms may occur, including, for example, fatigue, headaches or nasal, ocular and respiratory irritations. Moreover, air quality is also affected by unpleasant and/or intense odors, causing nausea, insomnia, discomfort and asthma [25]. The main sources of VOCs/odors are plastics, textiles, leathers, coatings and glues used to make vehicle interior parts. In the literature, some studies have shown that the lignocellulosic fibers used in NFPP composites are sources of both VOCs and odors. These VOCs/odors depend on the pretreatment of lignocellulosic fibers. Fischer et al. [26] established that the odor concentration of lignocellulosic fibers is related to the fineness of the fibers. They also showed that raw fibers contaminated with molds exhibit a higher odor concentration compared to uncontaminated ones. Finally, they demonstrated that the wide range of textile finishes greatly increases the odor concentration of the fibers. Moreover, Savary et al. [27] showed that the chemical composition and the olfactory profile of flax fibers change as a function of temperature. At 80 °C, the olfactory profile is characterized by fatty, green, acid, flowery and sweet notes, and beyond 215 and 230 °C, odors of the roasted, burnt and phenolic type appear, which can be perceived as unpleasant. Other researchers and original equipment manufacturers (OEMs) have been interested in reducing the emission of VOCs/odors emitted by NFPP composites based on lignocellulosic fibers by using different VOCs/odors removal techniques.

The easier VOCs/odors removal technique consists of decreasing the number of lignocellulosic fibers in composites, often to the detriment of the mechanical properties. Bledzki et al. [28] showed that the higher the abaca fiber content, the higher the odor concentration emitted by PP-abaca fiber composites. Nevertheless, the identification of the VOCs sources for the corresponding odors was not carried out. Badji et al. [24] showed that the higher the hemp fiber content in a composite, the greater the quantity of VOCs— in particular, VOCs from the alcohols, carboxylic acids, aldehydes, ketones and azines families. Unfortunately, no odor analysis was carried out to establish a link between VOCs and odors. In order to less significantly deteriorate or increase the mechanical properties of the composites, the second VOCs/odors removal technique consists of substituting lignocellulosic fibers with similar fibers emitting less VOCs/odors. Bledzki et al. [28] also showed that substituting abaca fibers in composites (made by injection molding) with jute or flax fibers allows for a much lower odor concentration. Fischer et al. [27] have also shown this in the case of composites made by the compression molding of nonwovens. Indeed, they showed that substituting contaminated lignocellulosic fibers with non-contaminated ones (or with contaminated fibers treated with soda and enzymatic treatment) in NFPP composites can reduce the odor concentration, whether for composites made in a laboratory or on an industrial scale. Fischer et al. [26] also showed that substituting contaminated lignocellulosic fibers with the same fibers treated with soda and enzymatic treatment allows for reducing the odor concentration threefold. Again, for composites made by compression molding, Li et al. [29] established that substituting hemp fiber with natural freezing-mechanical degumming hemp fiber decreases the VOCs monitored in vehicle interiors in China. Morin et al. [30] showed that substituting flax fibers with flax fibers pre-treated with deep eutectic solvent (DES) followed by ultrasound treatment efficiently allowed for reducing the odor intensity, as established through olfactory quality evaluation, as assessed via a panel following the D42 3109-C standard; in the same study, the authors were able to demonstrate a decrease in the number of VOCs. Nevertheless, in most of these studies [26,28–30], the VOCs causing the odor were not identified. The third VOCs/odors removal technique is to add VOCs/odors adsorbents. As an illustration, Courgneau et al. [31] prepared a low-odor-emissive cellulose fibers/PLA composite with PMPS as an adsorbent agent. The results indicate that the odor concentration decreases twofold by adding the adsorbents, while the odor intensity is only very slightly decreased. Again, no study of VOCs has been carried out in parallel to identify odorous VOCs. Additionally, Kim et al. [32] showed that porous inorganic materials were able to reduce the amount of Furfural, 5-methyl Furfural and hexanal emitted by composite materials based on PBS and PLA reinforced with bamboo flour or wood floor. Contrary to the previous study, the odor emitted by these composites was not investigated. The fourth VOCs/odors removal technique consists of changing the process of shaping the composite materials based on lignocellulosic fibers. This remediation strategy has the advantage of not requiring changing the constituents of the composite or adding any compounds such as odor adsorbents, unlike the previous removal techniques. This is a preventive method for limiting the formation of odorous or non-odorous VOCs. Faruk et al. [33] investigated the odor concentration of PP-abaca fiber composites as a function of three shaping processes. The study shows that the odor concentration emitted by PP-abaca fiber composites shaped by compression molding is significantly lower than that in the case of injection molding. Fischer et al. [26] showed that, in moving from a laboratory scale to an industrial scale, the hot-pressing process reduced the odor concentration of NFPP nonwoven composites. The laboratory scale thermocompression parameters have been defined (15 min and 180 °C). but those used on the industrial scale have not, which are probably at higher temperatures (around 200–220 °C) and a shorter time (no more than a few minutes) to ensure a good production rate. One can observe that only very few studies are interested in both odors and VOCs emitted by NFPP composite materials and the potential link that could unite them. Moreover, although studies have shown that the odor concentration or odor intensity of NFPP composite materials could decrease by using different VOCs/odors removal techniques, no one was interested in optimizing the shaping process of NFPP composites, especially from NFPP nonwovens.

In this study, the effect of the compressing molding temperature on both the VOCs and odors released by the NFPP nonwoven composites was examined by heating nonwoven preforms from 200 °C to 240 °C. This temperature range corresponds to the temperatures of the industrial shaping processes for fiber-reinforced composite materials. The heat press time has been defined according to the time needed to reach very specific temperatures at the core of the composite during the hot pressing. The first core temperature, 180 °C, corresponds to a temperature above the melting point (163–170 °C) of the polypropylene fibers (PP), allowing it to have the necessary minimum viscosity for the PP to coat and correctly bind the natural fibers. The second core temperature, 200 °C, is a temperature well above the melting point of PP fibers, ensuring (for sure) the necessary viscosity to produce composite materials corresponding to automotive specifications. Further core temperatures correspond to one of a hot plate whose temperature is between 200 and 240 °C. In each configuration, the heat press time is similar to the time of the industrial shaping processes;

therefore, the effect of the exposure temperature on the NFC materials could be evaluated. In each configuration, the VOCs were identified and quantified, and the odor intensity was assessed using a panel of expert olfaction assessors. Finally, a link was established between the odor intensity and the VOCs emitted by the composites for all configurations.

2. Materials and Methods

2.1. Materials

Flax (*Linum usitatissimum* L.) tows and scutched fiber from France were used in this study. The flax was cultivated under normal weather conditions and then dew-retted in fields and, finally, mechanically scutched. Flax fibers were used as reinforcement in the nonwoven material combined with PP fibers. Industrial flax/PP commingled nonwovens were manufactured according to the carding-overlapping-needle-punching technology of Ecotechnilin SAS (Valliquerville, France). For the nonwovens, the flax/PP ratio is about 50/50 w%, with a mass per unit area around 1500 g/m^2.

2.2. Methods

2.2.1. Composite Manufacturing

Flax/PP nonwovens were processed by hot-compression molding on a Dolouets (Soustons, France) laboratory hydraulic press, with a system of double plates, to obtain 21 × 30 cm^2 and 7 × 7 cm^2 plates. This press allows for a pressing of the material in two successive cycles—the first at a high temperature (up to 250 °C) and the second at a low temperature (room temperature). The 7 × 7 cm^2 Flax/PP nonwovens were used to determine the time required to obtain the core temperatures of the different configurations given in Table 1. Once this time was determined, 21 × 30 cm^2 Flax/PP nonwovens were thermocompressed under the same conditions, without a thermocouple. The sample size is larger in order to have enough samples to perform further analysis. In order to be consistent with industrial molding cycles and obtain mechanical properties in accordance with the specifications of car interior parts, nonwovens were hot pressed at 200, 210, 220, 230 and 240 °C under the pressure of 100 bars, with Teflon™ fabric on each side, using a set of shims whose thickness is 3 mm. The hot pressing is carried out until well-defined temperatures are obtained at the core of the composite materials. The temperature is measured using a type K thermocouple whose characteristics are described in Section 2.2.2. The target core temperatures for nonwovens, hot pressed at 200, 210, 220, 230 or 240 °C, are 180 °C, 200 °C and that at which the temperature at the core of the composite material is equal to the temperature of the heating plates of the hot press compression molding. The temperature parameters and the different configurations are presented in Table 1. Then, the hot-pressed nonwovens were placed directly between two cooled plates at room temperature under the same pressure for 40s, with a set of shims whose thickness is 2 mm.

Table 1. Temperature parameters and the different configurations of the natural fiber composite (NFC) manufactured.

Configuration	Molding Temperature (°C)	Core Temperature (°C)
NFC 200-180	200	180
NFC 210-180	210	180
NFC 220-180	220	180
NFC 230-180	230	180
NFC 240-180	240	180
NFC 200-200	200	200
NFC 210-200	210	200
NFC 220-200	220	200
NFC 230-200	230	200
NFC 240-200	240	200

Table 1. *Cont.*

Configuration	Molding Temperature (°C)	Core Temperature (°C)
NFC 210-210	210	210
NFC 220-220	220	220
NFC 230-230	230	230
NFC 240-240	240	240

2.2.2. Temperature Measurement System

During the hot-pressing process, real-time and in situ monitoring of the composite materials' core temperature was carried out using a thermocouple coupled to an HH74K Digital Thermometer (Oméga, Manchester, UK). The Type K thermocouple is widely used in industry due to the robustness and the simple working principle based on the Seebeck effect. The temperature measurement range is from −75 °C to 250 °C. The small size of the wire allows it to be a non-invasive sensor and induces a very low disturbance of the manufacture of the composite. The probe has been embedded in the center of nonwoven samples of 7×7 cm^2 (Figure 1). In order to measure and display the evolution of the average core temperature of the nonwovens in real time during the hot-pressing process, three core temperature measurements are carried out on three different samples for each configuration.

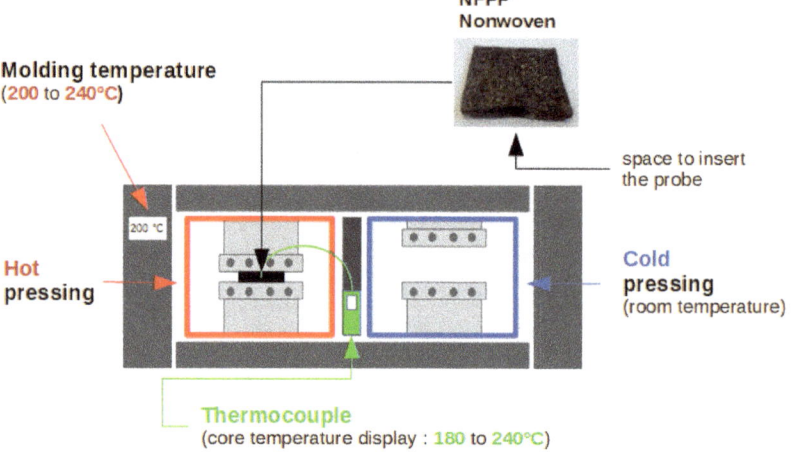

Figure 1. Composite manufacturing process by thermocompression and the placement of the thermocouple in the NFPP nonwoven to measure the core temperature in the nonwoven during the hot-pressing process.

2.2.3. Samples Preparation and Conditioning

From 21×30 cm^2 plates of flax/PP nonwoven composites, samples of around 1×1 cm^2 and 150 ± 1 mg were prepared. For the chemical and odor analyses, these specimens were stored in a climatic oven (Memmert IPP 110, Schwabach, Germany) at 23 °C/50% RH for 24 h.

2.2.4. Extraction of Volatile Organic Compounds by the HS-SPME Method

The extraction of volatile organic compounds (VOCs) was performed using the headspace–solid phase microextraction (HS-SPME) technique for all samples, as described in Section 2.2.3. Before the GC-MS analysis, the samples were placed into a 20 mL headspace-vial (HS-vial) and closed with caps integrating a PTFE/Silicone septum (Chromoptic, Courteboeuf, France). HS-vials were placed into a sampler tray of the MultiPurpose

Sampler (MPS) robot (GERSTEL GmbH & Co. KG, Mülheim an der Ruhr, Germany). By a robotic arm, the vial is picked up from the sampler tray and moved into the MPS Agitators/Incubators (GERSTEL GmbH & Co. KG, Mülheim an der Ruhr, Germany). The HS-Vial was heated to 80 °C for an incubation period of 15 min—to release VOCs into the headspace—and for an extraction period of 30 min—to extract VOCs into the headspace—with SPME fiber inserted into the vial. For the extraction period, a silica-based SPME fiber coated with a 50/30 thickness film of DVB/CAR/PDMS (50/30 µm film thickness, needle size 23 ga, 2 cm length, Stableflex, Supelco Chromoptic, Courteboeuf, France) was used. Before using the DVB/CAR/PDMS SPME fiber, the latter was conditioned in the injector port of the GC at 270 °C for 1 h. For each kind of sample described in Section 2.2.3, the VOCs analysis was carried out in triplicate for three distinct samples.

2.2.5. GC-MS Analysis

VOCs coming from the flax/PP nonwoven composite sample and extracted on the SPME fiber were desorbed at 250 °C for 5 min into a Gestel CIS inlet in a splitless configuration. Then VOCs were analyzed—i.e., separated, identified and quantified—by gas chromatography (GC-2010 Plus, Shimadzu, Kyoto, Japan) coupled with mass spectrometry with a simple quadrupole mass analyzer (GCMS-QP2010 SE, Shimadzu, Kyoto, Japan). The gas chromatography oven was equipped with an SLB-5MS fused silica capillary column: 5% diphenyl/95% dimethylpolysiloxane phase, 30 m × 0.25 mm i.d., 0.25 µm film thickness (Supelco, Sigma-Aldrich, St. Louis, MO, USA). The oven temperature was programmed as follows: initial hold at 40 °C for 3 min, ramp at 10 °C/min to 100 °C, at 4 °C/min to 190 °C and, finally, at 15 °C/min to 250 °C, held for 5 min. Thus, the total program time was 40.5 min. The carrier gas was helium gas at a constant flow rate of 1.0 mL/min. Mass spectrometry (MS) was operated in the electron impact positive ionization (ionization energy 70 eV; source temperature 200 °C). Full-scan data acquisition was registered over a mass range of 35–550 a.m.u. The instrument was calibrated using standard solutions of C6 through C16 n-alkanes diluted in ethanol. The standard solution was injected using a syringe with the same operating conditions. This calibration allows for identifying VOCs by comparing the MS spectra to the mass spectral coming from the NIST library and also by calculating the Kovats indices. On the chromatogram, only the peaks with a mass spectra similarity (similarity index) greater than or equal to 90% and a retention (Kovats) index difference of less than 50 were retained for identification. For each sample, three GC-MS analyses were conducted. The quantification of each VOC is expressed as the peak area units from the chromatograms.

2.2.6. Odor Analysis

For the odor analysis, a panel of six voluntary formed assessors (5 women/1 man) from 23 to 57 years old (average age: 31) was composed. The sensory analysis took place in a sensory laboratory composed of a sensory analysis cabin providing optimum analysis conditions. The assessors evaluated the odor intensity with an n-butanol odor intensity referencing scale of six levels (from 0 to 5), as described in Table 2 and inspired by the NF ISO 12219-7 standard. Moreover, the assessors could score with half-levels. Before the evaluation, the odor intensity scale was smelled by dipping an odorless sniffing paper stick in each level. The odor intensity scale could be smelled several times by the assessor during the evaluation.

Once the results of all the assessors are collected, the average value of the odor intensity is calculated. Since the average odor intensity value does not give any information on the distribution of values, the results are also presented in a box-plot diagram.

Table 2. n-butanol odor intensity referencing scale.

Level	Odor Intensity	n-butanol Aqueous Solution (g/L)
0	No odor	0
1	Very weak	1×10^{-2}
2	Weak	5×10^{-2}
3	Strong	5×10^{-1}
4	Very Strong	2.5
5	Very strong and insupportable	10

Regarding the sample preparation of the flax/PP nonwoven composite sample, the preparation and conditioning were described in Section 2.2.3. Afterwards, the samples were placed into a 20 mL amber glass-vial and heated in an oven at 80 °C for 45 min; the same temperature and time compared to the samples were used for the chemical analysis. Once removed from the oven, the samples were cooled to room temperature for one hour. Then, the assessors could evaluate them. The vials containing the flax/PP nonwoven composite samples were presented in different random orders to each assessor. On each vial, a three-digit number is inscribed. Each sample is smelled twice, on two different days, in order to characterize the repeatability of the evaluation of the odor intensity by the assessors.

2.2.7. Statistical Analysis

The experimental results of odor intensity are expressed by the mean value and box plots representation. Three-way analysis of variance (ANOVA)—for the factor materials, judge and session—was performed with XLSTAT 2016 to examine the difference between the NFC materials produced from different heat press temperatures and core temperatures regarding odor intensity. Each mean was compared with each other, and two results were considered statistically different if a p-value of less than 0.05 was observed. As ANOVA is a linear model, assumptions about the residual are verified or assumed: independence: no obvious relationship between measurements (assumed); normality: verified by Shapiro–Wilk's test and the Q-Q plot; equal variance: verified by Levene's test and not too many outliers observed among the residual value, i.e., 95% of the residuals are in an interval (-1.96, $+1.96$). Finally, the three-way ANOVA was followed by Fisher's post hoc test, which is used to determine significant differences between group means in an ANOVA. To help identify the significant differences between group means, letters are used. If the means share at least one letter in common, the means are not significantly different. Conversely, if the means have no letter in common, the means are significantly different.

3. Results and Discussion

3.1. Evolution of the Temperature of the Nonwoven In Situ and in Real Time during the Compression Molding Process

The evolution of the temperature of the nonwoven in situ and in real time during the hot-pressing process is presented in Figure 2. Depending on the configuration, the nonwoven is hot pressed between two heating plates at 200, 210, 220, 230 or 240 °C until the core reaches the same temperature as that of the hot plates. In Figure 2a, for each configuration, one can observe that the evolution of the temperature is linear for the first four seconds, with a slope around 20 °C/s. Once a temperature of 100 °C at the core of the nonwoven is reached, an inflection of the curves is observed, which is more or less pronounced depending on the configuration. The higher the heat press temperature, the lower the inflection of the curve. This inflection point around 100 °C may be explained by the residual water contained in the lignocellulosic fibers [34] and, more generally, in the lignocellulosic fiber nonwovens [35] that passes from a liquid state to a gaseous state. This phenomenon is more commonly called the vaporization/evaporation stage and is also reported for other lignocellulosic materials during thermocompression [36]. This vapor slows down the increase in temperature. This point of inflection of the temperature evolution was reported elsewhere during the hot pressing of the board of the beech veneer produced at a temperature of 250 °C, 4 MPa and 280 s from the thermocouples in the core of the material [37].

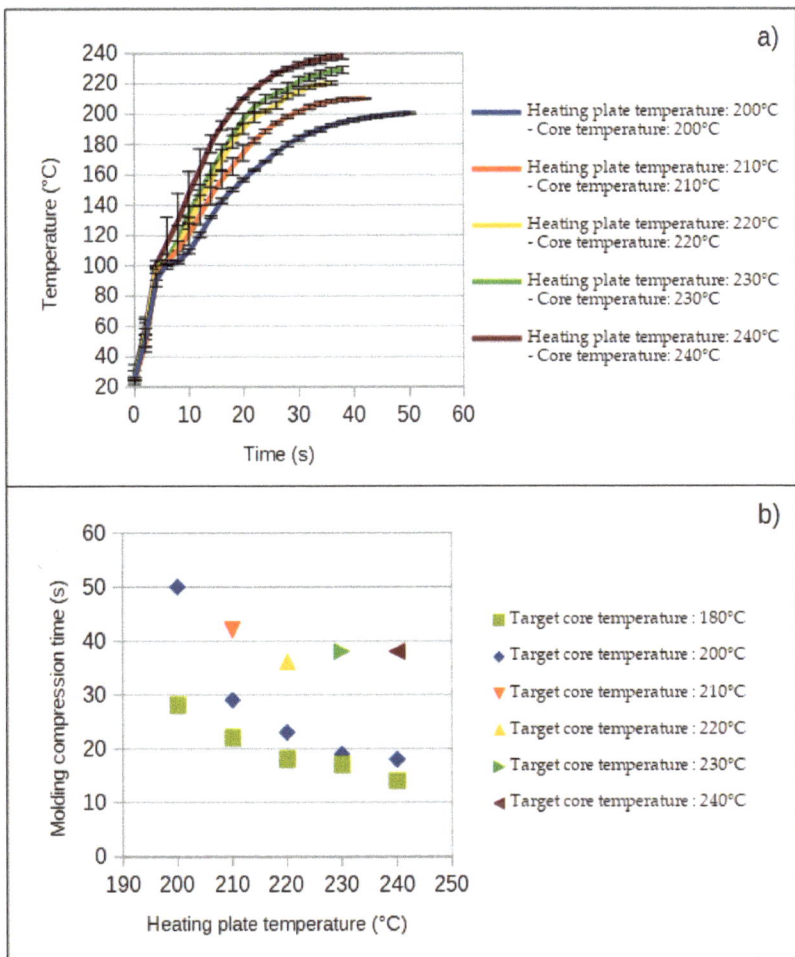

Figure 2. Evolution of the temperature of the non-woven in situ and in real time during the compression molding process: (a) temperature versus time and (b) molding compression time versus heating pate temperature for different target core temperatures.

The higher the heat press temperature, the faster this vapor forms and the quicker it escapes from the lignocellulosic fiber nonwoven, which would explain the less significant inflection of the curve when the temperature of the hot plates is increased during thermocompression. This phenomenon of vaporization is well known in the literature [38] and by the industrial sector. It is not uncommon in the industry to have one or more degassing cycles in order to prevent the water naturally contained in the lignocellulosic fibers from passing too quickly in the form of vapor and generating defects (blisters and/or cracks) within the material. Another way to limit vapor generation during the hot pressing of lignocellulosic matter consists of drying materials prior to hot pressing [39].

Figure 2b is obtained from the curves of Figure 2a. This figure represents the molding compression time that is necessary to obtain at core 180, 200, 210, 220, 230 and 240 °C, depending on the temperature of the hot plates. As expected, it is observed that the higher the heat press temperature, the faster the temperatures of 180 °C and 200 °C are reached at the core. To reach 180 °C at the core of the nonwoven, a time of 28, 22, 18, 17 and 14 s is needed for heat press temperatures of 200, 210, 220, 230 and 240 °C, respectively. Then, to reach 200 °C at the core of the nonwoven, a time of 50, 29, 23, 19 and 18 s for the same heat press temperatures is needed, respectively. It should be noted that the time difference in reaching 180 or 200 °C at the core decreases sharply from 22 s to 4 s when going from an initial temperature of the hot plates of 200 °C to 240 °C. Finally, to reach core temperatures of 200, 210, 220, 230 and 240 °C for identical heat press temperatures, it takes 50, 42, 36, 38 and 38 s, respectively. In addition, to see the heat press time decrease, it is also observed that, from 220 °C, the heat press time stabilizes around 36/38 s. In all the configurations, once 180 °C is reached at the core, the more the temperature increases, the more the slope of the curve decreases and tends to be zero.

As a first conclusion, the results make it possible to accurately control the time of thermocompression necessary to produce an NFC composite according to the heat press temperature and the target core temperature. Overall, the results show that the higher the heat press temperature and the lower the target core temperature, the shorter the heat press time. Whatever the configuration, the thermocompression time remains less than 60 s.

3.2. Evolution of the Odor Intensity Coming from the NFC According to the Molding Temperature and Core Temperature of the Nonwoven In Situ during the Compression Molding Process

The evolution of the odor intensity coming from the NFC according to the molding temperature and core temperature of the nonwoven in situ during the compression molding process is presented with box-pots in Figure 3. A three-way analysis of variances was performed on the scores given by the assessors. According to Table 3, the "NFC material" factor has a highly significant effect ($p < 0.0001$). Therefore, the assessors smelled marked differences between the intensities of the samples. Notably, the "Judge" factor also has a significant effect ($p < 0.0001$). Although formed and trained, it is common to have a "Judge" effect in sensory analysis. Finally, the "Session" factor has no significant effect ($p = 0.50$); this means that there is no difference between the evaluations in session 1 and in session 2. Still, according to Table 3, the interaction "NFC material × Judge" has a p-value close to 0.05, whereas the "NFC material × session" and "Judge × session" interactions have no significant effect ($p > 0.05$).

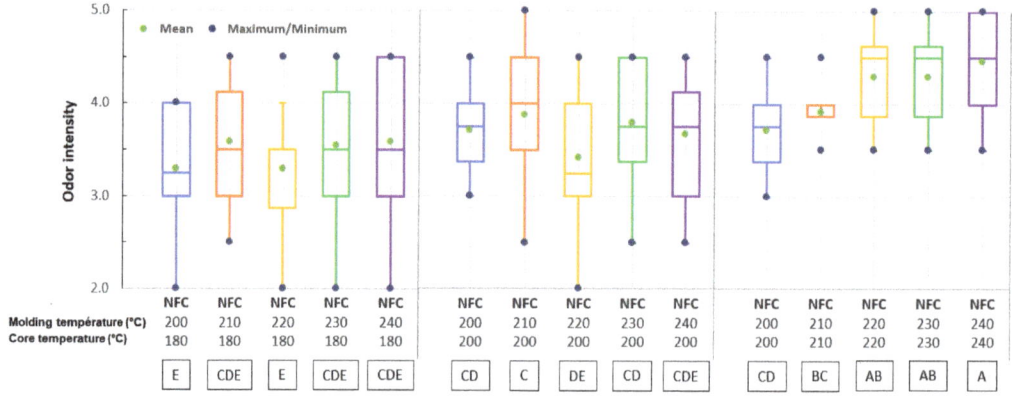

Figure 3. Box-plots of odor intensity coming from the NFC according to the molding temperature and the core temperature of the nonwoven in situ during the compression molding process. Letters A, B, C, D and E on the box-plot diagrams are the result of ANOVA analyses (Fisher's test). NB: Two means with at least one letter in common are not significantly different/two means with no letter in common are significantly different.

Table 3. Results of the three-way analysis of variance (ANOVA) with interaction.

Factors of Variation and Interaction	dF	MS	F	p-Value (Pr > F)
Natural Fiber Composite (NFC) material	13	1.63	6.26	<0.0001
Judge	5	3.88	14.87	<0.0001
Session	1	0.12	0.46	0.50
NFC material × Judge	65	0.42	1.62	0.03
Session × NFC	13	0.17	0.66	0.80
Judge × Session	5	0.31	1.19	0.33

dF = degree of freedom; MS = mean square; F = MS factor/MS residual; p-value = statistical significance.

As a result, significant differences were perceived between NFC materials. The assessors were repeatable between both sessions, even if some differences are observed in the use of the intensity scale due to various sensibilities. This may explain the standard deviations in Figure 3. Fisher's post hoc test was used to determine which samples were different from each other from distinct groups. The results from Figure 3 show that NFC materials with a core temperature of 180 and 200 °C (regardless of the heat press temperature between 200 and 240 °C) do not have significantly different means, with odor intensity means between 3.3 and 3.8. The lowest average odor intensity was obtained for the material hot pressed at 200 °C, with a core temperature of 180 °C. In contrast, NFC 220-220, NFC 230-230 and NFC 240-240 were perceived with the highest intensities, with means between 4.3 and 4.5. NFC 210-210 showed an intermediary odor intensity.

Consequently, to avoid a high odor intensity (>4), the heat press temperature and targeted core temperature should not exceed 210 °C.

3.3. Evolution of VOCs Emission Coming from Fibers of NFC According to the Molding Temperature and Core Temperature

A large number of VOCs of different natures coming from NFC were detected, identified and quantified by HS-SPME-GC-MS analysis. Both Tables 4 and A1 and Table A2 (tables in Appendix A, which is more detailed) summarize the main identified VOCs coming from flax fibers-reinforced hot-pressed NFC, which have also been identified in other scientific works [24,27]. The study specifically focuses on VOCs from the thermal degradation of flax fibers, as numerous studies have shown that the odors of composite materials based on lignocellulosic fibers mainly come from the latter [24,26–28,30].

It is well known that flax fibers consist of cellulose, hemicellulose, lignin and also a small quantity of lipophilic extractives [40–42], which are mainly present on the fibers' surface [43,44]. First of all, the evolution of VOCs from the thermal degradation of lipophilic extractives located mainly on the surface of flax fibers (Section 3.3.1) is presented as a function of the molding temperature and the targeted core temperature. Then, the evolution of VOCs from the thermal degradation of holocelluloses, i.e., cellulose and hemicelluloses, (Section 3.3.2) and lignin (Section 3.3.3), which constitute the main chemical constituents of flax fibers, was again evaluated depending on the molding temperature and the targeted core temperature.

Table 4. Volatile organic compounds coming from flax fibers of NFC shaped for heat press temperatures between 200 and 240 °C for different core temperature.

VOCs Families	Compounds [a,b]	LRI exp.	Odor Descriptors
Aldehyde	Hexenal	806	Green, fatty, aldehyde, herb [c,d,e,f,g,h,i]
	Furfural	831	Almond, baked bread, woody, nut [c,j,k,l,i]
	Heptanal	905	Fatty, citrus, rancid, green [c,g,h]
	2-Heptenal, (Z)-	913	Fatty, fruit/mushroom, soapy [h,i]
	5-Methylfurfural	920	Caramel, almond, burnt sugar [c]
	Benzaldehyde	982	Almond, burnt sugar, sweet, caramel [c,h,k,i]
	Octanal	1005	Fatty, lemon, green, rancid, soap [c,d,e,j,f,g,h]
	2-Nonenal, (E)-	1112	Cucumber, green, fatty [c,d,j,f,g,h]
	Decanal	1204	Fatty, soapy, orange peel, sugar [c,e,g,h]
	Vanillin	1392	Vanilla [c,e,f,l]
Alcohol	2-Furanmethanol	885	Burnt, sweet, caramel, bread, coffee [c,k,l]
Carboxylic acid	Acetic acid	656	Sour, acid, sunflower seed, cheesy [c,e,k]
	Octanoic acid	1183	Sweat, cheese, rancid [c,k]
Furan	Furan, 2-pentyl-	992	Fruity, green, earthy, orange [c,g]

Compounds identified by: (a) Mass Spectra Similarity ≥ 90%; (b) Retention Index obtained experimentally not exceeding ± 50 of the NIST library standard. Odor description was obtained from: (c) [27], (d) [45], (e) [46], (f) [47], (g) [48], (h) [49], (i) [50], (j) [51], (k) [52] and (l) [53], respectively.

3.3.1. VOCs Coming from Lipophilic Extractives of Flax Fibers

Among the 15 identified compounds (Table 4) coming from the flax fibers of hot-pressed NFC, most are aldehydes, including: aliphatic aldehydes (hexanal, heptanal, octanal and decanal), unsaturated aldehydes (2-heptenal, 2-nonenal) and aromatic aldehydes (benzaldehyde, vanillin) and others (furfural and 5-methylfurfural). According to Figure 4a, for a targeted core temperature of 180 °C, the molding temperatures ranging from 200 °C to 240 °C do not drastically change the quantity of aliphatic and unsaturated aldehydes released by NFC. The same observation was made in the case of thermocompression, in which the targeted core temperature was 200 °C (Figure 4b). Nevertheless, the change in the core temperature from 180 °C to 200 °C induced a greater release of each aldehyde, except for hexanal and 2-heptanal. Finally, when the core temperature and the heat press temperature are the same, going from 200 °C to 240 °C (Figure 4c), no change was observed in the quantity of heptanal, 2-heptanal and decanal. Nevertheless, in the case of octanal and 2-nonenal, the quantity in the headspace increased with the heat press temperature. In the case of hexanal, the quantity unexpectedly decreased when the heat press temperatures increased. It can therefore be assumed that a greater quantity of hexanal was released at the moment of the hot-pressing process; thus, they are not found in greater quantities in the final composite. This can be justified because hexanal is the more volatile compound among the aldehydes of this study.

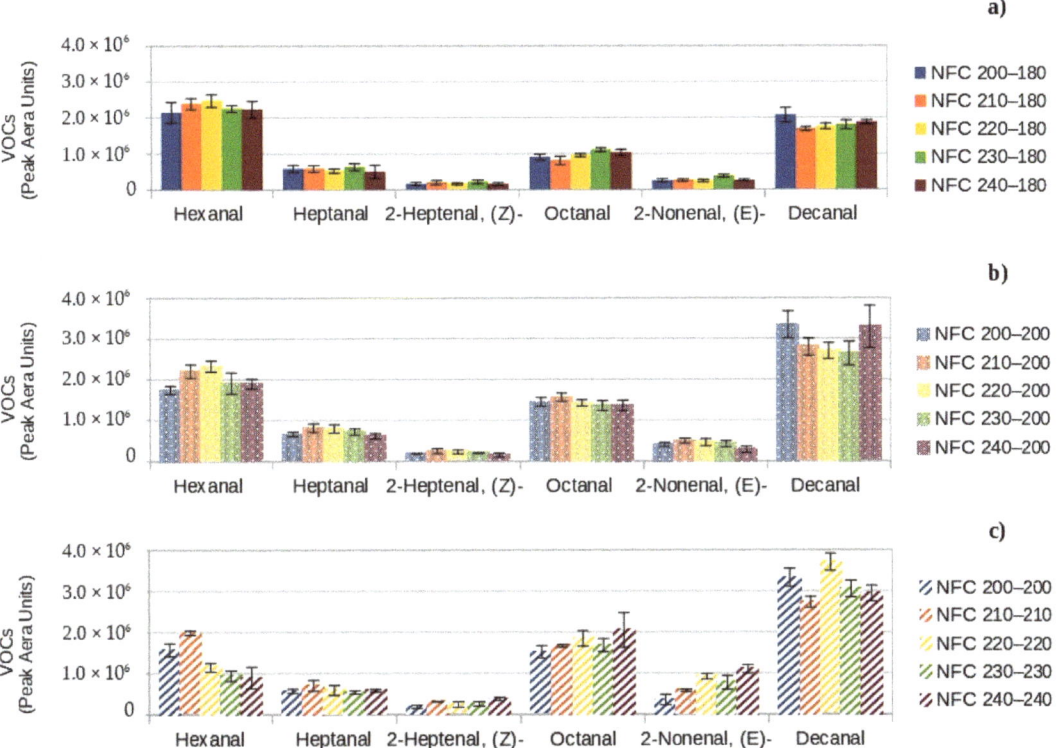

Figure 4. VOCs coming from lipophilic extractives of flax fibers: aliphatic and unsaturated aldehydes as a function of molding temperature conditions for a core temperature of (**a**) 180 °C, (**b**) 200 °C and (**c**) going from 200 °C to 240 °C.

Aliphatic aldehydes (hexanal, heptanal, octanal and decanal), unsaturated aldehydes (2-heptenal, 2-nonenal) and 2-pentylfuran are well known as secondary lipid oxidation/degradation products, especially for unsaturated fatty acid. Flax fibers mostly consist of cellulose, hemicellulose, lignin and also lipophilic extractives up to 1.8% [40–42], mainly on the surface of the fibers [43,44]. The main constituents of lipophilic extractives (i.e., lipid) are long chain fatty acids, long chain aldehydes, fatty alcohols and wax esters. Among the fatty acids, unsaturated fatty acids are found, such as 9-octadecenoic acid (oleic acid) and 9,12-octadecadienoic acid (linoleic acid) [41,42]. Hexanal is a secondary degradation product of linoleic acid generally formed due to the β-cleavage of the first degradation product (hydroperoxides) [54–56]. Hexanal can be used as an indicator of the degradation of lipid oxidation [54]. Hexanal can be described by green, fatty, aldehyde odorous facets (Table 4). Heptanal is also well known to be a secondary degradation product of fatty acid [57,58]. Nevertheless, the fatty acid source of heptanal has not been clearly identified. Some authors suggested that it is a breakdown product of oleic acid [59]. Heptanal was also detected as a volatile oxidation compound from linoleic acid [60,61], as in the case of 2-pentylfuran [60,62,63]. Grebenteuch et al. [57] showed that 2-pentylfuran and 2-nonenal can be precursors of heptanal. These two compounds have been identified in this article. Heptanal can be described by fatty, citrus, rancid odorous facets (Table 4). 2-nonenal is derived from linoleic acid acyl group oxidation [64], whose fragrant facets can be described as cucumber and green. Octanal, nonanal (not quantified in this study because its peak was co-eluted with a peak of VOC from PP) and decanal can be produced from oleic acid oxidation [55,64]. The odor descriptors of octanal, nonanal and decanal are in Table 4.

Therefore, aliphatic and unsaturated aldehyde compounds from lipophilic extractives of flax fibers globally increase with the temperature and give globally fatty, green, aldehyde odor facets of the NFC odor profile.

3.3.2. VOCs Coming from the Dehydration of the Primary Holocellulose Decomposition Products

Among the identified compounds (Table 4) coming from flax fibers of hot-pressed NFC, furan compounds such as furfural, 2-furanmethanol and 5-methylfurfural were identified. Such compounds have already been identified elsewhere as decomposition products of lignocellulosic fibers in NFC [24,27,65]. According to Figure 5a, in the case of thermocompression in which the targeted core temperature is 180 °C, a switch from a temperature of 200 °C to 240 °C does not change the emitted quantity of furfural and 2-furanmethanol. The same observation was made in the case of thermocompression in which the targeted core temperature is 200 °C (Figure 5b). However, one can note the appearance of 5-methylfurfural in small quantities. It should also be emphasized that furfural and 2-furanmethanol were released 1.5 to 2 times more when the temperature increased from 180 °C to 200 °C.

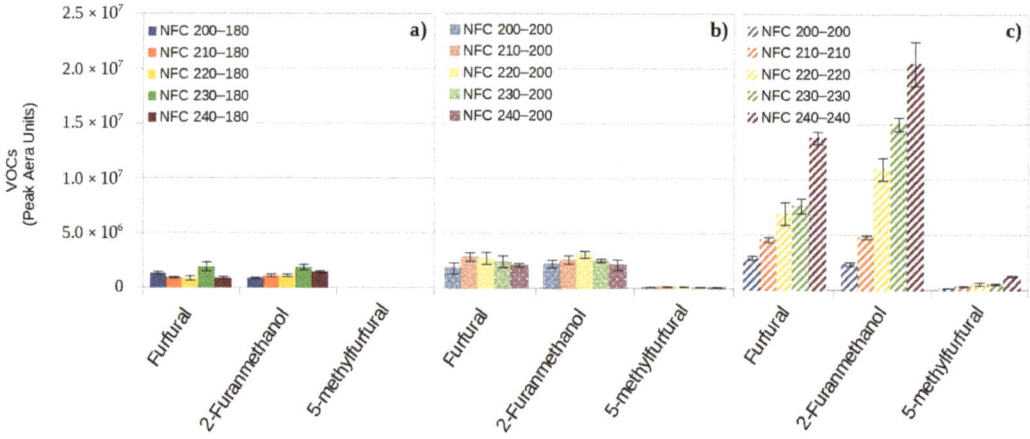

Figure 5. VOCs coming from the dehydration of the primary holocellulose decomposition products, Furfural, 2-Furanmethanol and 5-methylfurfural, as a function of molding temperature conditions for a core temperature of (**a**) 180 °C, (**b**) 200 °C and (**c**) going from 200 °C to 240 °C.

Finally, when the core temperature is at the heat press temperatures, going from a temperature of 200 °C to 240 °C, the results in Figure 5c show that the higher the heat press temperatures, the greater the presence of 2-methylfurfural, 2-furanmethanol and furfural. Between a thermocompression at 200 °C and one at 240 °C, the quantities of 2-methylfurfural, 2-furanmethanol and furfural emitted by the NFC were multiplied by 8.7, 8.8 and 4.8, respectively. One can also notice that the amounts of furfural and 2-furanmethanol at 210 °C are close to the amount of decanal, which is the aliphatic aldehyde in the largest quantity. Then, from 220 °C, the amount of furfural and 2-furanmethanol is well beyond the decanal. Qualitatively, the odor of 5-methylfurfural is described by notes of caramel, almond or even burnt sugar. Furfural is described by notes of almond, baked bread, sweet, woody and nut. Finally, 2-furanmethanol is described by sweet, burnt, caramel, baked bread and coffee notes (Table 4). Consequently, the increase in these VOCs may cause an increase in these different olfactory notes. The odor profile of hot-compressed NFC is more complex, with a pyrogenic (burnt smell) and phenolic smell. This evolution of odor has also been observed in the case of flax fibers heated between 200 and 230 °C [30]. Therefore, there seems to be a strong correlation between the increase in odor intensity and the increase in VOCs coming from the degradation of holocelluloses when NFPP nonwovens are thermocompressed at temperatures above 200 °C and up to 240 °C.

Flax fibers are lignocellulosic fibers composed of 60 to 85% cellulose, 14 to 20.6% hemicelluloses and 1 to 3% lignin [66]. It is well established that the thermal degradation of holocelluloses [67], i.e., cellulose [67–71] and hemicelluloses [68–70,72,73], genders thermal decomposition products such as furfural, 2-furanmethanol and 5-methylfurfural. A great difference was shown between the pyrolysis behaviors of hemicellulose, cellulose and lignin [70,73,74]. The pyrolysis TGA and DTG curves [70] show that the thermal degradation of hemicelluloses mainly occurred in the temperature range from 200 to 340 °C, with a maximum degradation temperature from 240 to 285 °C, while cellulose degradation is more pronounced between 300 and 400 °C, with a maximum degradation temperature at 355 °C. Finally, the thermal degradation of lignin mainly occurred between 138 and 780 °C and takes place in three stages (first step: <138 °C; second step: from 138 to 285 °C; third step: from 285 to 780 °C), with very low thermal degradation compared to hemicelluloses and cellulose. The primary products issued from the thermal decomposition of hemicelluloses and cellulose are considered here, whereas the products from lignin will be studied in the next section. The cell walls of flax fibers are made up of different types of hemicelluloses depending on the cell walls. The middle lamella and primary cell wall, lying on the surface, contain xyloglucans and xylans hemicelluloses, while the secondary/G-layer cell walls contain mannans/heteromannans hemicelluloses [75]. Xyloglucans (hexosans) are polysaccharides made up of a D-glucose main chain with pendant xylose residues. Xylans (pentosans) are also polysaccharides whose main chain consists of D-xylose monomers linked to each other by the β-(1,4)-glycosidic bonds, on which short-side chains are grafted. Furfural formation can be described in the literature as a simple, successive two-step reaction. The first step is the xylans hemicelluloses (acid-catalyzed) hydrolysis into xyloses (pentose sugars), and the second step is the dehydration of xyloses [76–78]. The evaporation of water (moisture content in NFC) in the beginning of thermocompression must be attributed to the xylans (contained in hemicelluloses) hydrolysis into xylose [78] and the byproduct acetic acid [74,77,79]. Under uncontrolled conditions such as pH and temperature, water can act as a weak acid and engender the acid-catalyzed hydrolysis of xylans into xylose. In parallel, water reacts with acetyl groups of hemicelluloses (xylans) and product acetic acid, which is present in our study [77]. The dehydration reaction within xylans takes place between 150 and 240 °C (breaking of a less stable linkage) and becomes significant from 200 °C [80]. Nitu et al. [81] showed, by the chemical analysis of a jute stick particleboard thermocompressed between 180 and 220 °C, that the higher the heat press temperatures, the greater the degradation of hemicellulose. Additionally, the decrease in pentosans and xylose may indicate their conversion to furfural. Cristescu et al. [37] showed that furfural appears after a thermocompression of pressed lignocellulosic boards at 200, 225 and 250 °C. Moreover, both 2-furanmetanol and 5-methylfurfural can be obtained by the thermal dehydration of cellulose [82]. Beyond heat press temperatures of 200 °C, the water (in a subcritical state) content in cellulose fibers migrates, and hydrolysis amorphizes cellulose. During thermocompression, compression (pressure) decreases the degree of polymerization (DP) of cellulose fibers due to the harsh friction between cellulose fibers and engenders the cleavage of the β-1,4-glycosidic bonds binding the D-glucopyranose monomer [77,82,83]. The dehydration of D-glucopyranose generates furfural and 2-furanmethanol, and two successive dehydrations generate 5-methyl-furfural [84].

Therefore, furan compounds (Furfural, 2-Furanmethanol and 5-methylfurfural) coming from the dehydration of the primary holocellulose decomposition products of flax fibers increase when the heat press temperature increases. Furan compounds give globally burnt sugar, baked bread, caramel and woody odor facets of the NFC odor profile.

3.3.3. VOCs Coming from the Primary Degradation of Lignin

Among the identified compounds (Table 4) coming from the flax fibers of hot-pressed NFC, aromatic aldehydes such benzaldehyde and vanillin were identified. These aromatic aldehydes have already been identified as lignocellulosic fibers decomposition products [24,27] in NFC. According to Figure 6, in the case of thermocompression with a targeted core temperature of 180 °C, increasing the temperature from 200 °C to 240 °C does not drastically change the emitted quantity of benzaldehyde and vanillin. The same result was observed in the case of thermocompression with a targeted core temperature of 200 °C. However, it should be underlined that the switch from a core temperature of 180 °C to 200 °C induced a more important release of benzaldehyde and vanillin from the composite in the headspace. As previously established in the case in which the core temperature is the same as the heat press temperature, going from 200 °C to 240 °C, the results reported in Figure 6 indicate that the higher the heat press temperature, the greater the released quantities of benzaldehyde and vanillin. Indeed, the amount of benzaldehyde and vanillin is almost four times higher for NFC 240-240 compared to that for NFC 200-200. On average, there is 2.6 times more benzaldehyde emitted compared to vanillin for each thermocompression temperature. A comparison of these results was made with the results of 2-furanmethanol (Figure 5). This compound is the most important VOC originating from the degradation of holocelluloses. This analysis underlined much smaller quantities of benzaldehyde and vanillin compared to 2-furanmethanol. Therefore, VOCs from the thermal degradation of holocelluloses may play a much more important role in the increase in NFC odor than VOCs issued from the thermal degradation of lignin. This may be explained by the amount of lignin in flax fibers being between 1 and 3% [66], with a complex aromatic polymer structure composed of three phenylpropane units (H, G and S) cross-linked to each other with a variety of chemical bonds. Therefore, lignin has a much wider thermal degradation temperature range of 138–780 °C, with a maximum degradation temperature around 350 °C [70], which is far from the heat press temperature range used in the present work. Additionally, it should be noted that, in the nonwoven, there always remains a residual quantity of flax shives (the central part of a flax stem) with a significant amount of lignin (around 25%) [39]. Thus, some of the VOCs from lignin degradation may also come from flax shives.

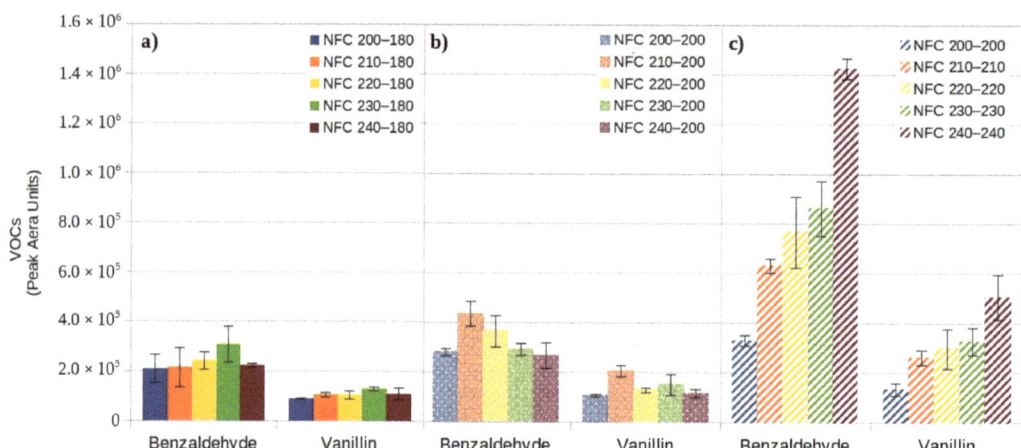

Figure 6. VOCs coming from the degradation of lignin: benzaldehyde and vanillin methylfurfural as a function of molding temperature conditions for a core temperature of (**a**) 180 °C, (**b**) 200 °C and (**c**) going from 200 °C to 240 °C.

Benzaldehyde and substituted benzaldehydes are commonly evidenced as degradation products from the pyrolysis of lignin [85]. Vanillin has already been identified as a compound emitted from composite materials reinforced with lignin and hot pressed at 235 °C [86]. According to Badji et al. [24], two distinct pathways could allow for the release of vanillin from ferulic acid. One pathway is the decarboxylation of ferulic acid, forming 4-vinylguaiacol, followed by the oxidation of 4-vinylguaiacol. Another pathway is the hydrolysis of ferulic acid, followed by deacetylation. According to Badji et al., the latter pathway seems to be more likely than the first one.

Therefore, aromatic aldehyde compounds, such as benzaldehyde and vanillin, coming from the dehydration of the primary holocellulose decomposition of lignin increase when the heat press temperature increases. Benzaldehyde and vanillin globally give the almond, burnt sugar and vanilla odor facets of the NFC odor profile.

4. Conclusions

In this study, the effects of compressing molding temperature on VOCs and odors released from the NFPP nonwoven composites were examined by heating nonwoven preforms from 180 °C to 240 °C. Moreover, during the thermocompression process, different core temperatures were targeted: 180 °C, 200 °C and until the temperature at the core of the composite material was equal to the temperature of the heating plates of hot press compression molding. The measurements of the core temperature of the nonwoven in situ and in real time during the thermocompression allowed for evidencing that the time required to reach 180 °C or 200 °C at the core decreased from 200 °C to 240 °C. The time difference needed to reach 180 or 200 °C at the core sharply decreased, from 22 s to 4 s, when going from an initial temperature of the hot plates of 200 °C to 240 °C. The results demonstrated a change in the odor intensity and VOCs composition with temperature. The odor approach based on the odor intensity scale rated by a trained panel allowed for establishing that:

- For a target core temperature of 180 and 200 °C, heat nonwovens from 200 °C to 240 °C did not drastically change the odor intensity of NFCs. It is noteworthy that the lowest average odor intensity was obtained for material hot pressed at 200 °C.
- For a target core temperature equal to the temperature of the heating plates of the hot press compression molding, heating nonwovens from 200 °C to 240 °C increased the odor intensity of NFCs. The lowest odor intensity is obtained for the heat press temperature of 200 °C and the targeted core temperature of 180 °C. Moreover, to avoid a high odor intensity (>4), the heat press temperature and targeted core temperature should not exceed 210 °C.

In addition, the chemical approach based on headspace solid-phase microextraction (HS-SPME) coupled with gas chromatography–mass spectrometry (GC-MS) revealed that:

- Some VOCs come from lipophilic extractives of flax fibers and are mainly aliphatic aldehydes (hexanal, heptanal, octanal and decanal) and unsaturated aldehydes (2-heptenal, 2-nonenal) associated with fatty and green odors.
- Some VOCs come from the dehydration of the primary holocellulose decomposition products and are furan compounds (2-furanmethanol, furfural and 5-methylfurfural) associated with burnt sugar, baked bread and burnt odors.
- Some VOCs come from the primary degradation of lignin, which are aromatic aldehydes (benzaldehyde and vanillin) associated with almond, burnt sugar and vanilla odors.
- Switching from a target core temperature of 180 to 200 °C and heat nonwovens from 200 °C to 240 °C slightly increases all VOCs coming from flax fibers of NFC.
- For a target core temperature equal to the temperature of the heating plates of hot press compression molding, heat nonwovens from 200 °C to 240 °C do not change the VOCs quantity coming from the lipophilic extractives of flax fibers but drastically increase the VOCs coming from the degradation of holocellulose and lignin.

Finally, on the basis of the original results obtained in the present work, a correlation between the increase in odor intensity and the increase in VOCs coming from the degradation of holocelluloses and lignin was established. To go further, analyses by gas chromatography–mass spectrometry (GC-MS) coupled with an olfactometer (O) would make it possible to clearly identify odorous VOCs and to follow their evolution according to the compression molding temperature.

Author Contributions: Conceptualization, K.B., M.G. and G.S.; methodology, B.B.-M., M.H., K.B., M.G. and G.S.; validation, B.B.-M., M.H., K.B., M.G. and G.S.; investigation, B.B.-M.; original draft preparation, B.B.-M.; writing—review and editing, B.B.-M., M.H., K.B., M.G. and G.S.; visualization, B.B.-M., M.H., K.B., M.G. and G.S.; supervision, K.B., M.G. and G.S.; funding acquisition, K.B., M.G. and G.S. All authors have read and agreed to the published version of the manuscript.

Funding: This research was funded by the National Research and Technology Association (ANRT).

Institutional Review Board Statement: The study was conducted in accordance with the Declaration of Helsinki and approved by the Ethics Committee of Université Le Havre Normandie (18 April 2022).

Informed Consent Statement: Informed consent was obtained from all subjects involved in the study.

Data Availability Statement: The data presented in this study are available on request from the corresponding author.

Acknowledgments: First of all, the authors would like to thank the National Research and Technology Association (ANRT) for funding Benjamin BARTHOD-MALAT's PhD thesis. The authors would also like to thank the formed panel of six voluntary assessors. Finally, the authors would also like to thank the laboratory technicians of URCOM for their invaluable assistance.

Conflicts of Interest: The authors declare no conflict of interest.

Appendix A

Table A1. Volatile organic compounds coming from flax fibers of NFC shaped for heat press temperatures between 200 and 240 °C for different core temperatures (1/2).

VOCs Families	Compounds [a,b]	MSS(%)	RT $_{exp.}$	LRI $_{exp.}$	LRI $_{lit.}$	Odor Descriptors	Quantification (Mean Peak Area Units $\times 10^3 \pm$ SD)				
							Core Temperature: 180 °C				
							NFC 200-180	NFC 210-180	NFC 220-180	NFC 230-180	NFC 240-180
Aldehyde	Hexenal	98	6.37	806	804	Green, fatty, aldehyde, herb [c,d,e,f,g,h,i]	2142 ± 288	2383 ± 154	2466 ± 174	2253 ± 92	2233 ± 229
	Furfural	98	7.00	831	836	Almond, baked bread, woody, nut [c,j,k,l,i]	1349 ± 116	1349 ± 116	834 ± 191	1912 ± 419	856 ± 115
	Heptanal	90	8.33	905	904	Fatty, citrus, rancid, green [c,g,h]	583 ± 98	581 ± 93	522 ± 56	631 ± 94	496 ± 185
	2-Heptenal, (Z)-	92	9.38	913	960	Fatty, fruit/mushroom, soapy [h,i]	161 ± 38	195 ± 59	162 ± 28	215 ± 53	154 ± 39
	5-Methylfurfural	94	9.46	920	964	Caramel, almond, burnt sugar [c]	/	/	/	/	/
	Benzaldehyde	96	9.52	982	967	Almond, burnt sugar, sweet, caramel [c,h,k,i]	209 ± 56	214 ± 79	241 ± 35	307 ± 72	223 ± 7
	Octanal	97	10.24	1005	1005	Fatty, lemon, green, rancid, soap [c,d,e,j,f,g,h]	908 ± 76	802 ± 115	950 ± 46	1105 ± 49	1031 ± 82
	2-Nonenal, (E)-	95	13.74	1112	1162	Cucumber, green, fatty [c,d,j,f,gh]	251 ± 42	254 ± 34	244 ± 39	384 ± 38	259 ± 29
	Decanal	96	14.88	1204	1208	Fatty, soapy, orange peel, sugar [c,e,g,h]	2071 ± 200	1680 ± 57	1750 ± 80	1801 ± 128	1873 ± 53
	Vanillin	91	20.23	1392	1402	Vanilla [c,e,f,l]	90 ± 2	105 ± 9	104 ± 16	129 ± 7	109 ± 23
Alcohol	2-Furanmethanol	98	7.43	885	858	Burnt, sweet, caramel, bread, coffee [c,k,l]	860 ± 59	860 ± 59	1118 ± 138	1873 ± 249	1466 ± 72

Table A1. Cont.

VOCs Families	Compounds [a,b]	MSS(%)	RT exp.	LRI exp.	LRI lit.	Odor Descriptors	Quantification (Mean Peak Area Units × 10³ ± SD)				
							Core Temperature: 180 °C				
							NFC 200-180	NFC 210-180	NFC 220-180	NFC 230-180	NFC 240-180
Carboxylic acid	Acetic acid	98	3.44	656	576	Sour, acid, sunflower seed, cheesy [c,e,k]	7545 ± 677	7480 ± 1119	7584 ± 977	8800 ± 217	71761 ± 214
	Octanoic acid	94	14.27	1183	1173	Sweat, cheese, rancid [c,k]	3318 ± 397	3106 ± 590	3585 ± 632	4555 ± 81	4099 ± 450
Furan	Furan, 2-pentyl-	97	9.99	992	1040	Fruity, green, earthy, orange [c,g]	3006 ± 70	3136 ± 354	3137 ± 461	3605 ± 203	3215 ± 130

Compounds identified by: [a] Mass Spectra Similarity ≥ 90%; [b] Retention Index obtained experimentally not exceeding ± 50 of the NIST library standard. "/" signifies no quantified compounds. Odor description was obtained from: [c] [28], [d] [46], [e] [47], [f] [48], [g] [49], [h] [50], [i] [51], [j] [52], [k] [53] and [l] [54], respectively.

Table A2. Volatile organic compounds coming from flax fibers of NFC shaped for heat press temperatures between 200 and 240 °C for different core temperature (2/2).

Compounds [a,b]	Quantification (Mean Peak Area Units × 10³ ± SD)								
	Core Temperature: 200 °C					Core Temperature: From 210 to 240 °C			
	NFC 200-200	NFC 210-200	NFC 220-200	NFC 230-200	NFC 240-200	NFC 210-210	NFC 220-220	NFC 230-230	NFC 240-240
Hexenal	1735 ± 102	2198 ± 165	2319 ± 132	1910 ± 261	1891 ± 117	1983 ± 52	1148 ± 104	942 ± 128	898 ± 261
Furfural	1814 ± 507	2847 ± 404	2731 ± 553	2419 ± 531	2118 ± 173	4518 ± 232	6898 ± 1025	7586 ± 672	13,777 ± 572
Heptanal	656 ± 52	809 ± 108	790 ± 102	712 ± 74	611 ± 65	707 ± 130	592 ± 122	543 ± 39	584 ± 34
2-Heptenal, (Z)-	170 ± 11	244 ± 65	225 ± 50	196 ± 10	150 ± 48	314 ± 5	247 ± 68	260 ± 53	376 ± 36
5-Methylfurfural	104 ± 9	138 ± 33	131 ± 47	119 ± 24	109 ± 15	296 ± 55	536 ± 132	507 ± 66	1230 ± 65
Benzaldehyde	279 ± 14	434 ± 50	434 ± 50	291 ± 25	267 ± 51	630 ± 29	765 ± 143	861 ± 111	1424 ± 41
Octanal	1445 ± 106	1557 ± 102	1420 ± 79	1353 ± 119	1355 ± 131	1663 ± 33	1843 ± 187	1678 ± 158	2045 ± 422
2-Nonenal, (E)-	398 ± 51	490 ± 61	452 ± 83	415 ± 78	279 ± 83	580 ± 29	923 ± 67	772 ± 169	1104 ± 102
Decanal	3343 ± 332	2788 ± 213	2692 ± 195	2629 ± 289	3288 ± 519	2726 ± 136	3702 ± 208	3054 ± 201	944 ± 194
Vanillin	107 ± 6	205 ± 23	205 ± 23	151 ± 42	116 ± 16	261 ± 30	296 ± 80	326 ± 55	505 ± 90
2-Furanmethanol	2214 ± 352	2596 ± 377	3063 ± 312	2531 ± 163	2123 ± 482	4744 ± 159	10,918 ± 1024	15,019 ± 598	20,480 ± 2001
Acetic acid	8523 ± 1719	9880 ± 219	8895 ± 1845	7893 ± 761	6458 ± 682	8579 ± 275	10,664 ± 1240	8,748 ± 1297	13,552 ± 1545
Octanoic acid	5986 ± 637	5821 ± 422	5078 ± 1188	5590 ± 2052	5678 ± 782	7059 ± 275	8378 ± 1031	7228 ± 995	14,051 ± 2381
Furan, 2-pentyl-	3375 ± 249	4148 ± 246	40,153 ± 81	3908 ± 121	3618 ± 348	4675 ± 149	4096 ± 530	4263 ± 619	4639 ± 324

Compounds identified by: (a) Mass Spectra Similarity ≥90%; (b) Retention Index obtained experimentally not exceeding ± 50 of the NIST library standard. "/" signifies no quantified compounds.

References

1. Paris Agreement, United Nations. 2015. Available online: https://unfccc.int/sites/default/files/english_paris_agreement.pdf (accessed on 16 April 2020).
2. CO2 Emission Performance Standards for Cars and Vans. Available online: https://climate.ec.europa.eu/eu-action/transport-emissions/road-transport-reducing-co2-emissions-vehicles/co2-emission-performance-standards-cars-and-vans_en (accessed on 7 November 2022).
3. Proposal for a Regulation of the European Parliament and of the Council Amending Regulation (EU) 2019/631 as Regards Strengthening the CO_2 Emission Performance Standards for New Passenger Cars and New Light Commercial Vehicles in Line with the Union's Increased Climate Ambition; European Commission: Brussels, Belgium, 2021.
4. La Transition Bas Carbone: Une Opportunité Pour l'industrie Automobile Française? The Shift Project: Paris, France, 2021; Rapport Final.
5. Meilhan, N. Comment faire enfin baisser les émissions de CO2 des voitures. La Note D'analyse 2019, 78, 1–12. [CrossRef]
6. Li, Y.; Ha, N.; Li, T. Research on Carbon Emissions of Electric Vehicles throughout the Life Cycle Assessment Taking into Vehicle Weight and Grid Mix Composition. Energies 2019, 12, 3612. [CrossRef]
7. Ducreux, B.O.; Dore, N. AVIS de l'ADEME: Voitures Électriques et Bornes de Recharges. ADEME 2022, 1–10.
8. Huda, M.K.; Widiastuti, I. Natural Fiber Reinforced Polymer in Automotive Application: A Systematic Literature Review. J. Phys.: Conf. Ser. 2021, 1808, 012015. [CrossRef]
9. Issa, A.; Salihi, M.K.; Aliyu, A.B. Automotive Applications of Animal and Plant Fiber Based Thermoplastic Composite: A Review. Am. J. Eng. Res. 2022, 12.
10. Thakur, V.K.; Thakur, M.K.; Gupta, R.K. Review: Raw Natural Fiber–Based Polymer Composites. Int. J. Polym. Anal. Charact. 2014, 19, 256–271. [CrossRef]
11. Meirhaeghe, C.; Bewa, H. Evaluation de la disponibilité et de l'accessibilité de fibres végétales à usages matériaux en France. ADEME-FRD 2011, 84.

12. Le Duigou, A.; Davies, P.; Baley, C. Environmental Impact Analysis of the Production of Flax Fibres to Be Used as Composite Material Reinforcement. *J. Biobased Mater. Bioenergy* **2011**, *5*, 153–165. [CrossRef]
13. de Beus, N.; Barth, M.; Carus, M. Carbon Footprint and Sustainability of Different Natural Fibres for Biocomposites and Insulation Material. *Nova Inst.* **2019**, 56.
14. Gueudet, A. Analyse Du Cycle de Vie Comparative de Panneaux de Porte Automobiles Biosourcé (PP/Fibres de Lin et de Chanvre) et Petrosource (ABS). *Quantis Frs Ecotechnilin* **2016**, 119.
15. Merotte, J.; Le Duigou, A.; Bourmaud, A.; Behlouli, K.; Baley, C. Mechanical and Acoustic Behaviour of Porosity Controlled Randomly Dispersed Flax/PP Biocomposite. *Polym. Test.* **2016**, *51*, 174–180. [CrossRef]
16. Zhang, J.; Khatibi, A.A.; Castanet, E.; Baum, T.; Komeily-Nia, Z.; Vroman, P.; Wang, X. Effect of Natural Fibre Reinforcement on the Sound and Vibration Damping Properties of Bio-Composites Compression Moulded by Nonwoven Mats. *Compos. Commun.* **2019**, *13*, 12–17. [CrossRef]
17. Hadiji, H.; Assarar, M.; Zouari, W.; Pierre, F.; Behlouli, K.; Zouari, B.; Ayad, R. Damping Analysis of Nonwoven Natural Fibre-Reinforced Polypropylene Composites Used in Automotive Interior Parts. *Polym. Test.* **2020**, *89*, 106692. [CrossRef]
18. Renouard, N.; Mérotte, J.; Kervoëlen, A.; Behlouli, K.; Baley, C.; Bourmaud, A. Exploring Two Innovative Recycling Ways for Poly-(Propylene)-Flax Non Wovens Wastes. *Polym. Degrad. Stab.* **2017**, *142*, 89–101. [CrossRef]
19. Bourmaud, A.; Fazzini, M.; Renouard, N.; Behlouli, K.; Ouagne, P. Innovating Routes for the Reused of PP-Flax and PP-Glass Non Woven Composites: A Comparative Study. *Polym. Degrad. Stab.* **2018**, *152*, 259–271. [CrossRef]
20. Shah, D.U. Developing Plant Fibre Composites for Structural Applications by Optimising Composite Parameters: A Critical Review. *J. Mater. Sci.* **2013**, *48*, 6083–6107. [CrossRef]
21. Engel, L. Biocomposites Performing Great—Not Only for Lightweight Construction. Nova-Institute. *Renew. Carbon News* **2019**.
22. Chen, X.; Zhang, G.; Chen, H. Controlling Strategies and Technologies of Volatile Organic Compounds Pollution in Interior Air of Cars. In Proceedings of the 2010 International Conference on Digital Manufacturing & Automation, Changcha, China, 18–20 December 2010; Volume 1, pp. 450–453.
23. Faber, J. Air Quality inside Passenger Cars. *AIMS Environ. Sci.* **2017**, *4*, 112–133. [CrossRef]
24. Badji, C.; Beigbeder, J.; Garay, H.; Bergeret, A.; Bénézet, J.-C.; Desauziers, V. Under Glass Weathering of Hemp Fibers Reinforced Polypropylene Biocomposites: Impact of Volatile Organic Compounds Emissions on Indoor Air Quality. *Polym. Degrad. Stab.* **2018**, *149*, 85–95. [CrossRef]
25. Guidelines on Odour Pollution and Its Control. *Central Pollution Control Board*; Ministry of Environment & Forests, Govt. of India: New Delhi, India, 2008; 57p.
26. Fischer, H.; Knittel, D.; Opwis, K. *Verbesserung Der Geruchseigenschaften von Naturfasern Zur Öffnung Neuer Märkte in Den Bereichen Technische Textilien Und Verbundwerkstoffe.*; University of Bremen: Bremen, Germany, 2008; ISBN 978-3-8370-7213-6.
27. Savary, G.; Morel, A.; Picard, C.; Grisel, M. Effect of Temperature on the Release of Volatile and Odorous Compounds in Flax Fibers. *J. Appl. Polym. Sci.* **2016**, *133*, 43497. [CrossRef]
28. Bledzki, A.; Al-Mamun, A.; Faruk, O. Abaca Fibre Reinforced PP Composites and Comparison with Jute and Flax Fibre PP Composites. *Express Polym. Lett. —Express Polym Lett.* **2007**, *1*, 755–762. [CrossRef]
29. Li, Z.; Wei, X.; Liu, J.; Han, H.; Jia, H.; Song, J. Mechanical Properties and VOC Emission of Hemp Fibre Reinforced Polypropylene Composites: Natural Freezing-Mechanical Treatment and Interface Modification. *Fibers Polym.* **2021**, *22*, 1050–1062. [CrossRef]
30. Morin, S.; Richel, A. *Study of Chemical and Enzymatic Functionalization of Lignocellulosic Natural Fibers: Designing Natural Fibers for Biocomposites*; Liège Université Gembloux Agro-Bio Tech: Liège, Belgium, 2021.
31. Courgneau, C.; Rusu, D.; Henneuse, C.; Ducruet, V.; Lacrampe, M.F.; Krawczak, P. Characterisation of Low-Odour Emissive Polylactide/Cellulose Fibre Biocomposites for Car Interior. *Express Polym. Lett.* **2013**, *7*, 787. [CrossRef]
32. Kim, H.-S.; Lee, B.-H.; Kim, H.-J.; Yang, H.-S. Mechanical–Thermal Properties and VOC Emissions of Natural-Flour-Filled Biodegradable Polymer Hybrid Bio-Composites. *J. Polym. Environ.* **2011**, *19*, 628–636. [CrossRef]
33. Faruk, O.; Bledzki, A.; Al-Mamun, A. Influence of Compounding Processes and Fibre Length on the Mechanical Properties of Abaca Fibre-Polypropylene Composites. *Polymery* **2008**, *53*, 35–42. [CrossRef]
34. Mahir, F.; Keya, K.N.; Sarker, B.; Nahiun, K.; Khan, R. A Brief Review on Natural Fiber Used as a Replacement of Synthetic Fiber in Polymer Composites. *Mater. Eng. Res.* **2019**, *1*, 88–99. [CrossRef]
35. Radkar, S.; Amiri, A.; Ulven, C. Tensile Behavior and Diffusion of Moisture through Flax Fibers by Desorption Method. *Sustainability* **2019**, *11*, 3558. [CrossRef]
36. Pintiaux, T.; Viet, D.; Vandenbossche, V.; Rigal, L.; Rouilly, A. Binderless Materials Obtained by Thermo-Compressive Processing of Lignocellulosic Fibers: A Comprehensive Review. *BioResources* **2015**, *10*, 1915–1963.
37. Cristescu, C. *Self-Bonding of Beech Veneers*; Luleå Tekniska Universitet: Luleå, Sweden, 2015.
38. Gager, V.; Le Duigou, A.; Bourmaud, A.; Pierre, F.F.; Behlouli, K.; Baley, C. Comportement Hygromécanique Des Biocomposites Non-Tissés Soumis à Des Variations d'humidité. In Proceedings of the JNC 21: 21ème Journées Nationales sur les Composites 2019; École Nationale Supérieure d'Arts et Métiers (ENSAM)-Bordeaux, Bordeaux, France, 1–3 July 2019.
39. Evon, P.; Barthod-Malat, B.; Gregoire, M.; Vaca-Medina, G.; Labonne, L.; Ballas, S.; Véronèse, T.; Ouagne, P. Production of Fiberboards from Shives Collected after Continuous Fiber Mechanical Extraction from Oleaginous Flax. *J. Nat. Fibers* **2018**, *16*, 453–469. [CrossRef]

40. Acera Fernández, J.; Le Moigne, N.; Caro-Bretelle, A.S.; El Hage, R.; Le Duc, A.; Lozachmeur, M.; Bono, P.; Bergeret, A. Role of Flax Cell Wall Components on the Microstructure and Transverse Mechanical Behaviour of Flax Fabrics Reinforced Epoxy Biocomposites. *Ind. Crops Prod.* **2016**, *85*, 93–108. [CrossRef]
41. Gutiérrez, A.; del Río, J.C. Lipids from Flax Fibers and Their Fate in Alkaline Pulping. *J. Agric. Food Chem.* **2003**, *51*, 6911–6914. [CrossRef]
42. Marques, G.; del Río, J.C.; Gutiérrez, A. Lipophilic Extractives from Several Nonwoody Lignocellulosic Crops (Flax, Hemp, Sisal, Abaca) and Their Fate during Alkaline Pulping and TCF/ECF Bleaching. *Bioresour. Technol.* **2010**, *101*, 260. [CrossRef] [PubMed]
43. Csiszár, E.; Fekete, E.; Tóth, A.; Bandi, E.K.; Koczka, B.; Sajó, I. Effect of Particle Size on the Surface Properties and Morphology of Ground Flax. *Carbohydr. Polym.* **2013**, *94*, 927–933. [CrossRef]
44. Sachs, R.; Ihde, J.; Wilken, R.; Mayer, B. Treatment of Flax Fabric with AP-DBD in Parallel Plane Configuration. *Plasma* **2019**, *2*, 272–282. [CrossRef]
45. Majcher, M.A.; Scheibe, M.; Jeleń, H.H. Identification of Odor Active Compounds in *Physalis peruviana* L. *Molecules* **2020**, *25*, 245. [CrossRef]
46. Brattoli, M.; Cisternino, E.; Dambruoso, P.R.; de Gennaro, G.; Giungato, P.; Mazzone, A.; Palmisani, J.; Tutino, M. Gas Chromatography Analysis with Olfactometric Detection (GC-O) as a Useful Methodology for Chemical Characterization of Odorous Compounds. *Sensors* **2013**, *13*, 16759–16800. [CrossRef]
47. Schreiner, L.; Bauer, P.; Buettner, A. Resolving the Smell of Wood-Identification of Odour-Active Compounds in Scots Pine (*Pinus sylvestris* L.). *Sci. Rep.* **2018**, *8*, 8294. [CrossRef] [PubMed]
48. Ma, R.; Liu, X.; Tian, H.; Han, B.; Li, Y.; Tang, C.; Zhu, K.; Li, C.; Meng, Y. Odor-Active Volatile Compounds Profile of Triploid Rainbow Trout with Different Marketable Sizes. *Aquac. Rep.* **2020**, *17*, 100312. [CrossRef]
49. Villavicencio, J.D.; Zoffoli, J.P.; Plotto, A.; Contreras, C. Aroma Compounds Are Responsible for an Herbaceous Off-Flavor in the Sweet Cherry (*Prunus avium* L.) Cv. Regina during Fruit Development. *Agronomy* **2021**, *11*, 2020. [CrossRef]
50. Nie, C.; Gao, Y.; Du, X.; Bian, J.; Li, H.; Zhang, X.; Wang, C.; Li, S. Characterization of the Effect of Cis-3-Hexen-1-Ol on Green Tea Aroma. *Sci. Rep.* **2020**, *10*, 15506. [CrossRef]
51. Lin, H.; Liu, Y.; He, Q.; Liu, P.; Che, Z.; Wang, X.; Huang, J. Characterization of Odor Components of Pixian Douban (Broad Bean Paste) by Aroma Extract Dilute Analysis and Odor Activity Values. *Int. J. Food Prop.* **2019**, *22*, 1223–1234. [CrossRef]
52. Wang, H.; Yang, P.; Liu, C.; Song, H.; Pan, W.; Gong, L. Characterization of Key Odor-Active Compounds in Thermal Reaction Beef Flavoring by SGC×GC-O-MS, AEDA, DHDA, OAV and Quantitative Measurements. *J. Food Compos. Anal.* **2022**, *114*, 104805. [CrossRef]
53. Casassa, L.; Ceja, G.; Vega-Osorno, A.; Fresne, F.; Llodrá, D. Detailed Chemical Composition of Cabernet Sauvignon Wines Aged in French Oak Barrels Coopered with Three Different Stave Bending Techniques. *Food Chem.* **2020**, *340*, 127573. [CrossRef]
54. Fereidoon, S.; Pegg, R.B. Hexanal as an Indicator of Meat Flavor Deterioration. *J. Food Lipids* **1994**, *1*, 177–186. [CrossRef]
55. Clarke, H.J.; McCarthy, W.P.; O'Sullivan, M.G.; Kerry, J.P.; Kilcawley, K.N. Oxidative Quality of Dairy Powders: Influencing Factors and Analysis. *Foods* **2021**, *10*, 2315. [CrossRef]
56. Grebenteuch, S.; Kroh, L.W.; Drusch, S.; Rohn, S. Formation of Secondary and Tertiary Volatile Compounds Resulting from the Lipid Oxidation of Rapeseed Oil. *Foods* **2021**, *10*, 2417. [CrossRef] [PubMed]
57. Kastrup Dalsgaard, T.; Sørensen, J.; Bakman, M.; Vognsen, L.; Nebel, C.; Albrechtsen, R.; Nielsen, J.H. Light-Induced Protein and Lipid Oxidation in Cheese: Dependence on Fat Content and Packaging Conditions. *Dairy Sci. Technol.* **2010**, *90*, 565–577. [CrossRef]
58. Snyder, J.M.; Frankel, E.N.; Selke, E.; Warner, K. Comparison of Gas Chromatographic Methods for Volatile Lipid Oxidation Compounds in Soybean Oil. *J. Amer. Oil Chem Soc.* **1988**, *65*, 1617–1620. [CrossRef]
59. Lipid Oxidation-2nd Edition. Available online: https://www.elsevier.com/books/lipid-oxidation/frankel/978-0-9531949-8-8 (accessed on 11 November 2022).
60. Yang, S.; Lee, J.; Lee, J.; Lee, J. Effects of Riboflavin-Photosensitization on the Formation of Volatiles in Linoleic Acid Model Systems with Sodium Azide or D_2O. *Food Chem.* **2007**, *105*, 1375–1381. [CrossRef]
61. García-Martínez, M.C.; Márquez-Ruiz, G.; Fontecha, J.; Gordon, M.H. Volatile Oxidation Compounds in a Conjugated Linoleic Acid-Rich Oil. *Food Chem.* **2009**, *113*, 926–931. [CrossRef]
62. Min, D.B.; Callison, A.L.; Lee, H.O. Singlet Oxygen Oxidation for 2-Pentylfuran and 2-Pentenyfuran Formation in Soybean Oil. *J. Food Sci.* **2003**, *68*, 1175–1178. [CrossRef]
63. Krishnamurthy, R.G.; Smouse, T.H.; Mookherjee, B.D.; Reddy, B.R.; Chang, S.S. Identification of 2-Pentyl Furan in Fats and Oils and Its Relationship to the Reversion Flavor of Soybean Oil. *J. Food Sci.* **1967**, *32*, 372–374. [CrossRef]
64. Cao, J.; Deng, L.; Zhu, X.-M.; Fan, Y.; Hu, J.-N.; Li, J.; Deng, Z.-Y. Novel Approach to Evaluate the Oxidation State of Vegetable Oils Using Characteristic Oxidation Indicators. *J. Agric. Food Chem.* **2014**, *62*, 12545–12552. [CrossRef] [PubMed]
65. Domenek, S.; Berzin, F.; Ducruet, V.; Plessis, C.; Dhakal, H.; Richaud, E.; Beaugrand, J. Extrusion and Injection Moulding Induced Degradation of Date Palm Fibre-Polypropylene Composites. *Polym. Degrad. Stab.* **2021**, *190*, 109641. [CrossRef]
66. Bourmaud, A.; Beaugrand, J.; Shah, D.; Placet, V.; Baley, C. Towards the Design of High-Performance Plant Fibre Composites. *Prog. Mater. Sci.* **2018**, *97*, 347–408. [CrossRef]

67. Paczkowski, S.; Paczkowska, M.; Dippel, S.; Schulze, N.; Schütz, S.; Sauerwald, T.; Weiß, A.; Bauer, M.; Gottschald, J.; Kohl, C.-D. The Olfaction of a Fire Beetle Leads to New Concepts for Early Fire Warning Systems. *Sens. Actuators B: Chem.* **2013**, *183*, 273–282. [CrossRef]
68. González Martínez, M.; Anca Couce, A.; Dupont, C.; da Silva Perez, D.; Thiéry, S.; Meyer, X.; Gourdon, C. Torrefaction of Cellulose, Hemicelluloses and Lignin Extracted from Woody and Agricultural Biomass in TGA-GC/MS: Linking Production Profiles of Volatile Species to Biomass Type and Macromolecular Composition. *Ind. Crops Prod.* **2022**, *176*, 114350. [CrossRef]
69. Zhao, C.; Jiang, E.; Chen, A. Volatile Production from Pyrolysis of Cellulose, Hemicellulose and Lignin. *J. Energy Inst.* **2017**, *90*, 902–913. [CrossRef]
70. Chen, W.-H.; Wang, C.-W.; Ong, H.C.; Show, P.L.; Hsieh, T.-H. Torrefaction, Pyrolysis and Two-Stage Thermodegradation of Hemicellulose, Cellulose and Lignin. *Fuel* **2019**, *258*, 116168. [CrossRef]
71. Scheirs, J.; Camino, G.; Avidano, M.; Tumiatti, W. Origin of Furanic Compounds in Thermal Degradation of Cellulosic Insulating Paper. *J. Appl. Polym. Sci.* **1998**, *69*, 2541–2547. [CrossRef]
72. Shen, D.K.; Gu, S.; Bridgwater, A.V. Study on the Pyrolytic Behaviour of Xylan-Based Hemicellulose Using TG–FTIR and Py–GC–FTIR. *J. Anal. Appl. Pyrolysis* **2010**, *87*, 199–206. [CrossRef]
73. Liu, Q.; Zhong, Z.; Wang, S.; Luo, Z. Interactions of Biomass Components during Pyrolysis: A TG-FTIR Study. *J. Anal. Appl. Pyrolysis* **2011**, *90*, 213–218. [CrossRef]
74. Dorez, G.; Ferry, L.; Sonnier, R.; Taguet, A.; Lopez-Cuesta, J.-M. Effect of Cellulose, Hemicellulose and Lignin Contents on Pyrolysis and Combustion of Natural Fibers. *J. Anal. Appl. Pyrolysis* **2014**, *107*, 323–331. [CrossRef]
75. Chabi, M.; Goulas, E.; Leclercq, C.C.; de Waele, I.; Rihouey, C.; Cenci, U.; Day, A.; Blervacq, A.-S.; Neutelings, G.; Duponchel, L.; et al. A Cell Wall Proteome and Targeted Cell Wall Analyses Provide Novel Information on Hemicellulose Metabolism in Flax*. *Mol. Cell. Proteom.* **2017**, *16*, 1634–1651. [CrossRef] [PubMed]
76. Luo, Y.; Li, Z.; Li, X.; Liu, X.; Fan, J.; Clark, J.H.; Hu, C. The Production of Furfural Directly from Hemicellulose in Lignocellulosic Biomass: A Review. *Catal. Today* **2019**, *319*, 14–24. [CrossRef]
77. Weil, J.; Dien, B.; Bothast, R.; Hendrickson, R.; Mosier, N.; Ladisch, M. Removal of Fermentation Inhibitors Formed during Pretreatment of Biomass by Polymeric Adsorbents. *Ind. Eng. Chem. Res.* **2002**, *41*, 132–6138. [CrossRef]
78. Fahm, T.Y.A.; Mobarak, F. Advanced Binderless Board-like Green Nanocomposites from Undebarked Cotton Stalks and Mechanism of Self-Bonding. *Cellulose* **2013**, *20*, 1453. [CrossRef]
79. Wakushie, N.; Woldeyes, B.; Demsash, H.; Jabasingh, S. An Insight into the Valorization of Hemicellulose Fraction of Biomass into Furfural: Catalytic Conversion and Product Separation. *Waste Biomass Valorization* **2021**, *12*, 531–552. [CrossRef]
80. Collard, F.-X.; Blin, J. A Review on Pyrolysis of Biomass Constituents: Mechanisms and Composition of the Products Obtained from the Conversion of Cellulose, Hemicelluloses and Lignin. *Renew. Sustain. Energy Rev.* **2014**, *38*, 594–608. [CrossRef]
81. Nitu, I.P.; Islam, M.N.; Ashaduzzaman, M.; Amin, M.K.; Shams, M.I. Optimization of Processing Parameters for the Manufacturing of Jute Stick Binderless Particleboard. *J. Wood Sci.* **2020**, *66*, 65. [CrossRef]
82. Wang, S.; Dai, G.; Yang, H.; Luo, Z. Lignocellulosic Biomass Pyrolysis Mechanism: A State-of-the-Art Review. *Prog. Energy Combust. Sci.* **2017**, *62*, 33–86. [CrossRef]
83. Pintiaux, T.; Heuls, M.; Vandenbossche Maréchal, V.; Murphy, T.; Wuhrer, R.; Castignolles, P.; Gaborieau, M.; Rouilly, A. Cellulose Consolidation under High-Pressure and High-Temperature Uniaxial Compression. *Cellulose* **2019**, *26*, 2941–2954. [CrossRef]
84. Badji, C.; Sotiropoulos, J.-M.; Beigbeder, J.; Garay, H.; Bergeret, A.; Bénézet, J.-C.; Desauziers, V. Under Glass Weathering of Hemp Fibers Reinforced Polypropylene Biocomposites: Degradation Mechanisms Based on Emitted Volatile Organic Compounds. *Front. Mater.* **2020**, *7*, 162. [CrossRef]
85. Nowakowski, D.J.; Bridgwater, A.V.; Elliott, D.C.; Meier, D.; de Wild, P. Lignin Fast Pyrolysis: Results from an International Collaboration. *J. Anal. Appl. Pyrolysis* **2010**, *88*, 53–72. [CrossRef]
86. Sallem-Idrissi, N.; Vanderghem, C.; Pacary, T.; Richel, A.; Debecker, D.P.; Devaux, J.; Sclavons, M. Lignin Degradation and Stability: Volatile Organic Compounds (VOCs) Analysis throughout Processing. *Polym. Degrad. Stab.* **2016**, *130*, 30–37. [CrossRef]

Disclaimer/Publisher's Note: The statements, opinions and data contained in all publications are solely those of the individual author(s) and contributor(s) and not of MDPI and/or the editor(s). MDPI and/or the editor(s) disclaim responsibility for any injury to people or property resulting from any ideas, methods, instructions or products referred to in the content.

Nanocellulose and Cellulose Making with Bio-Enzymes from Different Particle Sizes of Neosinocalamus Affinis

Jiaxin Zhao [1,†], Xiaoxiao Wu [1,†], Xushuo Yuan [1,†], Xinjie Yang [1], Haiyang Guo [2,*], Wentao Yao [1], Decai Ji [1], Xiaoping Li [1,*] and Lianpeng Zhang [1,*]

1. Yunnan Provincial Key Laboratory of Wood Adhesives and Glued Products, Southwest Forestry University, Kunming 650224, China
2. College of Biological, Chemical Sciences and Engineering, Jiaxing University, Jiaxing 314001, China
* Correspondence: guohy@zjxu.edu.cn (H.G.); lxp810525@163.com (X.L.); lpz@zju.edu.cn (L.Z.)
† These authors contributed equally to this work.

Abstract: Cellulose is one of the most abundant, widely distributed and abundant polysaccharides on earth, and is the most valuable natural renewable resource for human beings. In this study, three different particle sizes (250, 178, and 150 μm) of Neosinocalamus affinis cellulose were extracted from Neosinocalamus affinis powder using bio-enzyme digestion and prepared into nanocellulose (CNMs). The cellulose contents of 250, 178, and 150 μm particle sizes were 53.44%, 63.38%, and 74.08%, respectively; the crystallinity was 54.21%, 56.03% and 63.58%, respectively. The thermal stability of cellulose increased gradually with smaller particle sizes. The yields of CNMs for 250, 178, and 150 μm particle sizes were 14.27%, 15.44%, and 16.38%, respectively. The results showed that the Neosinocalamus affinis powder was successfully removed from lignin, hemicellulose, and impurities (pectin, resin, etc.) by the treatment of bio-enzyme A (ligninase:hemicellulose:pectinase = 1:1:1) combined with $NH_3 \cdot H_2O$ and H_2O_2/CH_3COOH. Extraction of cellulose from Neosinocalamus affinis using bio-enzyme A, the smaller the particle size of Neosinocalamus affinis powder, the more cellulose content extracted, the higher the crystallinity, the better the thermal stability, and the higher the purity. Subsequently, nanocellulose (CNMs) were prepared by using bio-enzyme B (cellulase:pectinase = 1:1). The CNMs prepared by bio-enzyme B showed a network structure and fibrous bundle shape. Therefore, the ones prepared in this study belong to cellulose nanofibrils (CNFs). This study provides a reference in the extraction of cellulose from bamboo using bio-enzymes and the preparation of nanocellulose. To a certain extent, the utilization of bamboo as a biomass material was improved.

Keywords: Neosinocalamus affinis; cellulose; nanocellulose; bio-enzymes; particle size

1. Introduction

Bamboo is a woody perennial plant of the Gramineae family. It is a potential sustainable biomass energy feedstock. Bamboo includes 75 genera and 1250 species, most of which have a short growth period, reaching maturity in about 5 years [1]. There are about 1000 species of bamboo in Asia. China alone has 44 genera and about 300 species, covering about 33,000 km, accounting for 2%–3% of the total forest area of the country. Neosinocalamus affinis is one of the most planted and cultivated bamboo species in southwest China. It has long internodes, a high cellulose content (close to that of wood), and a large fiber length. The growth cycle is much shorter than that of wood. Scientific management of bamboo forests can be used forever without damaging the ecological environment. This is a characteristic that no other gramine has.

Cellulose is the main component of the cell wall of higher plants. It is the most abundant and important organic polymer, naturally renewable. It is a very high-demand alternative energy source worldwide, with the molecular formula $(C_6H_{10}O_5)_n$. It is a polysaccharide, consisting of hundreds to thousands of glucose unit chains, and is found in abundance on the earth, with varying amounts of cellulose in different plants. Research shows that cellulose is a promising renewable and sustainable chemical raw material [2,3]. It is widely used in paper, pharmaceuticals, packaging, textiles, automobiles, environmental management, and other fields [4,5].

With globalization and sustainable development, there is an increasing demand for biodegradable nanomaterials for nanotechnology. Pathak et al. prepared polycrystalline silver indium selenide films on Si (100) substrates by vacuum evaporation at high temperatures using chemometric powders. AIS nanorods were successfully synthesized by infrared radiation with incident 200 MeV Ag^+ ions at a flux of 5×10 ions/cm^2 on Si (100) substrates [6]. Sagadevan et al. completed the controlled synthesis of $TiO_2/SiO_2/CdS$-nanocomposites by the hydrothermal-assisted method. Optical properties were obtained by UV-Vis absorption spectroscopy. The optical band gap of the nanocomposites was found to be 3.31 eV [7]. Pathak et al. prepared organic semiconductor films on glass substrates and annealed them at 55 °C. The engineered band gaps of the synthesized films were 2.18, 2.35, 2.36, 2.52 and 2.65 eV. This indicates that it can be used as a new lightweight and environmentally friendly carbon-based material for photovoltaic devices such as solar cells [8]. Hosseini synthesized $MgAl_2O_4/NiTiO_3$ nanocomposites for the removal of MeO dyes from water by the sol-gel method. The prepared nanocomposites showed excellent photodegradation of MeO under UV light. The photodegradation efficiency of MeO was 84% [9].

Biomass is considered an ideal alternative to fossil energy due to its low price, wide range of sources, and renewable and sustainable characteristics [10]. Therefore, research and development on the use of biomass energy have become a priority in many countries around the world. The global demand for new sustainable materials has grown rapidly in recent years, as evidenced by the UN 2030 Agenda [11]. A raw material with the potential for many new materials is wood and cellulose pulp fibers derived from wood. By separating the nanostructures that make up the fibers, it is possible to obtain materials with highly interesting properties—such as nanocellulose. As the most promising biomass nanomaterials. Nanocellulose has many advantages: biodegradable, low composite energy consumption, good processable type, good heat resistance, strong tensile properties, high specific surface area, high adsorption, non-cytotoxic and biocompatible. It is widely used in biomedical, energy storage, biosorption, food packaging, electronic devices and other fields [12,13]. There are many kinds of nanocellulose preparation methods, such as acid hydrolysis, physical-mechanical method, enzymatic method and solvent method. Seta et al. used a ball mill to pretreat the bamboo fibers and added maleic acid to hydrolyze the bamboo fibers to obtain CNC. A control sample without maleic acid was also prepared. The results showed that the CNC yields were 10.55%–24.50%. It was much higher than the control 2.80% [14]. Mahmud et al. prepared different forms and multiple morphologies of CNC by acid hydrolysis of medical cotton in sulfuric, hydrochloric and phosphoric acids [15]. OKsman et al. added DMF to 0.5 wt% lithium chloride solution as a swelling agent for cellulose, placed microcrystalline cellulose into the solution, stirred at 70 °C for 12 h then ultrasonically separated. The final nanocellulose with a width of about 10 nm was obtained [16]. Chen et al. used cellulase to hydrolyze cotton pulp fibers to prepare ribbon-shaped CNC. The results showed that when the concentration of cellulase was low, ribbon-shaped CNCs with a length of about 250–900 nm were prepared. When the concentration of cellulase increased at 300 µ/mL, the prepared CNCs were all granular [17]. Zhang et al. used the steam blasting method to pretreat poplar wood and then prepared nanocellulose by bio-enzymatic-assisted ultrasonication. Its width ranged from 20 to 50 nm and had a high aspect ratio and network entanglement structure [18]. Hayashi et al. enzymatically digested the microcrystalline cellulose of the algae of setae in a solution of cellulase at 48 °C and pH 4.8. Nanocellulose with an average length of 350 nm was obtained after 2–3 days [19]. Beltramino et al. investigated the effect of different cellulase doses and reaction times on NCC. The results showed that the yield of NCC exceeded 80% under optimal enzymatic conditions (20 U, 2 h). Moreover, the different intensities of enzyme treatment resulted in a decrease in fiber length and viscosity [20]. Currently, there are few studies related to the effect of particle size on the extraction of cellulose and the preparation of nanocellulose. However, in other fields, the particle size is equally important for the performance of the samples. Huang et al. prepared four different particle sizes of beet pulp powder by ultra-fine grinding and ordinary crushing methods. The effect of beet pulp particle size on the extraction and properties of pectin was investigated. The results showed that as the particle size of beet pulp decreased, the extraction rate and content of pectin gradually increased. Moreover, the viscosity showed a negative correlation with particle size [21]. Meng et al. investigated the effect of different particle sizes of nano-SiO_2 on the properties and microstructure of cement paste. The results showed that 50 nm nano-SiO_2 provided higher compressive strength than 15 nm nano-SiO_2. The 15 nm nano-SiO_2 provided a denser microstructure than the 50 nm nano-SiO_2 [22]. Cheng et al. prepared nano, micron and micron/nano ZnO/LDPE by melt blending using LDPE as matrix polymer and ZnO particles with diameters of 30 nm and 1 µ as inorganic fillers. The experimental results showed a tendency to reduce the AC breakdown field strength of all composites. However, the reduction in micron ZnO/LDPE is lower than that in nano ZnO/LDPE [23].

There are many reports on the preparation of nanocellulose by bio-enzymatic methods. However, there are fewer reports on the extraction of cellulose from bamboo using bio-enzymatic methods. Moreover, the range of cellulosic raw materials selected for the preparation of nanocellulose is relatively narrow, mainly including wood, pulp fibers and bacterial cellulose [24]. In comparison with other traditional methods, biological enzymes have specificity, mild enzymatic process conditions, and enzyme reagents are renewable resources. The bio-enzymatic method greatly reduces energy consumption, reduces the use of chemicals, and avoids problems such as pollution of the environment during the experimental process. It is of great significance to ecologically sustainable development. However, the reaction conditions (amount of enzyme, reaction time and reaction temperature, etc.) for the extraction of cellulose by bio-enzymatic methods and the preparation of nanocellulose are more demanding. The preparation efficiency is relatively slow. The presence of these factors hinders the industrialization of cellulose extraction by bio-enzymatic methods and the preparation of nanocellulose.

In this study, three different particle sizes (250, 178, and 150 µm) of Neosinocalamus affinis powder were selected, following the single variable principle. Cellulose was extracted by using bio-enzyme A (ligninase:hemicellulose:pectinase = 1:1:1) in combination with $NH_3·H_2O$ and H_2O_2/CH_3COOH, and characterized and analyzed using Fourier Transform infrared spectroscopy (FTIR), X-ray Diffraction (XRD), Thermogravimetric analysis (TG) and Carbon nuclear magnetic resonance spectra (^{13}C-NMR). The nanocellulose was prepared by bio-enzyme B (cellulase:pectinase = 1:1), and characterized and analyzed for morphology and structural dimensions. The results of this study will help to improve the utilization of bamboo materials, expand the sources of cellulose as well as nanocellulose raw materials and explore greener methods of cellulose extraction and preparation of nanocellulose.

2. Materials and Methods

2.1. Materials and Reagents

Material: Neosinocalamus affinis (Kunming, China). Neosinocalamus affinis was crushed through a high-speed multifunctional crusher, passed through 60 mesh sieve (250 µm particle size), 80 mesh sieve (178 µm particle size) and 100 mesh sieve (150 µm particle size). 105 ± 2 °C oven dried for 24 h and stored for spare.

Chemical reagents: Ammonia solution ($NH_3·H_2O$), Hydrogen peroxide (H_2O_2) and Glacial acetic acid (CH_3COOH) were purchased from Yunnan Shuoyang Biological Company, Ltd, Shuoyang, China. Pectinase (CAS: 9032-75-1), hemicellulose (CAS: 9025-56-3), ligninase (CAS: 80498-15-3) and cellulase (CAS: 9012-54-8) were purchased from Aladdin Biotechnology Co., Ltd. (Shanghai, China). The pectinase activity was 100,000 U/g, hemicellulase activity was 30,000 U/g, ligninase activity was 500 U/g and cellulase activity was 10,000 U/g. Analytical purity-grade reagents were used in all experiments.

2.2. Experiment Equipment

Constant temperature magnetic stirrer (C-MAG HS 7, Hangzhou Aipu Instruments Company Limited, Hangzhou, China). High-speed Multi-functional Crusher (XY-100, Yongkang Songqing Hardware Factory, Yongkang, China). Electric Heat Blast Dryer (DHG-9003, Shanghai Yiheng Scientific Instruments Company Limited, Shanghai, China). Electronic balance (BSM220, Shanghai Zhuojing Electronic Technology Company Limited, Shanghai, China). Thermostatic water bath (B220, Shanghai Yarong Biochemical Instrument Factory, Shanghai, China).

2.3. Cellulose Extraction

The method of extracting cellulose is shown in Figure 1. Take the above particle size of 250, 178, and 125 μm of Neosinocalamus affinis powder (10 g) first add the appropriate amount of $NH_3 \cdot H_2O$ pretreatment 25 °C soaked for 24 h, using distilled water rinse filter to neutral, 105 ± 2 °C constant temperature drying. The purpose of this is to soften the Neosinocalamus affinis pellets and to treat the sample with alkali to remove some of the hemicellulose and lignin. Then, an appropriate amount of H_2O_2/CH_3COOH (1:1) was added to the pretreated cichlid pellets for bleaching and partial removal of lignin in a constant temperature water bath at 70 °C for 10 h. After acid treatment, the samples were filtered and washed with distilled water until the filtrate was neutral and dried at 105 °C $\pm$ 2 °C. Finally, 10% ligninase solution, 5% hemicellulase solution and 5% pectinase solution were configured and calibrated. Add an appropriate amount of 10% ligninase solution, 5% hemicellulase solution, and 5% pectinase solution (1:1:1) to the sample after ammonia pretreatment and acid treatment and put it into magnetic stirring. The samples were treated at a constant temperature of 50 °C and stirring speed of 800–1200 r/min for 240 min to remove residual lignin, hemicellulose and impurities (e.g., resins, pectin, etc.). Afterward, the filtered samples were rinsed with plenty of distilled water until clean and dried at a constant temperature of 105 ± 2 °C.

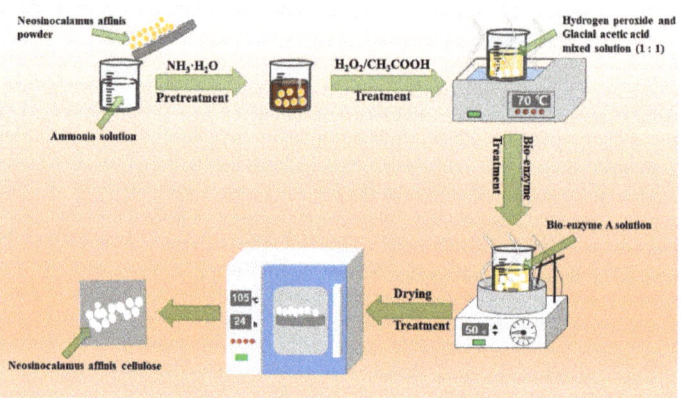

Figure 1. Flowchart of cellulose extraction.

2.4. Preparation of Nanocellulose (CNMs)

Nanocellulose was prepared using a bio-enzyme method [24] with slight modifications, as shown in Figure 2. Take the above cellulose (0.5 g) extracted from different particle sizes into a beaker and add 20 g of zirconia grinding beads (0.1 mm diameter). Configure and fix the volume of 5% cellulase solution and 5% pectinase solution, and add 15 mL each of cellulase solution and pectinase solution to the beaker. Then place it in a 100 °C water bath for 20 min and sonicate for 30 min. The purpose of this is to inactivate bio-enzymes. Afterward, the precipitate was collected and diluted with distilled water by centrifugation at 8000 r/min for 10 min to wash away the biological enzyme impurities and repeated five times. Finally, stir (800–1200 r/min) at room temperature for 30 min, sonicate for 30 min and repeat several times until the suspension is formed. The suspension was freeze-dried into a powder to obtain Neosinocalamus affinis nanocellulose (CNMs).

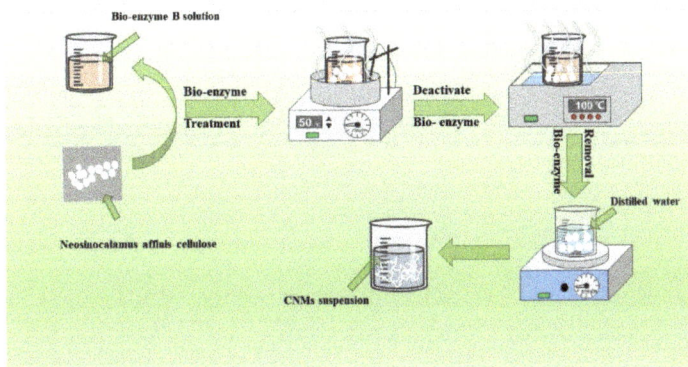

Figure 2. Flowchart of preparing CNMs.

2.5. Material Characterization

The chemical composition was determined and the extracted cellulose was characterized using Fourier transform infrared spectroscopy (FTIR), X-ray diffractometry (XRD), thermogravimetric analysis (TGA), and ^{13}C-NMR. Morphological characterization of nanocellulose (CNMs) was performed using field emission scanning electron microscopy (SEM) and transmission electron microscopy (TEM).

2.5.1. Determination of Chemical Composition

Lignin content was determined according to the Chinese standard GB/T2677. 8-94 (Determination of acid-insoluble lignin content of paper raw materials). Holo-cellulose content was determined according to the Chinese standard GB/T2677. 10-1995 (Determination of Holo-cellulose content of paper raw materials). α-cellulose content was determined according to the Chinese standard GB/T744-1989 (Determination of α-cellulose of pulp). The hemicellulose content was calculated according to Equation (1), and the experiment was repeated three times, and the results were averaged.

Hemicellulose content (%) = Holo-cellulose content (%) − α-cellulose content (%) (1)

2.5.2. FTIR

The samples were dried with KBr, mixed at a mass ratio of 1:100 and then ground and pressed. The samples were examined by a Varian 1000 Fourier transform infrared spectrometer (FTIR, Varian, Palo Alto, CA, USA) with a scan range of 400–4000 cm^{-1} at a resolution of 4 cm^{-1} and the spectra were collected for an average of 32 scans. The main absorption peaks were determined using ORIGIN (2018) software.

2.5.3. XRD

XRD was used to study the crystallinity of the samples. The samples were scanned using a LabX6000 X-ray diffractometer from Shimadzu, Kyoto, Japan, at 40 kV and 30 mA current and in the diffraction angle range of 10°–60°. Equation (2) [25] was used to calculate the crystallinity index (CrI) of the samples:

$$CrI\ (\%) = \frac{I_{200} - I_{Am}}{I_{200}} \times 100\% \quad (2)$$

CrI is the crystallinity index. I_{200} is the intensity of the diffraction peak near $2\theta = 22°$ on the 200-crystal plane, which is the diffraction intensity of the crystalline region. I_{Am} is the intensity of the lattice diffraction peak near $2\theta = 15.7°$, which is the diffraction intensity of the amorphous region.

2.5.4. TGA

The samples were analyzed by TGA using a Pyris Diamond thermogravimetric analyzer from Pekin-Elmer, Waltham, MA, USA. The thermogravimetric (TG) curves and derivative thermogravimetric (DTG) curves were recorded from 30 to 600 °C at a constant heating rate of 10 °C/min under nitrogen gas.

2.5.5. ^{13}C-NMR

A small amount of cellulose was dissolved in deuterated dimethyl sulfoxide, and ^{13}C-NMR spectra of the samples were recorded using Bruker (Billerica, MA, USA) MSI400 spectroscopy at 25 °C and 62.9 MHz with a MAS rate of 3 kHz. Each spectrum was obtained by accumulating 5000 scans. The delay time was 60 s with a proton 90 (pulse width of 9 mm and contact time of 2 ms for cross-polarization).

2.5.6. Calculation of CNMs Yield and Viscosity Determination

The CNMs suspension was prepared by stirring, sonication and centrifugation by homogeneous dispersion of CNMs in water. The CNMs yield was determined by weight analysis and Yield (%) was calculated according to Equation (3) [26]:

$$Y\ (\%) = \frac{(m_1 - m_2)V_1}{m_3 V_2} \quad (3)$$

In Equation (3), Y-yield of CNMs, %; m_1-total mass of the sample and sealed bag after freeze-drying, g; m_2-mass of the sealed bag, g; m_3-mass of Neosinocalamus affinis cellulose, g; V_1-total volume of CNMs suspension, mL; V_2-volume of CNMs suspension used for freeze-drying, mL.

The viscosity of CNMs suspension was measured at 25 °C using an SNB-2 digital viscometer (Shanghai Heng ping Instrument Factory, Shanghai, China). The measurement was performed according to Chinese standard GB/T 14074-2017 (Test method for adhesives and their resins for the wood industry). The experiment was repeated three times, and the results were averaged.

2.5.7. TEM

The microstructure of nanocellulose was studied by transmission electron microscopy (TEM, JEM 2100, Japan Electronics Co., Ltd., Tokyo, Japan) at an accelerating voltage of 200 kV. A drop of nanocellulose suspension (0.1 wt%) was placed on a copper grid and dried at room temperature. CNMs powder was prepared from the CNMs suspension and homogeneously mixed with KBr.

2.5.8. SEM

The samples were placed on a copper grid, sprayed with gold by an ion sputterer, and then analyzed, observed and photographed by field emission scanning electron microscopy for sample morphology images. The samples were characterized using an SEM instrument (JEOL JSM-7100 F, Tokyo, Japan) at a voltage of 15 kV.

3. Analysis and Discussion

3.1. Chemical Composition

Table 1 shows the cellulose content extracted from three different particle sizes of Neosinocalamus affinis powder. Table 1 shows that the cellulose content of 250, 178 and 150 µm of Neosinocalamus affinis was 53.44%, 63.38% and 74.08%, respectively. The results showed that the highest cellulose content was extracted from 150 µm of Neosinocalamus affinis powder. This is because the smaller particle size has a larger surface area and thus a larger surface area is exposed. This allows smaller particle sizes to react more readily with bio-enzymes than larger particle sizes, resulting in the easier removal of lignin and hemicellulose [27].

Table 1. Content of cellulose extracted from different particle sizes of Neosinocalamus affinis powder.

Chemical Compound	250 µm	178 µm	150 µm
Celluloses (%)	53.44	63.38	74.08
Hemicellulose (%)	14.72	10.62	7.59
Lignin (%)	7.60	4.35	3.55

3.2. FTIR Spectra of Cellulose

FTIR spectroscopy can visually determine the structural changes of the sample material. The FTIR spectra of cellulose extracted from three different particle sizes of Neosinocalamus affinis powder are shown in Figure 3. The attribution of some peaks of FTIR absorption is summarized in Table 2. The infrared spectra showed that almost all absorption peaks of the Neosinocalamus affinis powder were present in the spectrum of the extracted cellulose. There are wider and stronger absorption bands near 3370 cm^{-1} in the spectra of cellulose extracted from different particle sizes in Figure 3, which is the stretching vibration peak of O–H in cellulose. The C–H induced stretching vibration peak near 2900 cm^{-1} [28,29]. Near 1640 cm^{-1} is the stretching vibration peak caused by –OH, which is the signal absorption peak of water molecules in cellulose [30]. The absorption peak near 1500 cm^{-1} is the stretching vibration peak caused by C=C in lignin [31]. The absorption peaks of curves b, c, and d here gradually weaken and disappear. The absorption peak near 1114 cm^{-1} is a stretching vibration peak caused by C–O–C in pyranose. The peak near 890 cm^{-1} is the absorption peak of the β-D glucosidic bond in cellulose, which is a characteristic peak of cellulose [32,33]. The cellulose characteristic peaks in curves b, c, and d were increased due to the removal of lignin and hemicellulose compared with curve a. The cellulose content of Neosinocalamus affinis was increased due to the removal of lignin and hemicellulose. The intensity of the cellulose characteristic peaks was relatively enhanced. FTIR spectroscopy showed that bio-enzyme A combined with $NH_3 \cdot H_2O$ and H_2O_2/CH_3COOH could effectively remove lignin and hemicellulose from Neosinocalamus affinis. It disintegrates the strong surface structure of Neosinocalamus affinis and releases cellulose. It is favorable for the resource utilization of Neosinocalamus affinis cellulose.

Table 2. Characteristic peaks of some groups in the infrared spectrum.

Wave Number (cm^{-1})	Attribution of Characteristic Absorption Peaks	References
3370	O–H stretching vibration (hydrogen bonding)	[28]
2290	CH$_2$ asymmetric stretching vibration	[29]
1640	O–H bending vibration (absorption of H$_2$O), conjugated to C=C	[30]
1500	C=C stretching vibration	[31]
1162	C (1)–O–C (4) symmetric stretching vibration (sugar ring linkage bond characteristic peak)	[34]
1114	C (1)–O–C (5) intra-face pyranose ring asymmetric stretching vibration	[32]
1087	C (3)–OH stretching vibration	[35]
1030	C (6)–OH stretching vibration	[35]
890	C (1)–H asymmetric stretching vibration outside the face β-D glucosidic bond characteristic	[33]

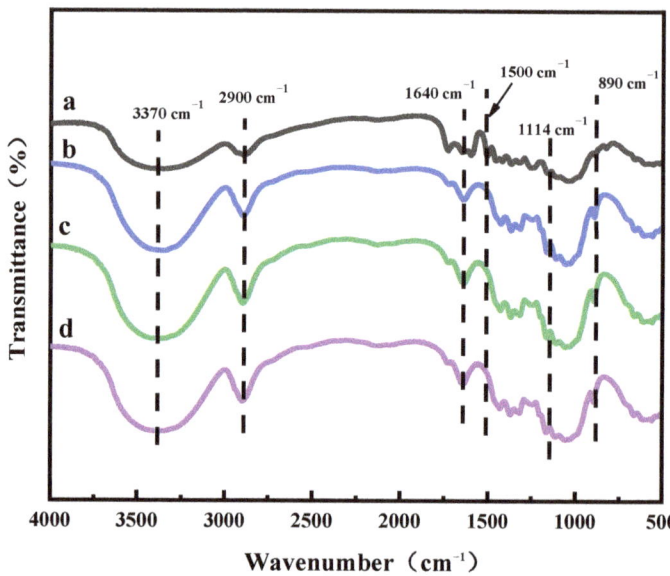

Figure 3. Infrared spectrum: (a) Neosinocalamus affinis powder; (b) 250 μm cellulose; (c) 178 μm cellulose; (d) 150 μm cellulose.

3.3. XRD Spectral Characteristics of Cellulose

Cellulose consists mainly of crystalline and amorphous regions formed by intra- or intermolecular (hydrogen bonding, van der Waals forces) interactions [36]. The crystallinity indices of cellulose extracted from three different particle sizes of Neosinocalamus affinis powder were analyzed by XRD. The crystallinity of cellulose affects its mechanical properties. Moreover, high crystallinity is more effective in enhancing the properties of cellulose in composites [37]. The XRD diffraction pattern is shown in Figure 4. Three distinct crystalline peaks are observed in the cellulose diffraction pattern, all samples have diffraction peaks at $2\theta = 15.7°$, $2\theta = 22.2°$, $2\theta = 34.7°$. The diffraction peaks out of 22.2° and 34.7° are attributed to (200) and (004) reflections. It is shown that all samples have a typical cellulose type I crystal structure [38,39]. Although a series of chemical treatments did not destroy the crystalline structure of the cellulose, its crystallinity was altered. In Figure 4, it can be seen that curve c has a steeper diffraction peak for I_{200} compared with curves a and b. After the calculation of Equation (2) and Table 3, it is known that the crystallinity of 250 μm is 54.21%, 178 μm crystallinity is 56.03%, and 150 μm crystallinity is 63.58%. XRD further demonstrated that the combination of bio-enzyme A with $NH_3·H_2O$ and H_2O_2/CH_3COOH successfully removed the most difficult to remove lignin from the biomass, and also removed hemicellulose and impurities (such as gum, resin, etc.). Thus, the crystallinity of cellulose is improved. The XRD results showed that the reaction contact area of the Neosinocalamus affinis powder became larger as the particle size became smaller. This has more degradation of cellulose non-crystalline region by using bio-enzymes, which makes cellulose crystallinity increase.

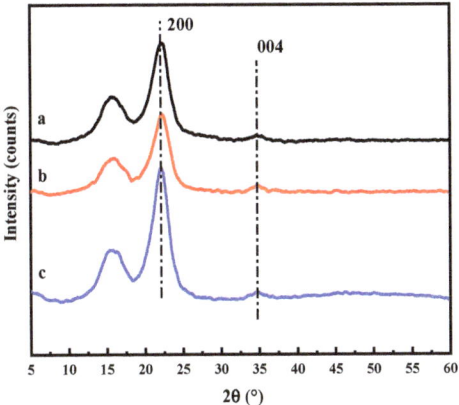

Figure 4. XRD spectra: (a) 250 μm cellulose; (b) 178 μm cellulose; (c) 150 μm cellulose.

Table 3. The crystallinity of cellulose extracted from different particle sizes of Neosinocalamus affinis powder.

Particle Size (μm)	CrI (%)
250	54.21%
178	56.03%
150	63.58%

3.4. Thermal Characterization of Cellulose

Thermal analysis was performed on cellulose extracted from different particle sizes of Neosinocalamus affinis. Figure 5 shows the TG and DTG curves of cellulose extracted from different particle sizes. Near the first stage (0~110 °C), 250, 178, and 150 μm have about 6.28%, 7.50%, and 4.66% thermal weight loss, respectively. This is due to the evaporation of water from the sample [40]. In the second stage, between approximately 200~400 °C, the glycosidic bonds of cellulose begin to break and a large amount of cellulose begins to be cleaved [41]. The heat loss rate of 250, 178 and 150 μm at this stage is about 70%, 75% and 68%, respectively. The DTG curve shows that the weight loss rate of 150 and 178 μm is less than that of 250 μm, so the heat resistance of 250 μm is worse. The third stage starts from about 400 °C to the completion of pyrolysis [42]. The residues decomposed slowly, and the residues were mainly non-decomposable ash, with residual amounts of 13.09%, 11.15%, and 14.03%, respectively. The above pyrolysis-related parameters can be combined to assign a composite index D to characterize the ease of pyrolysis of cellulose. The smaller the composite index D, the better its thermal stability [43].

$$D = \frac{(dw/dt)_{max}(dw/dt)_{mean}V_\infty}{T_s T_{max} \Delta T_{1/2}} \quad (4)$$

In addition, the following main parameters of the reaction pyrolysis characteristics commonly used are known from the TG/DTG curves and related calculations: pyrolysis initial temperature T_s; maximum pyrolysis weight loss rate $(dw/dt)_{max}$; peak temperature T_{max} corresponding to $(dw/dt)_{max}$; average pyrolysis weight loss rate $(dw/dt)_{mean}$; maximum pyrolysis weight loss rate V_∞; temperature interval $\Delta T_{1/2}$ corresponding to $(dw/dt)/(dw/dt)_{max} = 1/2$ of the temperature interval $\Delta T_{1/2}$. The D values are calculated by Equation (4) (Table 4) and the above data show that the D values of 250, 178, and 150 μm cellulose are 5.82×10^{-6}, 5.56×10^{-6}, and 4.87×10^{-6}, respectively. The TG/DTG curves further demonstrated the effective removal of hemicellulose and lignin by the treatment with bio-enzyme A combined with $NH_3 \cdot H_2O$ and H_2O_2/CH_3COOH. Cellulose content was increased. In other words, the thermal stability of cellulose is improved due to the removal of lignin and hemicellulose, which have poor thermal stability. The results show that as the particle size becomes smaller, the overall index D value also becomes smaller. It indicates that the purity and thermal stability of cellulose are higher and better.

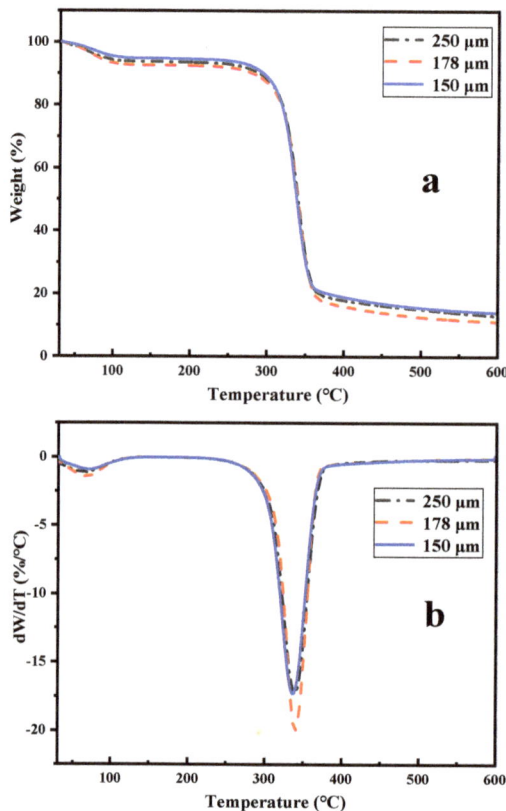

Figure 5. (a) Cellulose TG curve; (b) Cellulose DTG curve.

Table 4. Pyrolytic properties of cellulose.

Particle Size (μm)	$(dw/dt)_{max}$ (%/°C)	$(dw/dt)_{mean}$ (%/°C)	V_∞ (%)	T_S (°C)	T_{max} (°C)	$\Delta T_{(1/2)}$ (°C)	D
250	−17.30	−1.53	86.913	309.36	339.60	37.61	5.82×10^{-6}
178	−19.99	−1.57	88.850	312.58	340.30	47.18	5.56×10^{-6}
150	−17.31	−1.51	85.965	316.87	337.40	43.19	4.87×10^{-6}

3.5. ^{13}C-NMR Characterization of Cellulose

The cellulose was further analyzed in combination with ^{13}C-NMR because of its more complex structure. Based on the above series of characterization, the best Neosinocalamus affinis cellulose (150 μm particle size) was selected and characterized by ^{13}C-NMR. Figure 6 shows the ^{13}C-NMR spectra of 150 μm Neosinocalamus affinis cellulose. The peak at 63.11 ppm is caused by the C_6 glucopyranose repeat unit in cellulose. Peaks around 72.54, 73.91 and 75.35 ppm are caused by C_2, C_3 and C_5. The peak at 84.4 ppm is caused by C_4 in the non-crystalline region of cellulose. No significant characteristic peak appears at

84.44 ppm in Figure 6. It indicates that the 150 μm Neosinocalamus affinis cellulose was preferentially degraded in the non-crystalline region after the treatment. The resulting increase in crystallinity of the cellulose is in general agreement with the XRD characterization results. The absorption peak at 101.66 ppm belongs to C_1 of glucose in cellulose [44].

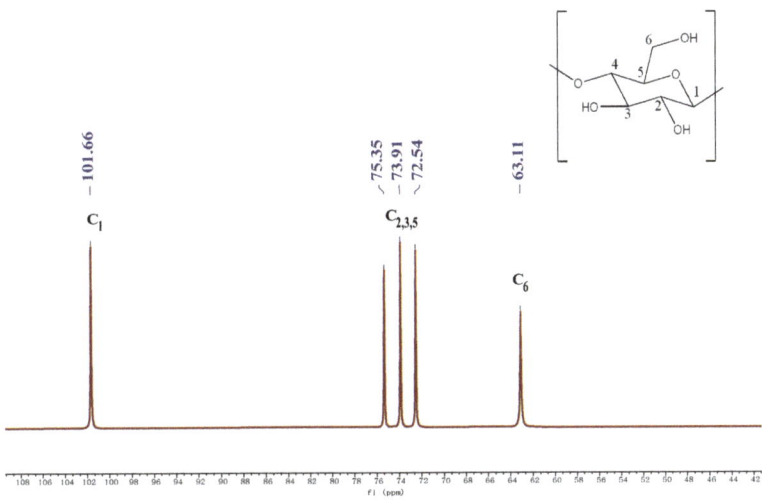

Figure 6. 150 μm cellulose ^{13}C-NMR.

3.6. CNMs Yield and Viscosity Analysis

To study in-depth the preparation of nanocellulose using bio-enzymes, the extracted cellulose was prepared into nanocellulose by bio-enzyme B. Figure 7 shows the preparation of CNMs from cellulose extracted from different particle sizes by bio-enzyme B. The yield of CNMs was calculated by Equation (3). The figure shows that the yield of CNMs was 14.27% for 250 μm, 15.44% for 178 μm and 16.38% for 150 μm, respectively. It was clearly shown that the yield of CNMs gradually increased as the particle size of Neosinocalamus affinis cellulose decreased and the contact area of the reaction became larger. Owing to the small particle size, the reaction contact area is large. By the synergistic action of pectinase and cellulase, the endonuclease in cellulase effectively promotes the breakage of molecular chains in the amorphous region of cellulose [45].

Figure 8 shows the CNMs suspensions prepared from three different particle sizes of Neosinocalamus affinis cellulose. The CNMs suspensions were in translucent colloidal form with concentrations of about 0.097 wt%, 0.102 wt%, and 0.119 wt%. The concentration of the nanocellulose suspension affects its dispersion state, from stable suspension to gel state as the concentration increases [46]. The CNMs suspensions prepared in this experiment were relatively low in concentration and presented a relatively stable suspension state.

From Table 5, it can be learned that the viscosities of the prepared CNMs suspensions were 18.79, 20.21, and 22.64 mPa·s. The viscosity of CNMs suspension was found to increase with the increase in concentration. Moreover, the viscosity was negatively correlated with the particle size. This may be due to the relatively high yield and concentration of CNMs with a particle size of 150 μm, which makes the viscosity of the suspension relatively large as well. Then the CNMs suspension was left at room temperature for a certain period. It was found that there was no significant change in the suspension of CNMs with 150 μm particle size (e.g., precipitation, stratification, etc.), and the suspension was still relatively turbid. The results showed that the suspension of CNMs with 150 μm particle size had better stability and the highest ratio of nano/microfibrils in the suspension [47,48].

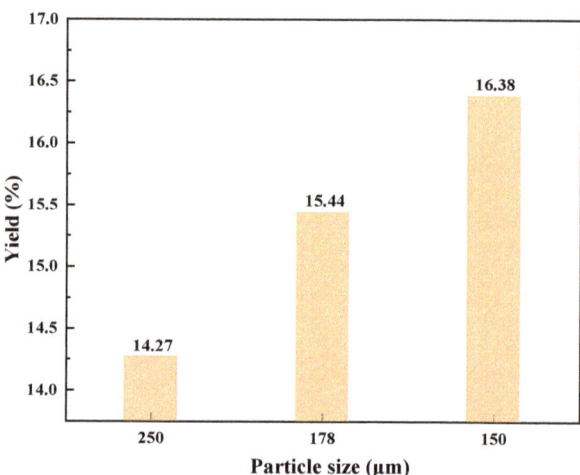

Figure 7. The yield of CNMs.

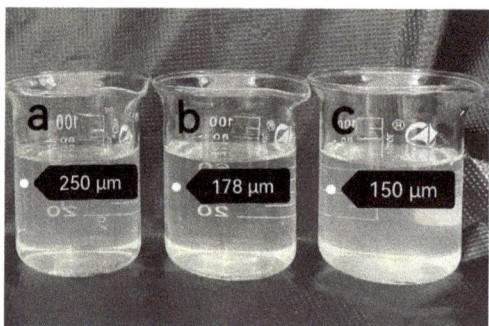

Figure 8. (a) 250 μm CNMs suspension; (b) 178 μm CNMs suspension; (c) 150 μm CNMs suspension.

Table 5. The viscosity of CNMs suspensions with different particle sizes.

Particle Size (μm)	Concentration (wt%)	Viscosity (mPa·s)
250	0.097	18.79
178	0.102	20.21
150	0.119	22.64

Due to the relatively high concentration, yield and viscosity of the CNMs prepared at 150 μm particle size, the CNMs suspension was more stable and had the highest ratio of nano/microfibrils. Therefore, the CNMs prepared at 150 μm particle size were further analyzed (TEM, SEM) for their morphological characteristics and dimensions.

3.7. TEM Analysis

Figure 9 shows the transmission electron microscopy images of CNMs prepared from 150 μm particle size of Neosinocalamus affinis powder using bio-enzyme B. Figure 9 shows that the CNMs prepared in this study were in the form of fibrous bundles and cross-twisted with each other. It is due to the amorphous region in the internal structure of cellulose, which causes CNMs to exhibit aggregation [49]. Figure 9 shows that the CNMs prepared in this study are cellulose nanofilaments (CNFs).

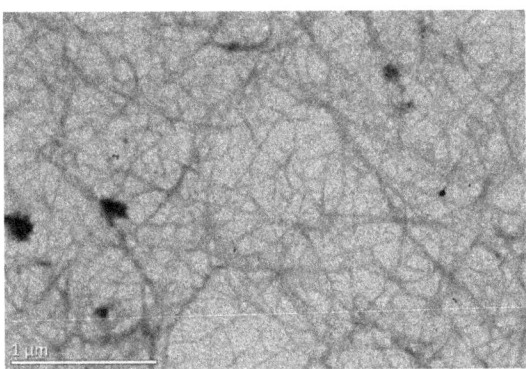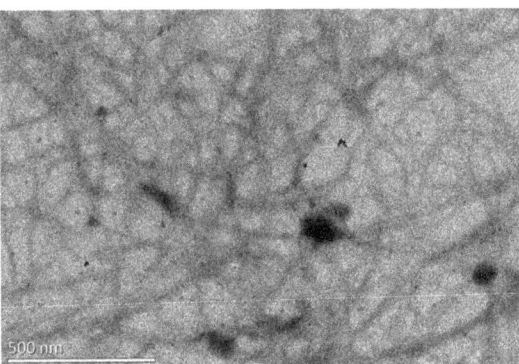

Figure 9. TEM of CNMs.

3.8. SEM Analysis

Figure 10 shows the electron micrographs of CNFs prepared from Neosinocalamus affinis powder with a particle size of 150 μm. CNFs in Figure 10 show irregular fibrous bundles. It is because the hemicellulose outside the microfiber inhibits the cleavage and cleavage of the microfiber by cellulase [50]. The average diameter of the CNFs measured and calculated by Image J software was 74 nm. The microstructure of the prepared CNFs can be seen in Figure 10 as a network structure. The nanofibers entangle the fibers in the network structure. It is caused by the interaction between the hydroxyl groups [51]. This network formation capability is an important feature of nanocellulose. This structure greatly increases the regional interface for bonding with the composite and plays a toughening role in the composite [52]. The analysis showed that the prepared CNFs were in the nanoscale range. They can be added to the composites as a toughening component to improve the properties of the composites.

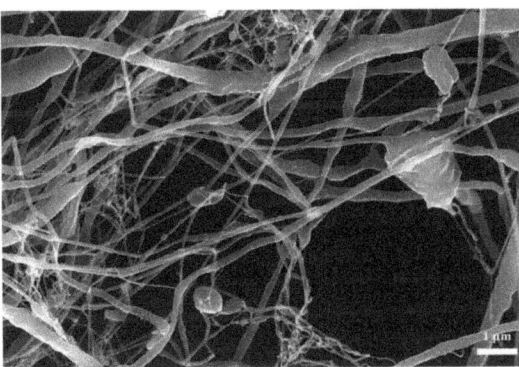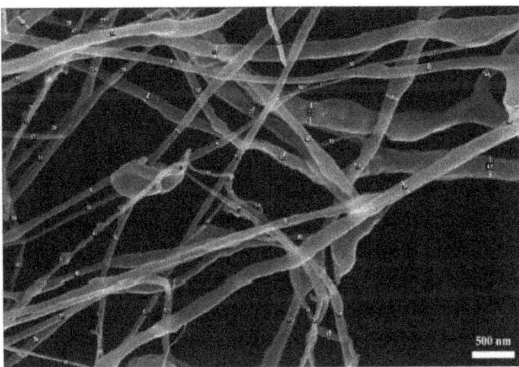

Figure 10. SEM of CNMs.

4. Conclusions

In this study, cellulose was extracted and nanocellulose was prepared from three different particle sizes (250, 178, and 150 μm) of Neosinocalamus affinis using a bio-enzyme method. The results showed that the highest cellulose content (74.08%), the highest crystallinity (63.58%), the highest purity and the best thermal stability were extracted from the Neosinocalamus affinis powder with a particle size of 150 μm. The highest yield (16.38%) of nanocellulose was prepared from Neosinocalamus affinis powder with a particle size of 150 μm. The concentration (0.119 wt%) and viscosity (22.64 mPa·s) of the nanocellulose suspension were relatively the highest. The nanocellulose exhibits a network structure. This network structure can play a toughening role. As a toughening component, it can be compounded with other materials to improve the toughness of the composite material.

Studies have shown that the smaller the particle size of Neosinocalamus affinis, the easier it is to react with bio-enzymes. The higher the content and purity of the extracted cellulose, the better the thermal stability and the higher the crystallinity. The smaller the particle size, the higher the yield of the prepared nanocellulose. The concentration and viscosity of the nanocellulose suspension are also relatively higher. The results illustrate the feasibility of cellulose extraction and preparation of nanocellulose from bamboo using the bio-enzyme method. The source of raw materials for cellulose extraction and preparation of nanocellulose using the bio-enzyme method was expanded to improve the utilization of bamboo as a biomass material.

Author Contributions: Conceptualization, J.Z. and L.Z.; Data curation, J.Z.; Formal analysis, J.Z., X.W. and X.Y. (Xushuo Yuan).; Funding acquisition, X.L. and L.Z.; Investigation, J.Z., X.W. and X.Y. (Xushuo Yuan); Methodology, X.L.; Project administration, X.L. and L.Z.; Supervision, W.Y. and D.J.; Validation, H.G. and X.L.; Visualization, X.Y. (Xinjie Yang); Writing—original draft, J.Z., X.W. and X.Y. (Xushuo Yuan); Writing—review and editing, L.Z. All authors have read and agreed to the published version of the manuscript.

Funding: This work is supported by the Natural Science Foundation of China (22161043, 31870551 and 21801096), Natural Science Foundation of Zhejiang Province (LY21B020009), Yunnan Fundamental Research Project (202201AU070071), "High-level Talent Introduction Program" project of Yunnan Province (YNQR-QNRC-2019-065), Innovation and Entrepreneurship Training Program for College Students in Yunnan Province (20201364003), and the Start Up Funding of Southwest Forestry University (112126), also supported by the 111 project (D21027).

Institutional Review Board Statement: Not applicable.

Informed Consent Statement: Not applicable.

Data Availability Statement: The data presented in this study are available on request from the corresponding author.

Conflicts of Interest: The authors declare that the research was conducted in the absence of any commercial or financial relationships that could be construed as a potential conflict of interest. There is no conflict to declare.

References

1. Scurlock, J.M.O.; Dayton, D.C.; Hames, B. Bamboo: An overlooked biomass resource? *Biomass Bioenergy* **2000**, *19*, 229–244. [CrossRef]
2. Bochek, A.M. Effect of hydrogen bonding on cellulose solubility in aqueous and nonaqueous solvents. *Russ. J. Appl. Chem.* **2003**, *76*, 1711–1719. [CrossRef]
3. Liang, X.; Liu, J.; Fu, Y.; Chang, J. Influence of anti-solvents on lignin fractionation of eucalyptus globulus via green solvent system pretreatment. *Sep. Purif. Technol.* **2016**, *163*, 258–266. [CrossRef]
4. Klemm, D.; Heublein, B.; Fink, H.P.; Bohn, A. Cellulose: Fascinating biopolymer and sustainable raw material. *Angew. Chem. Int. Ed.* **2005**, *44*, 3358–3393. [CrossRef] [PubMed]
5. Khalil, H.; Bhat, A.H.; Yusra, A.F.I. Green composites from sustainable cellulose nanofibrils: A review. *Carbohydr. Polym.* **2012**, *87*, 963–979. [CrossRef]
6. Pathak, D.; Bedi, R.; Kaur, D.; Kumar, R. Fabrication of densely distributed silver indium selenide nanorods by using Ag+ ion irradiation. *J. Korean Phys. Soc.* **2010**, *57*, 474–479. [CrossRef]
7. Sagadevan, S.; Pal, K. A facile synthesis of $TiO_2/SiO_2/CdS$-nanocomposites 'optical and electrical' investigations. *J. Mater. Sci. Mater. Electron.* **2017**, *28*, 9072–9080. [CrossRef]
8. Pathak, D.; Kumar, S.; Andotra, S.; Thomas, J.; Kaur, N.; Kumar, P.; Kumar, V. New tailored organic semiconductors thin films for optoelectronic applications. *Eur. Phys. J. Appl. Phys.* **2021**, *95*, 10201. [CrossRef]
9. Akbar Hosseini, S. Investigation of the structural, photocatalytic and magnetic properties of $MgAl2O4/NiTiO3$ nanocomposite synthesized via sol-gel method. *J. Mater. Sci. Mater. Electron.* **2017**, *28*, 10765–10771. [CrossRef]
10. Wu, M.; Liu, J.-K.; Yan, Z.-Y.; Wang, B.; Zhang, X.-M.; Xu, F.; Sun, R.-C. Efficient recovery and structural characterization of lignin from cotton stalk based on a biorefinery process using a γ-valerolactone/water system. *RSC Adv.* **2016**, *6*, 6196–6204. [CrossRef]

11. General Assembly of the United Nations. Transforming Our World: The 2030 Agenda for Sustainable Development. Geneva, Switzerland, 2015. Available online: https://sdgs.un.org/2030agenda (accessed on 13 October 2022).
12. Thomas, B.; Raj, M.C.; Athira, K.B.; Rubiyah, M.H.; Joy, J.; Moores, A.; Drisko, G.L.; Sanchez, C. Nanocellulose, a Versatile Green Platform: From Biosources to Materials and Their Applications. *Chem. Rev.* **2018**, *118*, 11575–11625. [CrossRef] [PubMed]
13. Heise, K.; Kontturi, E.; Allahverdiyeva, Y.; Tammelin, T.; Linder, M.B.; Nonappa; Ikkala, O. Nanocellulose: Recent Fundamental Advances and Emerging Biological and Biomimicking Applications. *Adv. Mater.* **2021**, *33*, 30. [CrossRef] [PubMed]
14. Seta, F.T.; An, X.Y.; Liu, L.Q.; Zhang, H.; Yang, J.; Zhang, W.; Nie, S.X.; Yao, S.Q.; Cao, H.B.; Xu, Q.L.; et al. Preparation and characterization of high yield cellulose nanocrystals (CNC) derived from ball mill pretreatment and maleic acid hydrolysis. *Carbohydr. Polym.* **2020**, *234*, 10. [CrossRef]
15. Mahmud, M.M.; Perveen, A.; Jahan, R.A.; Matin, M.A.; Wong, S.Y.; Li, X.; Arafat, M.T. Preparation of different polymorphs of cellulose from different acid hydrolysis medium. *Int. J. Biol. Macromol.* **2019**, *130*, 969–976. [CrossRef] [PubMed]
16. Oksman, K.; Mathew, A.P.; Bondeson, D.; Kvien, I. Manufacturing process of cellulose whiskers/polylactic acid nanocomposites. *Compos. Sci. Technol.* **2006**, *66*, 2776–2784. [CrossRef]
17. Chen, X.Q.; Pang, G.X.; Shen, W.H.; Tong, X.; Jia, M.Y. Preparation and characterization of the ribbon-like cellulose nanocrystals by the cellulase enzymolysis of cotton pulp fibers. *Carbohydr. Polym.* **2019**, *207*, 713–719. [CrossRef]
18. Zhang, Y.; Chen, J.N.; Zhang, L.; Zhan, P.; Liu, N.; Wu, Z.P. Preparation of nanocellulose from steam exploded poplar wood by enzymolysis assisted sonication. *Mater. Res. Express* **2020**, *7*, 9. [CrossRef]
19. Hayashi, N.; Kondo, T.; Ishihara, M. Enzymatically produced nano-ordered short elements containing cellulose Iβ crystalline domains. *Carbohydr. Polym.* **2005**, *61*, 191–197. [CrossRef]
20. Beltramino, F.; Roncero, M.B.; Vidal, T.; Valls, C. A novel enzymatic approach to nanocrystalline cellulose preparation. *Carbohydr. Polym.* **2018**, *189*, 39–47. [CrossRef]
21. Huang, X.; Li, D.; Wang, L.-J. Effect of particle size of sugar beet pulp on the extraction and property of pectin. *J. Food Eng.* **2018**, *218*, 44–49. [CrossRef]
22. Meng, T.; Ying, K.; Hong, Y.; Xu, Q. Effect of different particle sizes of nano-SiO_2 on the properties and microstructure of cement paste. *Nanotechnol. Rev.* **2020**, *9*, 833–842. [CrossRef]
23. Cheng, Y.; Yu, G.; Duan, Z. Effect of different size ZnO particle doping on dielectric properties of polyethylene composites. *J. Nanomater.* **2019**, *2019*, 9481415. [CrossRef]
24. Tang, Z.J.; Yang, M.W.; Qiang, M.L.; Li, X.P.; Morrell, J.J.; Yao, Y.; Su, Y.W. Preparation of Cellulose Nanoparticles from Foliage by Bio-Enzyme Methods. *Materials* **2021**, *14*, 4557. [CrossRef] [PubMed]
25. French, A.D.; Cintron, M.S. Cellulose polymorphy, crystallite size, and the Segal Crystallinity Index. *Cellulose* **2013**, *20*, 583–588. [CrossRef]
26. Bondancia, T.J.; Florencio, C.; Baccarin, G.S.; Farinas, C.S. Cellulose nanostructures obtained using enzymatic cocktails with different compositions. *Int. J. Biol. Macromol.* **2022**, *207*, 299–307. [CrossRef] [PubMed]
27. Pirayesh, H.; Khazaeian, A. Using almond (*Prunus amygdalus* L.) shell as a bio-waste resource in wood based composite. *Compos. Part B Eng.* **2012**, *43*, 1475–1479. [CrossRef]
28. Ilangovan, M.; Guna, V.; Prajwal, B.; Jiang, Q.R.; Reddy, N. Extraction and characterisation of natural cellulose fibers from *Kigelia africana*. *Carbohydr. Polym.* **2020**, *236*, 7. [CrossRef]
29. Rastogi, A.; Banerjee, R. Production and characterization of cellulose from *Leifsonia* sp. *Process Biochem.* **2019**, *85*, 35–42. [CrossRef]
30. Pang, Z.Q.; Wang, P.Y.; Dong, C.H. Ultrasonic pretreatment of cellulose in ionic liquid for efficient preparation of cellulose nanocrystals. *Cellulose* **2018**, *25*, 7053–7064. [CrossRef]
31. Huang, Y.; Wang, Z.G.; Wang, L.S.; Chao, Y.S.; Akiyama, T.; Yokoyama, T.; Matsumoto, Y. Analysis of Lignin Aromatic Structure in Wood Fractions Based on IR Spectroscopy. *J. Wood Chem. Technol.* **2016**, *36*, 377–382. [CrossRef]
32. Alghooneh, A.; Amini, A.M.; Behrouzian, F.; Razavi, S.M.A. Characterisation of cellulose from coffee silverskin. *Int. J. Food Prop.* **2017**, *20*, 2830–2843. [CrossRef]
33. Chen, J.H.; Xu, J.K.; Wang, K.; Cao, X.F.; Sun, R.C. Cellulose acetate fibers prepared from different raw materials with rapid synthesis method. *Carbohydr. Polym.* **2016**, *137*, 685–692. [CrossRef] [PubMed]
34. Deepa, B.; Abraham, E.; Cordeiro, N.; Mozetic, M.; Mathew, A.P.; Oksman, K.; Faria, M.; Thomas, S.; Pothan, L.A. Utilization of various lignocellulosic biomass for the production of nanocellulose: A comparative study. *Cellulose* **2015**, *22*, 1075–1090. [CrossRef]
35. Dai, H.; Ou, S.; Huang, Y.; Huang, H. Utilization of pineapple peel for production of nanocellulose and film application. *Cellulose* **2018**, *25*, 1743–1756. [CrossRef]
36. Beroual, M.; Boumaza, L.; Mehelli, O.; Trache, D.; Tarchoun, A.F.; Khimeche, K. Physicochemical Properties and Thermal Stability of Microcrystalline Cellulose Isolated from Esparto Grass Using Different Delignification Approaches. *J. Polym. Environ.* **2021**, *29*, 130–142. [CrossRef]
37. Franco, T.S.; Potulski, D.C.; Viana, L.C.; Forville, E.; de Andrade, A.S.; de Muniz, G.I.B. Nanocellulose obtained from residues of peach palm extraction (*Bactris gasipaes*). *Carbohydr. Polym.* **2019**, *218*, 8–19. [CrossRef]
38. Martins, M.A.; Teixeira, E.M.; Correa, A.C.; Ferreira, M.; Mattoso, L.H.C. Extraction and characterization of cellulose whiskers from commercial cotton fibers. *J. Mater. Sci.* **2011**, *46*, 7858–7864. [CrossRef]
39. French, A.D. Idealized powder diffraction patterns for cellulose polymorphs. *Cellulose* **2014**, *21*, 885–896. [CrossRef]

40. NagarajaGanesh, B.; Muralikannan, R. Extraction and characterization of lignocellulosic fibers from Luffa cylindrica fruit. *Int. J. Polym. Anal. Charact.* **2016**, *21*, 259–266. [CrossRef]
41. Kim, H.S.; Kim, S.; Kim, H.J.; Yang, H.S. Thermal properties of bio-flour-filled polyolefin composites with different compatibilizing agent type and content. *Thermochim. Acta* **2006**, *451*, 181–188. [CrossRef]
42. Yang, H.P.; Yan, R.; Chen, H.P.; Zheng, C.G.; Lee, D.H.; Liang, D.T. In-depth investigation of biomass pyrolysis based on three major components: Hemicellulose, cellulose and lignin. *Energy Fuels* **2006**, *20*, 388–393. [CrossRef]
43. Sun, Z.B.; Tang, Z.J.; Li, X.P.; Li, X.B.; Morrell, J.J.; Beaugrand, J.; Yao, Y.; Zheng, Q.Z. The Improved Properties of Carboxymethyl Bacterial Cellulose Films with Thickening and Plasticizing. *Polymers* **2022**, *14*, 3286. [CrossRef] [PubMed]
44. Samuel, R.; Pu, Y.; Foston, M.; Ragauskas, A.J. Solid-state NMR characterization of switchgrass cellulose after dilute acid pretreatment. *Biofuels* **2010**, *1*, 85–90. [CrossRef]
45. Chen, X.Q.; Deng, X.Y.; Shen, W.H.; Jia, M.Y. Preparation and characterization of the spherical nanosized cellulose by the enzymatic hydrolysis of pulp fibers. *Carbohydr. Polym.* **2018**, *181*, 879–884. [CrossRef] [PubMed]
46. Vinogradov, M.I.; Makarov, I.S.; Golova, L.K.; Gromovykh, P.S.; Kulichikhin, V.G. Rheological properties of aqueous dispersions of bacterial cellulose. *Processes* **2020**, *8*, 423. [CrossRef]
47. Benini, K.; Voorwald, H.J.C.; Cioffi, M.O.H.; Rezende, M.C.; Arantes, V. Preparation of nanocellulose from Imperata brasiliensis grass using Taguchi method. *Carbohydr. Polym.* **2018**, *192*, 337–346. [CrossRef]
48. Rosa, S.M.L.; Rehman, N.; de Miranda, M.I.G.; Nachtigall, S.M.B.; Bica, C.I.D. Chlorine-free extraction of cellulose from rice husk and whisker isolation. *Carbohydr. Polym.* **2012**, *87*, 1131–1138. [CrossRef]
49. Chen, W.S.; Li, Q.; Cao, J.; Liu, Y.X.; Li, J.; Zhang, J.S.; Luo, S.Y.; Yu, H.P. Revealing the structures of cellulose nanofiber bundles obtained by mechanical nanofibrillation via TEM observation. *Carbohydr. Polym.* **2015**, *117*, 950–956. [CrossRef]
50. Xu, J.T.; Chen, X.Q. Preparation and characterization of spherical cellulose nanocrystals with high purity by the composite enzymolysis of pulp fibers. *Bioresour. Technol.* **2019**, *291*, 6. [CrossRef]
51. Abe, K.; Yano, H. Cellulose nanofiber-based hydrogels with high mechanical strength. *Cellulose* **2012**, *19*, 1907–1912. [CrossRef]
52. Tian, C.H.; Yi, J.A.; Wu, Y.Q.; Wu, Q.L.; Qing, Y.; Wang, L.J. Preparation of highly charged cellulose nanofibrils using high-pressure homogenization coupled with strong acid hydrolysis pretreatments. *Carbohydr. Polym.* **2016**, *136*, 485–492. [CrossRef] [PubMed]

Article

Design and Analysis of a 5-Degree of Freedom (DOF) Hybrid Three-Nozzle 3D Printer for Wood Fiber Gel Material

Jifei Chen [1,*,†], Qiansun Zhao [1,†], Guifeng Wu [1], Xiaotian Su [1], Wengang Chen [1] and Guanben Du [2,*]

1. School of Mechanical and Transportation, Southwest Forestry University, Kunming 650224, China; zhaoqiansun@swfu.edu.cn (Q.Z.); wuguifeng@swfu.edu.cn (G.W.); suxiaotian@swfu.edu.cn (X.S.); chenwengang999@swfu.edu.cn (W.C.)
2. School of Material Science and Engineering, Southwest Forestry University, Kunming 650224, China
* Correspondence: cjf100cjf@swfu.edu.cn (J.C.); gongben9@hotmail.com (G.D.)
† These authors contributed equally to this work.

Abstract: Wood is an organic renewable natural resource. Cellulose, hemicellulose and lignin in wood are used in tissue engineering, biomedicine and other fields because of their good properties. This paper reported that the possibility of wood fiber gel material molding and the preparing of gel material were researched based on the wood fiber gel material as a 3D printing material. A five-degree of freedom hybrid three nozzle 3D printer was designed. The structural analysis, static analysis, modal analysis and transient dynamic analysis of 3D printers were researched, and the theoretical basis of the 3D printer was confirmed as correct and structurally sound. The results showed that the 5-DOF hybrid 3-nozzle 3D printer achieved the 3D printing of wood fiber gel material and that the printer is capable of multi-material printing and multi-degree-of-freedom printing.

Keywords: gel material; wood material; 3D printer; design and analysis

Citation: Chen, J.; Zhao, Q.; Wu, G.; Su, X.; Chen, W.; Du, G. Design and Analysis of a 5-Degree of Freedom (DOF) Hybrid Three-Nozzle 3D Printer for Wood Fiber Gel Material. *Coatings* **2022**, *12*, 1061. https://doi.org/10.3390/coatings12081061

Academic Editor: Philippe Evon

Received: 6 June 2022
Accepted: 18 July 2022
Published: 27 July 2022

Publisher's Note: MDPI stays neutral with regard to jurisdictional claims in published maps and institutional affiliations.

Copyright: © 2022 by the authors. Licensee MDPI, Basel, Switzerland. This article is an open access article distributed under the terms and conditions of the Creative Commons Attribution (CC BY) license (https://creativecommons.org/licenses/by/4.0/).

1. Introduction

3D printing is an additive manufacturing technology used in aerospace, biomedicine, food, construction, industrial manufacturing, new materials and other fields [1]. As one of the technologies of the third industrial revolution, it has started to enter the consumer market after continuous development and research from the beginning of its development to the present day [2]. These 3D printers also have different structures and functions according to different materials. Currently, there are three main types of 3D printing materials, non-metallic materials, metallic materials and biomaterials. Among them, non-metallic materials mainly include photosensitive resins, polylactic acid, plastics, ceramics and so on. In pursuit of a green, low-carbon and sustainable development concept, researchers are also looking for natural materials for 3D printing. Wood, as a natural renewable resource, has attracted the attention of researchers. Through the rational use of tree resources, it can solve the material problem and add a way of wood utilization, which has important significance and good development prospects [3].

Wood is composed of primarily organic matter and a small amount of inorganic matter. The former includes mainly cellulose, hemicellulose and lignin, which may amount to over 90% of its biomass [4]. Cellulose, one of the main components of plants, is also the most naturally abundant and sustainable polymeric raw material. Due to its various excellent properties and wide range of sources, cellulose is already used in the commercial production of products. In the field of 3D printing, cellulose is also present in solid printing as a structural reinforcement [5], binder [6,7] and thickener [8].

Hemicellulose is a naturally occurring polysaccharide polymer, second only to cellulose, which is a branched amorphous heteropolysaccharide, consisting of five main types of xylan, glucomannan, arabinoglycan, galactan and dextran [9]. Despite being one of the most abundant natural polymers, hemicellulose is underutilised. Although hemicellulose has great potential in the field of 3D printed condensation, there is currently very little 3D printing of hemicellulose. The main hemicellulose materials used for 3D printing are in their derivative form with other polymers for 3D printing [10] and require extensive chemical modification and various post-printing processes [11].

Like cellulose and hemicellulose, lignin is a natural polymer with reproducibility, biodegradability and biocompatibility. Lignin, a major component of chemical pulp black liquor, is the world's second most abundant natural polymer and the only renewable raw resource that contains considerable amounts of aromatic hydrocarbons [12]. The structural properties of lignin facilitate the synthesis of useful chemicals, fuels and low mass molecules and are also consistent with sustainable development and a reduced carbon footprint. It is attracting increasing global attention and is showing great economic potential for a variety of high-value bioproducts, including the 3D printing of biomaterials [13]. In terms of current research on the use of lignin and its derivatives as biomass materials for 3D printing, lignin can be physically or chemically blended with other components, such as polylactic acid (PLA) [14], to take advantage of the strengths of each material component after material compounding to achieve excellent 3D printing. Through chemical cross-linking, physical cross-linking and interpenetrating polymer network methods, wood gel materials are prepared for direct ink 3D printing and the complete printing of wood fiber gel materials. This is a new means of utilizing wood resources, which not only extends the range of 3D printing materials but also provides a new way to diversify the use of wood [15].

Due to the different properties of wood fiber gel materials, the forming mechanism of 3D printing and the structure and function of 3D printers are also different. In this study, wood fiber extracted from a branch, waste wood and wood scraps of Cunninghamia lanceolate. Using wood fiber gel material as the base material for 3D printing, the basic characteristics of wood fiber gel material printing and forming are studied, and a 3D printer corresponding to wood fiber gel material is designed according to the material characteristics, providing a way to use the wood resources. The 3D printer also has the ability to print three materials on the same platform with different printheads and is complemented by five degrees of freedom of movement for print versatility. Finally, it also provides equipment support for the subsequent wood fiber composite printing of biomimetic wood.

2. Materials and Methods

2.1. Materials and Equipment

Cellulose, hemicellulose and lignin were extracted from cedar wood powder. Sodium hydroxide (AR analytical purity-flakes, purity $\geq$ 96.0%), Chengdu Jinshan Chemical Reagent Co. Chitosan (AR analytical purity—off-white powder or flakes, purity $\geq$ 96.0%), Sinopharm Chemical Co. Acetic acid (AR analytical purity, purity $\geq$ 99.5%), Shanghai Aladdin Biochemical Technology Co. Urea (AR analytical purity—pellets, purity $\geq$ 99.0%), Sinopharm Chemical Co. Epichlorohydrin (AR Analytical purity, purity $\geq$ 98.0%), Shanghai Aladdin Biochemical Technology Co. Hydrochloric acid (AR analytical purity—liquid, purity $\geq$ 36.0%~38.0%), Shan Dian Pharmaceutical Co. Colouring agent (green), Guangdong Timothy Colour Food Colour Technology Co. Distilled water, homemade.

Collective thermostatic heating magnetic stirrer, Type DF-101S, Henan Province Yuhua Instruments Co. Rheometer: MARS III type, Germany HAAKE company. Vacuum drying oven: type DZF, Beijing Kewei Yongxing Instruments Co.

2.2. Cellulose Gel Material Preparation

An aqueous base system of sodium hydroxide, urea and distilled water (7:12:81 by mass) was used as the solvent, to which cellulose was added while stirring with an electromagnetic centrifugal stirrer until all the cellulose powder was dispersed in the solution and a milky-white solution without visible powder aggregates was formed. The prepared cellulose suspension was stored in a refrigerator at −10 °C for 4–6 h and removed to obtain a clear cellulose gel [16,17] as shown in Figure 1a.

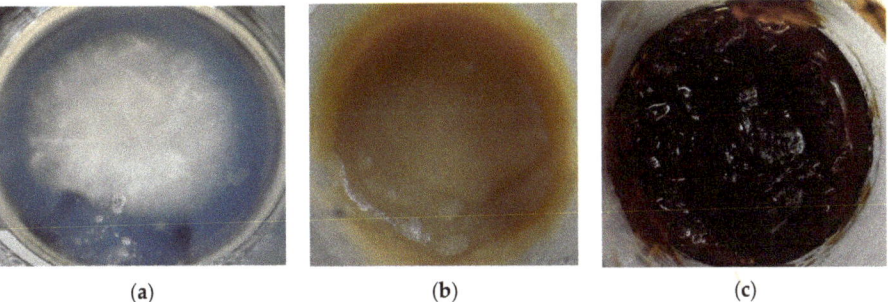

Figure 1. Wood fiber gel material. (**a**) Cellulose gel; (**b**) Hemicellulose gel; (**c**) Lignin gel.

2.3. Hemicellulose Gel Material Preparation

Mix hemicellulose and chitosan in a ratio of 6:4 in deionized water and add hydrochloric acid to adjust the pH value to 4. Heat the solution to 95 °C and keep it for 20 min to form a suspension of hemicellulose powder dissolved without cellulose suspension, then put it in the refrigerator at −12 °C for 4–6 h and take it out to prepare hemicellulose gel [18,19] as shown in Figure 1b.

2.4. Lignin Gel Material Preparation

Chitosan solution was prepared by dissolving 2.5 g of chitosan in 100 mL of 2.5 v/v % acetic acid, and lignin solution was prepared by dissolving 1 g of alkaline lignin in 10 mL of water. The chitosan solution and lignin solution were then mixed and chilled in a refrigerator at −10 °C for 5 h, and then removed and stored at 20 °C for half an hour to form a lignin gel [20] as shown in Figure 1c.

2.5. Rheology Tests

Using a HAAKE MARS III rheometer at a constant shear rate, cellulose gels (using a temperature ramp of 2 °C/min with a start temperature of 10 °C and measurements stopped at 80 °C) and hemicellulose gels (using a temperature ramp of 5 °C/min with a start temperature of 20 °C and measurements stopped at 140 °C), the pattern of viscosity change and maximum gelling temperature were measured. The gelling point was determined using Thermo Scientific HAAKE RheoWin software. (version 4.63.0004, Germany HAAKE company).

3. Results and Discussion

3.1. Gelation Point Analysis

The structural design of 3D printers is closely related to the forming materials, and the forming and curing mechanisms of different materials vary. For the analysis of wood gel material properties, the rheological properties of the material were experimented with the data obtained from the experiments and analyzed to obtain the characteristics of the wood fiber gel material, as shown in Figure 2.

Both cellulose and hemicellulose gels show a reduction in viscosity at the beginning of the test, which is a shear thinning characteristic and fits well with the printing characteristics of direct ink technology. This is a shear thinning characteristic that is very much in line with the printing characteristics of direct ink writing technology. It can also be seen that the wood fiber gel material shows a tendency to increase in viscosity with higher temperatures, and the graph shows that cellulose produced a mutation condition near 52 °C and hemicellulose produced a mutation near 80 °C. The reason for the mutation is due to the curing and molding of the wood fiber gel material [21,22]. Therefore, in the process of designing 3D printers of wood fiber gel materials, the molding process needs to be designed around the material properties.

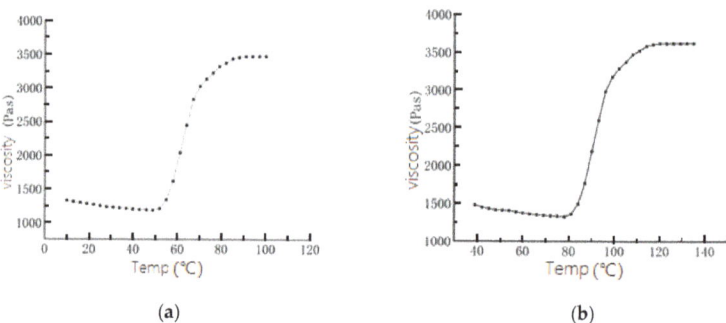

Figure 2. Properties of wood fiber gel material. (**a**) Cellulose gel; (**b**) Hemicellulose gel.

3.2. Printer Type Selection

Currently, the main technologies applicable to the additive manufacturing of wood materials are FDM technology, which is used for hot-melt cold-set wood–plastic composites [22], and DIW technology, which is used for direct forming or externally formed flowing materials [23]. As the wood component prepared the printing material like a gel while combining different printing consumables, the researchers explored the use of external effects to assist the 3D printing process, such as light [23], rotation [24]. In addition, it can enable the 3D printing of materials, such as cyclopentadiene titanium [23] and short carbon fiber epoxy composites [24]. Those can further increase the additive technology applicability [25]. In Figure 3, 3D printer-assisted devices are shown. Based on the experimental data in Figure 2, it is also known that the prepared wood fiber gel materials have shear thinning properties. It is suitable for the DIW mode of printing. At the same time, based on the advantages of the DIW type of printing technology, assisted by the external action and the characteristics of the printing materials that we have studied, this type of wood fiber gel material can be extruded using an external laser to accelerate the curing of the wood fiber gel material. In addition, according to the study, the design of 3D printers should be biased towards the DIW type. For the design of 3D printers, multiple printheads are required to meet the design goals for cellulose gel materials, hemicellulose gel materials and lignin gel materials printed on the same model and on the same platform. The more common five-degree-of-freedom configuration was chosen to meet the development needs of 3D printers with multiple degrees of freedom.

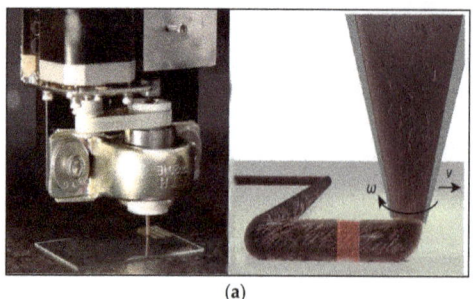
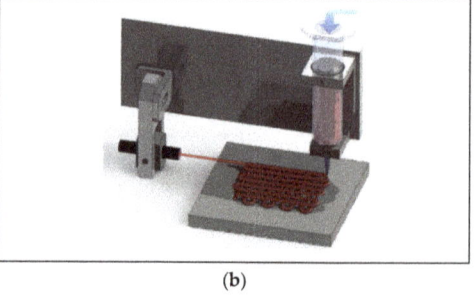

Figure 3. 3D printer auxiliary equipment. (**a**) Rotating auxiliary devices [24]; (**b**) Laser auxiliary devices [23].

4. Structural Design and Description

4.1. Overall Structure

A 5-Degree of Freedom (DOF) hybrid three-nozzle 3D printer was designed based on the performance feature of wood fiber gel material. Its structure mainly includes a printing nozzle mechanism, a nozzle moving mechanism, a z-axis moving mechanism, a printing platform and a support frame. Wood fiber gel material was printed with three nozzles. The 3D printers that have multiple degrees of freedom and multiple nozzles have attracted increasing attention to meet the printing needs of a variety of materials [26,27]. Multi-nozzle 3D printers should be designed in such a way to ensure a smooth nozzle switch, smooth and accurate plane movement, accurate positioning of the nozzle, accurate platform control, etc. The structure of the 5-DOF hybrid three-nozzle 3D printer is illustrated in Figure 4.

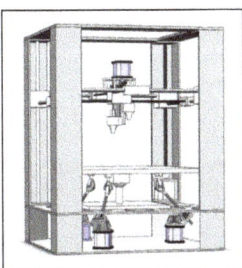 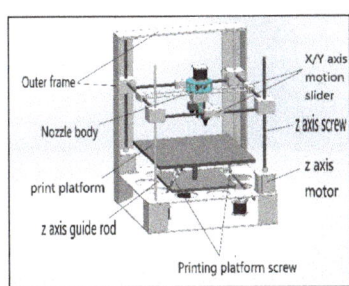

Figure 4. The 5-DOF hybrid three-nozzle 3D printer.

The movement of 3D printer nozzles mainly relies on the cross axis to realize the x and y axis movement in the plane, and the rotation of the screw rod is controlled by the stepper motor to drive the z-axis movement. Three nozzles are designed in the nozzle mechanism, which squeezes cellulose gel, Hemicellulose gel and lignin gel accurately, according to the computer structure diagram of the molding structure to a certain range; it can be a reduced material forming support structure by controlling the printer platform to move between +45° and −45°. The outer frame is the base that supports all mechanisms.

4.2. Design of Nozzle Mechanism

The nozzle mechanism is a core mechanism for the 3D printing of wood fiber gel material. Nozzle head mechanism includes nozzle head bracket, printer nozzle head, nozzle head switching motor and nozzle head switching gear group (Figure 5).

In the mechanism, the screw is driven by the motor and gear to move the three printer nozzles independently to make the print. The gear group of nozzle head switching is installed on the gearbox. The gear and shaft in the nozzle head switching gear group are made of non-magnetic materials. The output shaft of the nozzle switching motor is connected with the driving gear shaft through the telescopic shaft and matched with the spline. The electromagnet device comprises an electromagnet arranged on a driving gear and a magnet arranged in a gearbox. When the driving gear rotates, it drives the first gear to rotate and the second gear to rotate reversely, simultaneously, and then the screw connected to the two printer nozzles will rotate at the same time, but in the opposite direction. In addition, the two printer nozzles can move up and down in opposite directions. When the third print nozzle is needed, the first gear will rotate and the second switchgear will control the printer nozzle position to adjust to the central position. The electromagnet switch gear locates its axial position to make sure that the driving gear and the third switchgear will mesh, and starting the nozzle switch motor will engage the third printer head. By controlling the switch motor and electromagnetic mechanism of the nozzle, the three printer heads can be switched freely and can improve the printing efficiency and avoid the interference between the nozzles and the fixing frame while allowing the rapid switching of the three nozzles.

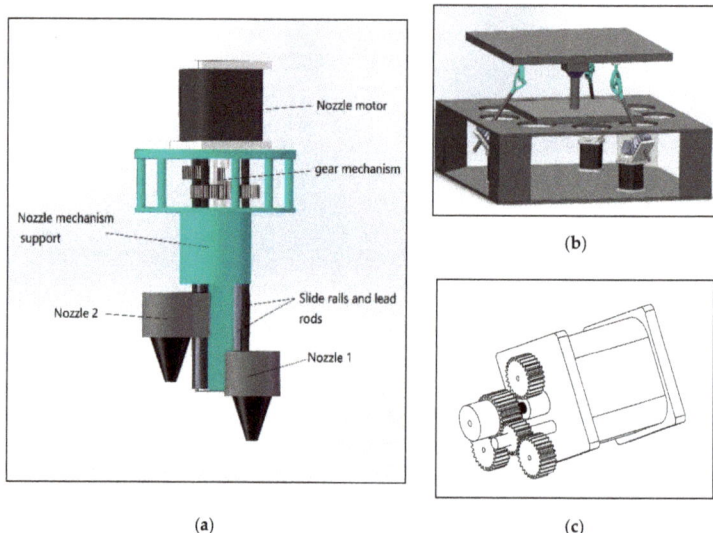

Figure 5. Printer nozzle mechanism. (**a**) Nozzle body; (**b**) Nozzle structure gear layout; (**c**) Printer platform mechanism.

4.3. Design of Printer Platform Mechanism

The printer platform mechanism makes the printing plate flip and tilt through a three-branch parallel mechanism and has two degrees of freedom of rotation in the x and y axis. It uses three stepping motors to drive the matching bevel gears, and then the bevel gear controls the platform screw to move, push and pull, and to drive the printer platform, so that the printing plate will complete the flip tilt. The parallel mechanism has three branches that are evenly distributed at 120° angle. Each branch of the parallel mechanism is driven by an independent motor that drives the screw to the required deflection of the printed platform. The printer platform mechanism can reduce the center of gravity of the 3D printer and can minimize the impact of vibration on the printing accuracy during motor operation, reduce the lateral force of the screw nut mechanism and effectively avoid screw deformation due to long-term stress, so that the printing accuracy is not affected (Figure 6).

The 3D printer is capable of increasing printing efficiency, reducing structural support and making prints of various types of materials.

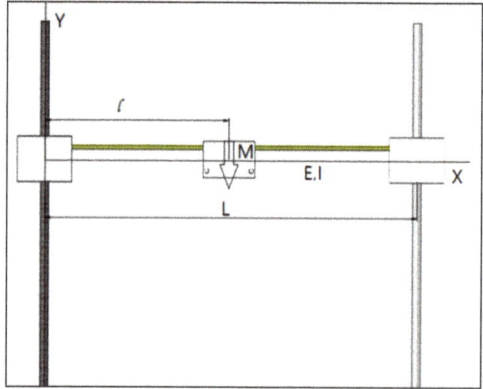

Figure 6. Schematic diagram of the simplified structure.

5. Structural and Mechanical Analysis

5.1. Degree of Freedom Analysis

The 3D printer is composed of the printing nozzle of a series of mechanisms and the printing platform of a parallel structure to form a 5-DoF motion. There are virtual constraints and local DoFs in its structure, and the modified CGK general DoF formula is used.

$$F = d(n - g - 1) + \sum_{i=1}^{g} f_i + \mu - \zeta \tag{1}$$

where F is the degree of freedom of the mechanism; d is the order of mechanism (space mechanism $d = 6$); n is the number of components including the frame (the base included); g is the number of motion pairs; f_i is the degree of freedom of the i-th motion pair; μ is the virtual constraint number of mechanism; ζ is the local degree of freedom.

The parameters of the 3D printer were put into the Formula (1), and the freedom degree of the motion mechanism of the 5-DOF hybrid three-nozzle 3D printer was calculated to be five. The five DoFs are the xy plane DoF of the nozzle part and the cross axis structure, the vertical DoF of z-axis direction (3T), and the x and y axis rotation movement DoF of the derivative structure of Delta parallel structure of the printer platform part (2R), respectively. The 5-DoF hybrid three-nozzle 3D printer has 5 DoFs (3T-2R).

5.2. Statics Analysis

The equation of the deflection curve [28] is obtained based on the structural form of simply supported beams, and the maximum deflection is determined. The structure is simplified as a mass–spring system, and the cross-arm slides are simplified as a uniform cross beam (Figure 6).

The differential equation of the flexural function equation is established according to the structural reaction force (Formulas (2)–(4)). The quadratic integration is carried out, and the value of the constant term is obtained according to the boundary conditions, and the maximum deflection is finally obtained, as follows.

$$\left. \begin{array}{l} EIw_1''(x) = -\frac{F(L-l)}{L} \\ EIw_2''(x) = -\frac{F(L-l)}{L}x + F(x-l)x \end{array} \right\} \tag{2}$$

$$\left. \begin{array}{l} EIw_1'(x) = -\frac{F(L-l)}{2L}x^2 + C_1 \\ EIw_2'(x) = -\frac{F(L-l)}{2L}x^2 + \frac{1}{2}F(x-l)^2 + C_2 \end{array} \right\} \tag{3}$$

$$\left. \begin{array}{l} EIw_1(x) = -\frac{F(L-l)}{2L}x^3 + C_1 x + D_1 \\ EIw_2(x) = -\frac{F(L-l)}{2L}x^3 + \frac{1}{6}F(x-l)^3 + C_2 x + D_2 \end{array} \right\} \tag{4}$$

where E is the elastic coefficient of the material, I is the moment of inertia of the section, L is the length of the transverse guide rod, l is the position of the nozzle at the guide rod, F is the force the rod received, w_i'' and w_i' are the second and first-order derivatives of the deflection, respectively.

The further calculation is made according to boundary conditions and continuous conditions. Because both ends of the guide rod are supported, the boundary conditions mainly consider both ends of the guide rod and its central position. Under continuous conditions, the left and right ends of the second and first order deflection differential equations are equal, respectively, in the central position, and the value of the constant is obtained.

Boundary condition,

$$l = 0; w = 0; l = \frac{1}{2}L; w_1 = w_2; l = L; w = 0;$$

Condition of continuity,

$$EIw_1'(l) = EIw_2'(l); \quad C_1 = C_2 = \frac{F(L-l)}{6L}(2Ll - l^2);$$
$$EIw_1(l) = EIw_2(l); \quad D_1 = D_2 = 0$$

$$\left.\begin{array}{l} EIw_1(x) = -\frac{F(L-l)}{2L}x^3 + \frac{F(L-l)}{6L}(2Ll - l^2)x \\ EIw_2(x) = -\frac{F(L-l)}{2L}x^3 + \frac{1}{6}F(x-l)^3 + \frac{F(L-l)}{6L}(2Ll - l^2)x \end{array}\right\} \quad (5)$$

Given the proportional relationship between l and L, the maximum deflection can be obtained. When $l = \frac{1}{2}L$, $W_{max} = \frac{FL^3}{48EI}$; When $l > \frac{1}{2}L$, $W_{max} = \frac{F(L-l)}{9EIL}\sqrt{\frac{2Ll-l^2}{3}}$.

5.3. Modal Analysis

ANSYS software modal analysis function is used to analyze the dynamic characteristics of the 3D printer system. The structure dynamic characteristics of the 3D printer were analyzed and evaluated to ensure printing accuracy. The use of different materials for different parts of the structure will further affect the rationality of the 3D printer. The 3D printer's support and printing platform uses a 6063 aluminum alloy, while 45 steel is used for the guide, slider and screw. The 3D model of the 3D printer was constructed in Solid Works, and the structural model was imported into ANSYS analysis software to divide the mesh and optimize the mesh quality. The average quality index of grid elements is greater than 0.80, and the average aspect ratio is 1.93, which meets the requirements of structural field analysis (Figure 7).

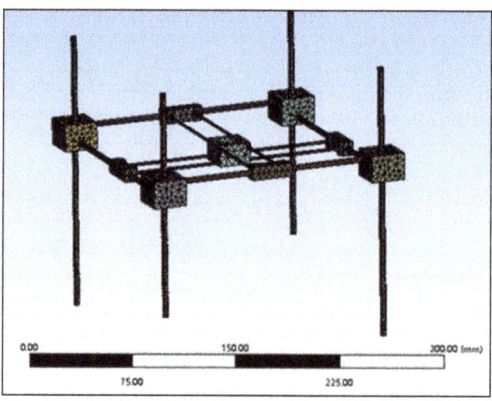

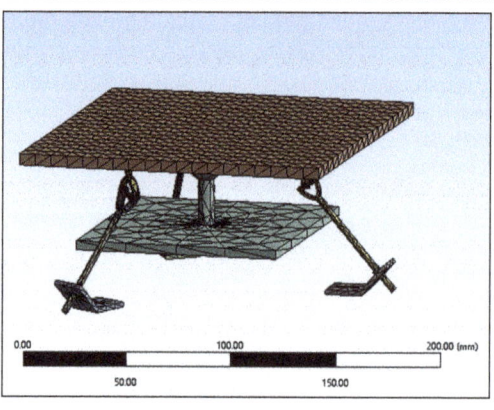

Figure 7. Meshing of kinematic mechanism of the ANSYS model.

The modal analysis of 3D printers is relatively fixed. The order of modal analysis is determined according to the modal calculation quality of each DoF and needs to be optimized and adjusted multiple times. The first eight order modal frequencies and the deformations of the 3D printer's motion mechanism are shown in Table 1.

Table 1. Modal frequency and deformation of the moving mechanism.

Order	Frequency/Hz	Deformation Behavior of XYZ Kinematic Mechanism Model	Order	Frequency/Hz	Printer Platform Mechanism Model Deformation
1	56.891	Pillar bent and deformed	1	74.119	Parallel branches move longitudinally
2	58.225	Pillar bent and deformed	2	74.317	Parallel branches move longitudinally
3	71.416	Guide rod deformed	3	74.327	Parallel branches move longitudinally
4	71.437	The guide rod deformed	4	74.602	Parallel branches move laterally
5	79.567	Overall torsion deformation occurred in the structure	5	74.982	Parallel branches move laterally
6	90.275	Guide rod deformed	6	75.162	Parallel branches move laterally
7	90.328	Guide rod deformed	7	103.57	Printer platform rotates along the z-axis
8	105.42	Guide rod deformed	8	133.76	Printer platform swings longitudinally

The results show that the deformation and natural frequency of models with different orders are different, which provides theoretical data support for subsequent design and manufacturing.

6. Transient Dynamics Analysis

3D printing is to drive each part to move by the motor, so as to drive the nozzle to move. In combination with the nozzle material extrusion and laser curing, the model printing molding is completed. In planning the printing paths and model slice processing, the reciprocating motion of the nozzle is inevitable, and the vibration generated therefrom will affect the printing accuracy. ANSYS Workbench software was used for the transient dynamics analysis of 3D printers, and structural deficiencies were improved to ensure structural stability and printing accuracy.

For nozzle movement, the influences of the x-axis and y-axis alone and the linkage of the two axes on the structure were analyzed. For z-axis, only the influence of its independent motion on the structure was analyzed. According to the motion characteristics, the total load step number was set to three in the software. The end times of the 1st, 2nd and 3rd load steps were 0.25 s, 0.75 s, and 1 s, respectively. The load was applied to the z-axis and analyzed in linkage with the x and y axis with added load. Since the motion of the x, y and z axes and the printing platform were independent, the single chain and multi-chain motion of the printer platform were simulated and the results analyzed.

(1) When z-axis moved, other motion axes were static or do not interfere with z-axis motion, and transient motion simulation was carried out when the z-axis moved independently. The results showed that the maximum displacement in the z-axis direction was close to the average displacement, and the motion accuracy in the z-axis direction was sufficient (Figure 8a). The movement displacement in the x and y axis directions was far less than the nozzle diameter, which secured accurate movement (Figure 8b,c). The motion displacement curve of the z-axis mechanism is shown in Figure 8.

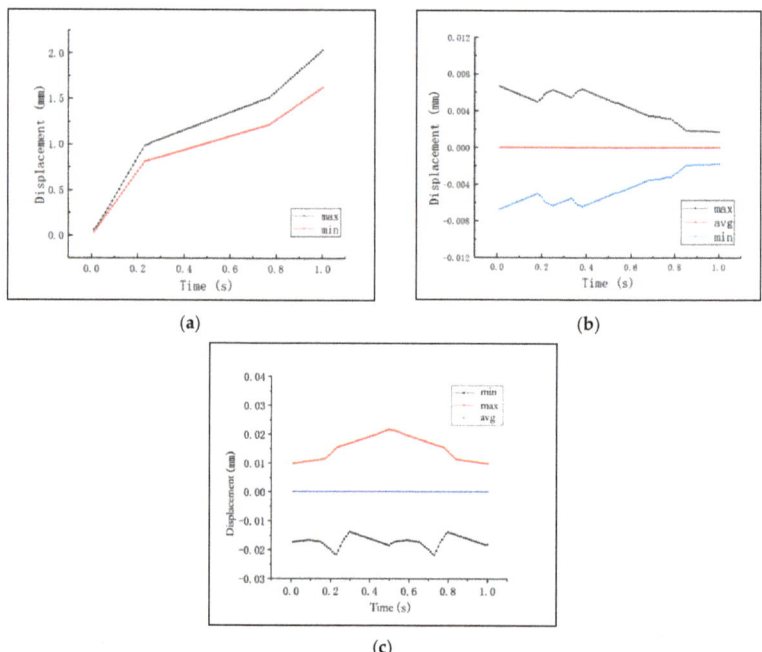

Figure 8. Motion displacement curve of z-axis mechanism; (**a**) the z-axis displacement; (**b**) the x-axis displacement; (**c**) the y-axis displacement.

(2) The nozzle head was linked with the x-axis and y-axis to analyze its motion precision and its influence on the z-axis when moving in the x and y directions. The nozzle moved along the x and y axes simultaneously. By moving 8 mm in the positive direction of the x-axis first, then moving 3 mm in the opposite direction and 10 mm in the positive direction of the y-axis, the final linkage effect was evaluated. The displacement curves of the x-axis and y-axis are shown in Figure 9a. The results show that the first two axes moved at the same time in 0.75 s and were relatively stable. At 0.75 s~1 s, there was an instantaneous reverse movement of x-axis, which had a certain influence on the interaction between the two axes. When the x and y axis was linked, the moving displacement of the nozzle in the z-axis direction is shown in Figure 9b. At the maximum displacement, it was still not sufficient to pose significant impact on the overall printing.

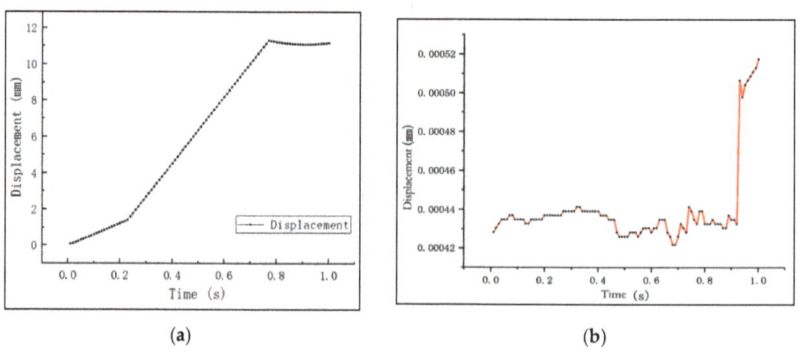

Figure 9. x-axis and y-axis linkage displacement curves. (**a**) the moving displacement of the x and y axis linked with the nozzle head; (**b**) the displacement of the z-axis.

(3) The printing platform drives the reversal of the platform through a parallel structure. Below the printer platform were three branches of the power sources distributed at 120°. Due to the limitation of freedom, it only rotated along the x or y axes. In transient analysis, the independent branch drive and multi-branch simultaneous drive were analyzed (Figure 10). The rotation and movement of the platform were transformed into the displacement represented at the x, y and z directions. The results showed that the printer platform ran smoothly without interference, and the moving displacement was consistent with the theoretical value.

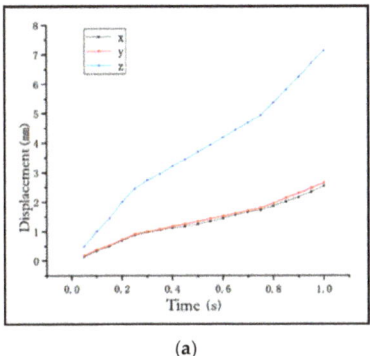
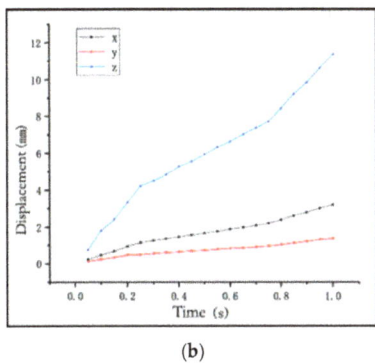

Figure 10. Print platform transient analysis curve. (**a**) The moving displacement of the platform driven by a single branch; (**b**) The moving displacement of the platform driven by a multiple branches.

7. Conclusions

Based on the 3D printing of the wood fiber gel material, cellulose, hemicellulose and lignin gel materials from a branch, waste wood and wood scraps of Cunninghamia lanceolata that can be used for 3D printing were analyzed and prepared. The results showed that cellulose produced a mutation condition near 52 °C and hemicellulose produced a mutation near 80 °C. The structure of a 5-DoF hybrid three-nozzle 3D printer was designed based on the material temperature of mutation condition, and a 3D solid model was established to analyze the structural properties, mechanical properties and motion mechanics. The results showed that the 5-DoF hybrid three-nozzle 3D printing mechanism can realize the 3D printing of wood fiber gel materials. Through the analysis of the structure performance and kinematics, the correctness of the theoretical basis was verified. At the same time, the motion simulation of the structure was realized in the transient motion analysis, which confirmed the rationality of the structure.

This study is a 3D printer design and analysis for wood gel material. The wood fibers mainly contain cellulose, hemicellulose and lignin, which correspond to the three printheads of the printer, and through subsequent studies aim to achieve the solid printing of bionic wood through this type of printer. Meanwhile, the design and implementation of purely multi-jet and multi-degree-of-freedom printers can provide a reference or a direction and idea for the 3D printing industry.

Author Contributions: Formal analysis, X.S.; Investigation, J.C., G.W. and W.C.; Software, X.S.; Supervision, J.C. and G.D.; Writing—original draft, J.C. and Q.Z.; Writing—review & editing, Q.Z. All authors have read and agreed to the published version of the manuscript.

Funding: This work was supported by the National Natural Science Foundation of China (No. 51865053) and the Scientific Research Fund Project of Yunnan Provincial Education Department (No. 2022J0501).

Institutional Review Board Statement: Not applicable.

Informed Consent Statement: Not applicable.

Data Availability Statement: Not applicable.

Conflicts of Interest: The authors declare no conflict of interest.

References

1. Kalkal, A.; Kumar, S.; Kumar, P.; Pradhan, R.; Willander, M.; Packirisamy, G.; Kumar, S.; Malhotra, B.D. Recent advances in 3D printing technologies for wearable (bio)sensors. *Addit. Manuf.* **2021**, *46*, 102088. [CrossRef]
2. Lu, B.H. Additive manufacturing technology—Surrent status and future. *China Mech. Eng.* **2020**, *31*, 19–23. (In Chinese)
3. Zhao, J.Y. Development and application trends of 3D printers. *Pap. Equip. Mater.* **2020**, *49*, 114. (In Chinese)
4. Liu, Y.X.; Zhao, G.J. *Wood Science*; China Forestry Press: Beijing, China, 2012. (In Chinese)
5. Wang, Q.; Ji, C.; Sun, L.; Sun, J.; Liu, J. Cellulose Nanofibrils Filled Poly(Lactic Acid) Bio composite Filament for FDM 3D Printing. *Molecules* **2020**, *25*, 2319. [CrossRef]
6. Cao, W.T.; Ma, C.; Mao, D.S.; Zhang, J.; Ma, M.G.; Chen, F. MXene-Reinforced Cellulose Nano fibril Inks for 3D-Printed Smart Fibers and Textiles. *Adv. Funct. Mater.* **2019**, *29*, 1905898. [CrossRef]
7. Tang, A.; Liu, Y.; Wang, Q.; Chen, R.; Liu, W.; Fang, Z.; Wang, L. A new photoelectric ink based on nanocellulose/CdS quantum dots for screen-printing. *Carbohydr. Polym.* **2016**, *148*, 29–35. [CrossRef] [PubMed]
8. Gunasekera, D.H.; Kuek, S.; Hasanaj, D.; He, Y.; Tuck, C.; Croft, A.K.; Wildman, R.D. Three dimensional ink-jet printing of biomaterials using ionic liquids and co-solvents. *Faraday Discuss.* **2016**, *190*, 509–523. [CrossRef]
9. Yang, J.; An, X.; Liu, L.; Tang, S.; Cao, H.; Xu, Q.; Liu, H. Cellulose, hemicellulose, lignin, and thei r derivatives as multi-components of bio-based feedstocks for 3D printing. *Carbohydr. Polym.* **2020**, *250*, 116881. [CrossRef]
10. Xu, W.; Zhang, X.; Yang, P.; Långvik, O.; Wang, X.; Zhang, Y.; Feng, F.; Österberg, M.; Willför, S.; Xu, C. Surface Engineered Biomimetic Inks Based on UV Cross-Linkable Wood Biopolymers for 3D Printing. *ACS Appl. Mater. Interfaces* **2019**, *11*, 12389–12400.
11. Markstedt, K.; Escalante, A.; Toriz, G.; Gatenholm, P. Biomimetic Inks Based on Cellulose Nano fibrils and Cross-Linkable Xylans for 3D Printing. *ACS Appl. Mater. Interfaces* **2017**, *9*, 40878–40886. [CrossRef] [PubMed]
12. Graichen, F.H.M.; Grigsby, W.J.; Hill, S.J.; Raymond, L.G.; Sanglard, M.; Smith, D.A.; Thorlby, G.J.; Torr, K.M.; Warnes, J.M. Yes, we can make money out of lignin and other bio-based resources. *Ind. Crops Prod.* **2017**, *12*, e431. [CrossRef]
13. Ebers, L.S.; Arya, A.; Bowland, C.C.; Glasser, W.G.; Chmely, S.C.; Naskar, A.K.; Laborie, M.P. 3D printing of lignin: Challenges, opportunities and roads onward. *Biopolymers* **2021**, *112*, e23431. [CrossRef]
14. Hong, S.H.; Park, J.H.; Kim, O.Y.; Hwang, S.H. Preparation of Chemically Modified Lignin-Reinforced PLA Bio composites and Their 3D Printing Performance. *Polymers* **2021**, *13*, 667. [CrossRef]
15. Gao, X. Current status and research ideas of cellulose materials in 3D printing. *Sci. Technol. Innov. Prod.* **2020**, *7*, 56–58. (In Chinese)
16. Feng, T.; Chen, J.F.; Wang, C.; Chen, W.G. Laser-cured cellulose gel for 3D printing. *Cellul. Sci. Technol.* **2020**, *28*, 42–47. (In Chinese)
17. Kam, D.; Braner, A.; Abouzglo, A.; Larush, L.; Chiappone, A.; Shoseyov, O.; Magdassi, S. 3D Printing of Cellulose Nanocrystal-Loaded Hydrogels through Rapid Fixation by Photopolymerization. *Langmuir* **2021**, *37*, 6451–6458. [CrossRef] [PubMed]
18. Gabrielii, I.; Gatenholm, P.; Glasser, W.G.; Jain, R.K.; Kenne, L. Separation, characterization and hydrogel-formation of hemicellulose from aspen wood. *Carbohyd. Polym.* **2000**, *43*, 367–374. [CrossRef]
19. Bahçegül, E.G.; Bahçegül, E.; Özkan, N. 3D Printing of Hemicellulosic Biopolymers Extracted from Lignocellulosic Agricultural Wastes. *ACS Appl. Polym. Mater.* **2020**, *2*, 2622–2632. [CrossRef]
20. Ravishankar, K.; Venkatesan, M.; Desingh, R.P.; Mahalingam, A.; Sadhasivam, B.; Subramaniyam, R.; Dhamodharan, R. Biocompatible hydrogels of chitosan-alkali lignin for potential wound healing applications. *Mater. Sci. Eng. C Mater. Biol. Appl.* **2019**, *102*, 447–457. [CrossRef]
21. Feng, T.; Chen, J.F.; Wang, C.; Liu, B.; Chen, W.G. Design and analysis of FDM-type multi-jet 3D printers. *Digit. Print.* **2020**, *5*, 53–61. (In Chinese)
22. Lu, Y.Z.; Xu, J.F.; Chen, Y.J.; Wang, Z.G.; Ma, J.X. Effect of KH550 modification on the properties of micro-nano-cellulose/PLA composite 3D printing materials. *Chin. J. Papermak.* **2019**, *34*, 14–19. (In Chinese)
23. Zhu, J.; Zhang, Q.; Yang, T.; Liu, Y.; Liu, R. 3D printing of multi-scalable structures via high penetration near-infrared photopolymerization. *Nat. Commun.* **2020**, *11*, 3462. [CrossRef]
24. Raney, J.R.; Compton, B.G.; Mueller, J.; Ober, T.J.; Shea, K.; Lewis, J.A. Rotational 3D printing of damage-tolerant composites with programmable mechanics. *Proc. Natl. Acad. Sci. USA* **2018**, *115*, 1198–1203. [CrossRef]
25. Jiang, Y.F.; Zhao, F.Q.; Li, H.; Zhang, M.; Jiang, Z.F. Ink direct writing additive manufacturing technology and its research progress in the field of energy-containing materials. *J. Pyrotech.* **2022**, *45*, 1–19. (In Chinese)
26. Wu, C.M.; Dai, C.K.; Wang, C.L.; Liu, Y.J. Research progress of multi-degree-of-freedom 3D printing technology. *Chin. J. Comput.* **2019**, *42*, 1918–1938. (In Chinese)
27. Yu, Y.; Chen, Z.Y.; Wang, Z.M. Structure design of free switching multi-color 3D printing with three nozzles. *Modern Salt Chem. Ind.* **2021**, *48*, 185–186. (In Chinese)
28. Li, W.M.; Luo, H.B.; Zhu, H.P.; Xia, Y. Effect of moving mass velocity on the dynamic deflection response of simply supported beams. *J. Huazhong Univ. Sci. Technol. (Nat. Sci. Ed.)* **2008**, *36*, 117–120. (In Chinese)

Article

Comparative Analysis of Polymer Composites Produced by FFF and PJM 3D Printing and Electrospinning Technologies for Possible Filter Applications

Tomasz Kozior [1,*], Al Mamun [2], Marah Trabelsi [2,3] and Lilia Sabantina [2,*]

1. Faculty of Mechatronics and Mechanical Engineering, Kielce University of Technology, 25-314 Kielce, Poland
2. Junior Research Group "Nanomaterials", Faculty of Engineering and Mathematics, Bielefeld University of Applied Sciences, 33619 Bielefeld, Germany; al.mamun@fh-bielefeld.de
3. Ecole Nationale d'Ingénieurs de Sfax, Sfax 3038, Tunisia; marah.trabelsi@enis.tn
* Correspondence: tkozior@tu.kielce.pl (T.K.); lilia.sabantina@fh-bielefeld.de (L.S.)

Citation: Kozior, T.; Mamun, A.; Trabelsi, M.; Sabantina, L. Comparative Analysis of Polymer Composites Produced by FFF and PJM 3D Printing and Electrospinning Technologies for Possible Filter Applications. *Coatings* **2022**, *12*, 48. https://doi.org/10.3390/coatings12010048

Academic Editor: Philippe Evon

Received: 27 November 2021
Accepted: 19 December 2021
Published: 1 January 2022

Publisher's Note: MDPI stays neutral with regard to jurisdictional claims in published maps and institutional affiliations.

Copyright: © 2022 by the authors. Licensee MDPI, Basel, Switzerland. This article is an open access article distributed under the terms and conditions of the Creative Commons Attribution (CC BY) license (https://creativecommons.org/licenses/by/4.0/).

Abstract: Three-dimensional printing technologies are mainly used to build objects with complex shapes and geometry, largely prototypes, and thanks to the possibility of building very thin layers of material with small pores, electrospinning technology allows for the creation of structures with filtration properties, in particular very small particles. The combination of these technologies creates new possibilities for building complex-shape composites that have not been comprehensively tested so far. The article describes the results of research on composites manufactured by combining samples prepared with two 3D printing technologies, Fused Filament Fabrication (FFF) and Photo-Curing of Liquid Polymer Resins (PJM) in combination with electrospinning (ES) technology. The surface morphology of composites manufactured from biocompatible materials was investigated using confocal laser scanning microscopy (CLSM) and contact angle measurements, and chemical composition analysis was studied using Fourier transform infrared spectroscopy (FTIR). This approach to creating composites appears to be an alternative to developing research for filtration applications. The article presents basic research illustrating the quality of composites produced by combining two unconventional technologies: 3D printing and electrospinning (ES). The analysis of the research results showed clear differences in the structure of composites produced with the use of various 3D printing technologies. The CLSM analysis showed a much better orientation of the fibers in the MED610 + PAN/gelatin composite, and the measurement of the contact angle and its indirect interpretation also for this composite allows for the conclusion that it will be characterized by a higher value of adhesion force. Moreover, such composites could be used in the future for the construction of filtering devices and in medical applications.

Keywords: quality; electrospinning; 3D printing; PJM; FFF; bio-materials; filtering

1. Introduction

The production of complex objects using conventional technologies requires the use of numerous technological processes and a large number of tools (injection molds, casting molds, dies, and cutting tools) [1] and thus the production of objects becomes time-consuming. The use of 3D printing technology to produce complex objects, e.g., freeform objects [2], especially prototypes, in the era of industrial transformation 4.0, is a natural choice and fits perfectly into the realities of such issues as lean manufacturing and LPPD (lean process and product development) [3], where the main objective is to produce structures with a minimum weight and maximum fulfillment of product quality requirements [4] while maintaining an optimal manufacturing process. These technologies are successfully used wherever the time of product implementation is a key competitive parameter. Unconventional technologies include 3D printing, but also electrospinning, electro-erosion, electrochemical, laser, and ultrasonic processing and modern coating methods [5–7].

Due to the necessity of reducing the mass of the manufactured objects or increasing the strength and utility properties, composite materials manufactured of plastics are increasingly used in industrial applications. In the case of conventional technologies, there are known polymeric materials with additives of carbon and glass fibers, with molybdenum disulfide, with additives improving lubricating properties, etc. [8]. Three dimensional printing technologies also have some possibilities of using composite materials [9,10], especially technologies such as Selective Laser Sintering—SLS, Fused Filament Fabrication—FFF, also commercially known as Fused Deposition Modeling—FDM® [11].

It seems that the SLS technology is well suited for manufacturing objects from composite materials, while the FFF technology is more appropriate for creating objects that are only made after combining with new material, e.g., using electrospinning technology. In many applications, however, the use of a composite material alone is not sufficient in order to achieve appropriate properties. The use of a composite material may increase strength and wear resistance, for example tribological [12] or rheological [13,14], but it does not necessarily provide the required properties, such as filtration.

An interesting solution may be a combination of two technologies, for example 3D printing and electrospinning [15,16], or 3D printing and textile object [17,18]. The combination of 3D printing and electrospinning allows for the production of objects with very complex geometric shapes and high strength, and thanks to the use of electrospinning technology and nanofibers, such objects can exhibit filtration properties. The production of objects from biocompatible materials using 3D printing and electrospinning opens up new possibilities in medicine, e.g., in the field of implantology, where the use of 3D printing is increasing. In addition, 3D printing of metals, in particular, is used because of their high mechanical properties, as they are far more durable than human bones. By using electrospinning technology, the functional properties of the 3D objects produced can be improved and new properties can be added.

The use of 3D printing is very widespread and concerns such industries as rapid prototyping, foundry [19], injection molding, architecture, medicine [20], aviation, automotive [21], food, and flexible (lean) production series. The most important 3D printing technologies include SLS (Selective Laser Sintering), SLM (Selective Laser Melting), FFF (Fused Filament Fabrication), SLA (Stereolithography), Photo-Curing Technologies: PJM (polyjet matrix), PJ (polyjet), DoDjet (Drop on Demand Jetting Technology), and MJP (multijet). Due to the low initial cost of the machines with FFF-technology in contrast to other 3D printing technologies such as Selective Laser Sintering (SLS) or Selective Laser Melting (SLM), FFF technology includes the advantages of very low material costs, easy handling, and a wide range of materials for the production of composite materials. When choosing the combination of 3D printing and electrospinning technology, FFF technology seems to offer the optimal solution due to the above-mentioned advantages. With this FFF-technology, many interesting materials are available such as acrylonitrile butadiene styrene (ABS), poly(lactic acid) (PLA), PLA + carbon fiber, PET, Nylon, TPU, and PC, which are also reinforced with additives. Although the FFF technology can use composite materials, they cannot be easily modified independently, as is the case with SLS technology. In FFF, these materials can be treated both thermally and chemically, which improves the quality of the manufactured objects [22,23]. PLA material, being the cheapest and the most important due to its very good functional features, is a good solution for the construction of prototypes with filtration properties [24] combined with electrospinning technology. Comparing both technologies, it can be clearly stated that the big disadvantage of the PJM technology is the high cost of the model material (liquid resin) compared to the filament in the FDM technology, which costs only a few dollars per kilogram. However, its great advantage is its high manufacturing accuracy, especially in the Z axis, due to the very small thickness of the layer being built (as small as 16 µm). Many studies have aimed at determining the influence of technological parameters on the properties of the manufactured objects [25–30] and the influence of measurement methods on the reliability of the test results [31].

The electrospinning technique is the most widely used, simplest, and most cost-effective method to produce nanofibers from a wide range of bio-based and synthetic polymers as well as admixtures, such as particles [24].

For nanofiber fabrication, it is possible to use nontoxic or low-toxicity solvents such as dimethyl sulfoxide (DMSO) and polymers soluble in it [32,33]. The polymer polyacrylonitrile (PAN) is often used for nanofiber mat fabrication because it can be spun from DMSO, has no cell toxicity and is water resistant [34,35]. Moreover, by adding other polymers, such as gelatin or magnetic particles to the PAN solution, nanofiber mats for defined purposes are easily prepared by simply adding these compounds to the polymer electrospinning solution [36,37]. PAN nanofiber mats are flexible and become even more flexible when wet, and can be easily stretched and draped without any problems [38]. In addition, PAN is often used for the production of carbon material due to its high carbon yield [39]. Gelatin is a biobased polymer which is often used in biotechnological fields due to its biocompatibility and is also used for the production of nanofibers.

With 3D printing technology, the available range of materials is wide, which allows for selection of the appropriate material depending on the field of application, for example in medicine. As a result, these technologies can be used, among others, as medical dressings, scaffolding in tissue engineering or controlled drug delivery systems. Despite the long time that has passed since the first papers on electrospinning technology were published, they are still in the testing phase due to the production of new electrospinning devices and new materials with improved properties. Research work largely focuses on the study of new materials [32,37,40], including composites [41], and the properties [42] of the produced objects, such as filtration, as well as composites obtained by combining electrospinning technology and, for example, 3D printing. Moreover, the electrospinning technology can also be used for the production of nanofibers and in the construction of composite materials, e.g., by conventional methods [43].

Our previous research works in this field [15,16] focused on composites using combination of FFF technology and needle-free electrospinning technology. The novelty of this study is that in the present work, two technologies were used to produce composites, namely PJM and FFF 3D printing technologies in combination with needle-based electrospinning technology for use in filter applications in medical filed, and the variable technological parameters were studied in more detail. For use in filtering areas, final composite consists of 3D printed structure that incorporates the robustness and thus supports the fine and fragile nanofiber mats, which has excellent filtering properties and requires mechanical protection. The printing direction in relation to the XZ build platform and the filling level with the model material were analyzed. In addition, 3D objects were fabricated from two biocompatible materials such as PLA and MED610, which show a high potential for use in the medical industry due to their biocompatibility. These 3D objects were electrospun using two different types of nanofiber mats, such as pure polyacrylonitrile (PAN) nanofibers and a combination of PAN and gelatin. PAN nanofiber mats are often used in medical and filter areas and the PAN polymer is valued for its water-resistant properties. In addition, it should be mentioned that the nanofiber mats were produced using the low-toxicity solvent dimethyl sulfoxide (DMSO) and there are few studies in this area, because mostly highly toxic solvents such as dimethylacetamide (DMAc) and dimethylformamid (DMF) are widely used in the production of nanofiber mats. The combination of 3D materials such as PLA and MED610 with nanofibers with PAN and gelatin show a high potential for effective application in the medical industry for filtration applications. The fabrication of such novel composite models by combining two different technologies (3D printing and electrospinning) is important for many reasons, but the most important is the one concerning the geometrical possibilities of the created models. In addition, the use of these two methods and indirect assessment of the diameter of the fibers, their orientation, and contact angle measurements will allow the initial assessment of the quality of the produced composites also in the context of filtration, where the distribution of the fibers, their diameter, and directionality depend on the material, 3D printing technology and parameters and material in the electrospinning technology.. This is why the purpose of this research is to conduct a preliminary comparative analysis. In the case of 3D printing, the use of biodegradable PLA material in FFF technology and PJM technology allows the use of MED610 material with biocompatible properties, which are beneficial for some medical applications. The preliminary research results presented below require further research, but they can provide a valid basis for further, expanded research that considers the body's response to such composites.

2. Materials and Methods

The test samples were designed in Solidworks re-SolidWorks (Dassault Systèmes SolidWorks Corp., Waltham, MA, USA), in the shape of a rectangular prism with the dimensions of $50 \times 50 \times 0.5$ mm^3. The SolidWorks objects were written as STL files (stereolithography language), mapping the objects as a mesh of triangles. The translation accuracy was in the so-called fine record mode, but in this case, due to the simplicity of the object, each form of notation maps the object in the same way, i.e., with 12 triangles. These designed objects were sent to internal 3D printer systems, manufactured using two 3D printing technologies—FFF [44] and PJM, and then removed from the platform and cleaned: FFF—manually, PJM—with water under pressure. After completing the printing process and cleaning, the objects were subjected to the electrospinning process, which, using high-voltage electrical discharges, applied a layer of material on a surface with the dimensions of 50×50 mm^2 (see Figure 1).

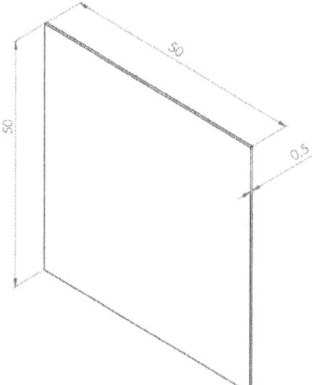

Figure 1. 3D CAD sample, dimensions in mm.

2.1. 3D printing Methods

Two 3D printing machines with FFF and PJM technologies were used to build the objects and later coated with nanofibers. In the case of the photo-curing technology of liquid polymer resins—PJM (Objet Geometries Ltd., Rehovot, Israel), a Connex 350 machine by Object (currently Stratasys, Ltd., Eden Prairie, MN, USA) was used, and in the case of the Fused Filament Fabrication—FFF (also known as FDM) technology, the Makerbot Replicator 5th generation machine was used. In the FFF technology, polylactide (PLA) material (trade name—EASY PLA, Fiberlogy company, Poland) was used to build the samples, and in the PJM technology, the biocompatible material MED610 (Stratasys company, Ltd., Eden Prairie, MN, USA). MED610 material is a material that is approved for contact with the human skin for a period of 30 days and for contact with the mouth for 24 h, making it very attractive in medicine, such as an element of medical devices, filtering masks, prototypes of orthoses, as well as in the field of dental interests, such as dental appliances [45]. MED610 has been evaluated and deemed acceptable for the following biological risk [45]: Cytotoxicity—EN ISO 10993-5: 2009 [46]; Irritation—EN ISO 10993-10: 2013 [46] Delayed-type hypersensitivity—EN ISO 10993-10: 2013 [46]; Genotoxicity—EN

ISO 10993-3: 2014 [46]; Chemical characterization—EN ISO 10993-18: 2009 [46]; USP Plastic Class VI—USP 34 <88> [47].

Table 1 and Figure 2 show the chemical composition of the materials used in the construction of samples with 3D printing technologies.

Table 1. Composition of MED610 material [45].

Information on Ingredients		
CAS#	**Component**	**Percent (%)**
5888-33-5	Isobornyl acrylate	15–30
Proprietary	Acrylic monomer	15–30
Proprietary	Urethane acrylate	10–30
Proprietary	Acrylic monomer	5–10; 10–15
Proprietary	Epoxy acrylate	5–10; 10–15
Proprietary	Acrylate oligomer	5–10; 10–15
Proprietary	Photoinitiator	0.1–1; 1–2

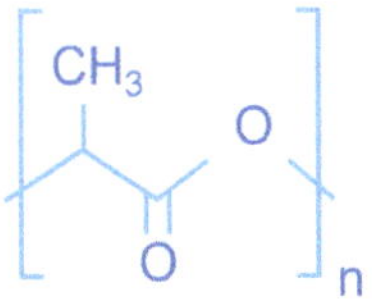

Figure 2. Structural model of PLA material [48].

The test specimens were manufactured in three predetermined orientations relative to the 0°, 45°, and 90° build platform. The overview of samples orientation on the build platform prepared for FFF process is shown in the Figure 3. Moreover, in the FFF technology, samples 1–9 (see Figure 3a) were manufactured with a layer thickness of 0.2 mm, samples 10–18 (not shown here) with a layer thickness of 0.3 mm and both types with 95% filling, and samples 19–27 (see Figure 3b) with a layer thickness of 0.2 mm and a filling of 50%.

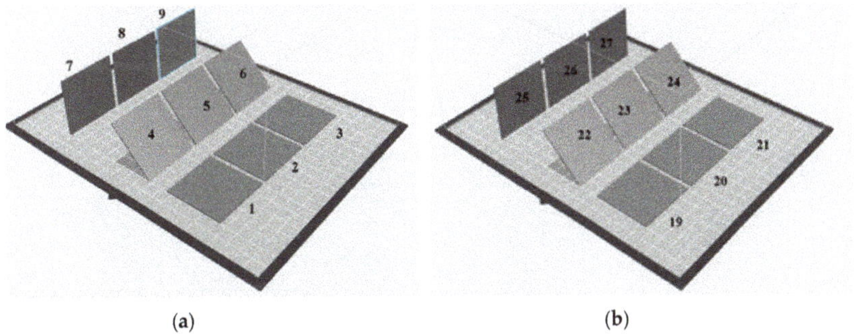

Figure 3. Overview of samples orientation on the build platform prepared for FFF process: (**a**) samples 1–9, (**b**) samples 19–27.

Samples in the PJM technology were manufactured in two variants: 28–36 (see Figure 4a) with a layer thickness of 0.032 mm, i.e., in the *High Speed* mode and 100% filling, and the samples 37–45 (see Figure 4b) were manufactured in the so-called *High Quality* mode with a set layer thickness of 0.016 mm and also 100% filling. A total of 45 samples were manufactured, but the total number of different types of samples is only 15, because three pieces of each type were made in order to statistically evaluate the repeatability of the measurement results, so there are no differences between samples 1, 2, and 3, and in subsequent batches after 3 pieces. Figures 3 and 4 show the location of samples on virtual platforms of working machines.

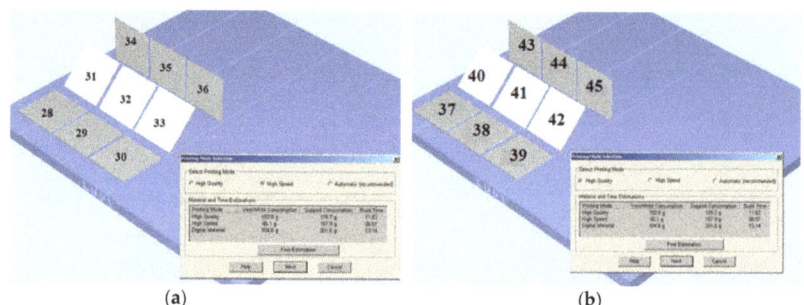

Figure 4. Overview of samples orientation on the build platform prepared for PJM process: (**a**) samples 28–36, (**b**) samples 37–45.

Figure 5a shows 3D-printed sample prepared from PLA using FFF technology and Figure 5b shows the sample prepared from MED610 using PJM technology.

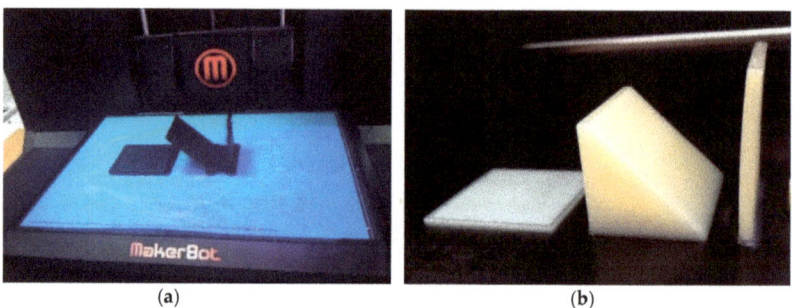

Figure 5. Overview of 3D printed samples: (**a**) 3D printed samples prepared from PLA by FFF technology and (**b**) from MED610 by PJM technology.

2.2. Electrospinning

The nanofibers were produced by the needle-based electrospinning machine (Spinbox—Bioinicia S.L., Paterna, Valencia, Spain) (see Figure 6).

Figure 6. Electrospun 3D printed sample in the electrospinning device.

The applied spinning parameters in the manufacturing process were as follows: high voltage 8–18 kV, temperature in the chamber 20–23 °C, relative humidity in the chamber 35–40%, a 20 cm distance from the needle tip to the nanofiber collector, and the flow rate of polymer solution 300–500 µL/h. The polymer nanofibers were produced using an

18-gauge needle (1.23 mm outer diameter). Spinning duration was approximately 15 min, which was also used as 3D printing material on non-woven polypropylene (PP) substrates (Elmarco, Czech Republic). Polyacrylonitrile (PAN) (X-PAN, Dralon, Dormagen, Germany) polymer and dimethyl sulfoxide (DMSO, min 99.9%, purchased from S3 Chemicals, Bad Oeynhausen, Germany) solvent were used for production of all nanofiber mats.

The polymer solutions were mixed by magnetic stirring for 2 h at room temperature. Two types of nanofiber mats were used based on polymer properties for biological/medical/chemical purposes to create composites with biocompatible materials as composites for filtration applications (see Table 2). The first spinning solution is the mixture of 13% PAN and 87% DMSO. The second spinning solution contained 13% PAN and 8% gelatin powder (Abtei OP Pharma GmbH, Marienmünster, Germany) and 79% DMSO.

Table 2. Overview of the 3D printed samples.

3D Printed Sample No.	3D Material	Nanofiber Mats
1–6	PLA	PAN 13% + DMSO 87%
7–18	PLA	PAN 13% + gelatin 8% + DMSO 79%
19–27	PLA	PAN 13% + gelatin 8% + DMSO 79%
28–36	MED610	PAN 13% + gelatin 8% + DMSO 79%
37–45	MED610	PAN 13% + gelatin 8% + DMSO 79%

The aim of this research was the preliminary assessment of polymer composites produced by 3D printing and electrospinning technologies in the application of filtering and medical purpose. For this purpose, several types of samples were selected for the comparative tests and are presented in Table 2.

The 3D printed objects were vertically attached and electrospun. Figure 7a shows PLA sample prepared using FFF and coated with PAN nanofiber mat and Figure 7b shows MED610 sample prepared using PJM and coated with PAN/gelatin nanofiber mat.

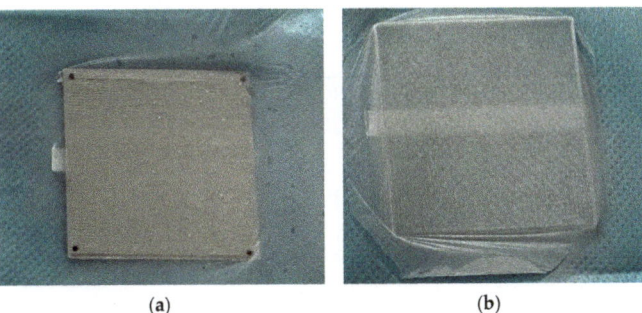

(a)   (b)

Figure 7. (a) Sample No. 3—prepared from PLA using FFF and coated with PAN nanofiber mat and (b) sample No. 29—prepared from MED610 using PJM and coated with PAN/gelatin nanofiber mat (b).

2.3. Research Methods

Fourier transform infrared spectroscopy (FTIR) is a tool that provides significant information about chemical bonding, molecular structures, and component miscibility. In this study, the chemical study was performed with an Excalibur 3100 fourier transform infrared spectroscope (FTIR) (Varian Inc., Palo Alto, CA, USA). The spectra were recorded between 4000 and 650 cm^{-1} frequency ranges.

The optical evaluation of the morphology of the samples was performed with the use of a confocal laser scanning microscope (CLSM) VK-8710 (Keyence, Osaka, Japan).

The water contact angle was measured using the contact angle goniometer OCA 15 Pro (Dataphysics, Filderstadt, Germany). The measurements were performed with 5 µL water droplets at the inner surface of the polystyrene well plates in ambient conditions.

Eight measurements were performed for each material. For contact angles below 90°, the surface is considered as hydrophilic. Contact angles greater than 90° are considered as hydrophobic.

The diameters of the nanofibers were evaluated using the ImageJ 1.51j8 program.

3. Results

The results of testing are presented below. In Section 3.1, the analysis of fourier transform infrared spectroscopy is investigated. Then, the results of surface measurements are described using confocal laser scanning microscopy (CLSM) in Section 3.2. The analysis of the contact angle is presented in Section 3.3.

3.1. FTIR Results

FTIR was used to show that the nanofiber mats were electrospun onto the printed samples. Figure 8a shows electrospun PAN/gelatin nanofiber mats on PAN nanofiber mats on PLA and Figure 8b shows MED610 and PAN/gelatin nanofiber mats electrospun on MED610. The electrospun nanofiber mats on PLA and MED610 are advantageous for cell growth and also for numerous filter applications, which presents new manufacturing possibilities for medical applications, especially in implant production.

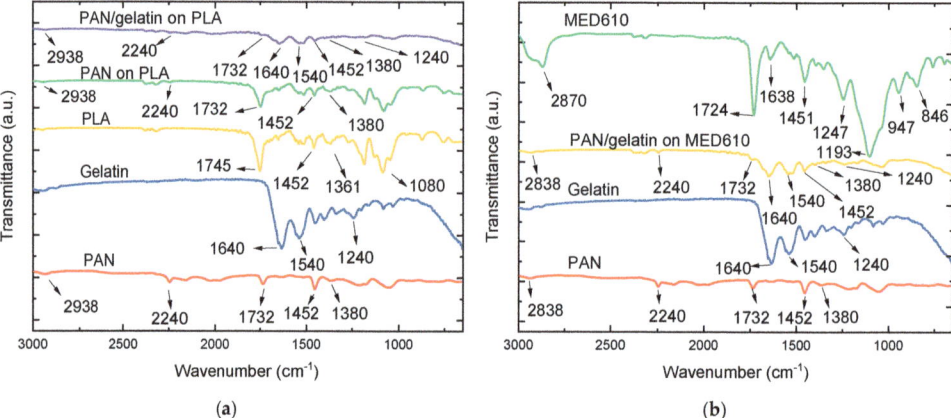

Figure 8. FTIR results: (a) PAN/gelatin nanofiber mats on PLA sample, PAN nanofiber mats on PLA sample, PLA sample without coating with nanofiber mat, pure gelatin, and PAN nanofiber mat; (b) MED610 sample without coating with nanofiber mat, PAN/gelatin nanofiber mats on MED610, pure gelatin, PAN nanofiber mat.

The overview of typical spectra of used materials can be seen in Table 3. In Figure 8a,b, typical and characteristic FTIR peaks of the materials used. In addition, these peaks are shown again in Table 3 for clarity. When looking at Figure 8a, it can be seen that some peaks occur continuously at the marked locations with a crossed-out line, and thus it is not easy to distinguish between the characteristic peaks for a specified material. However, some peaks, such as PLA of 1452 cm^{-1}, also occur in PAN and can be seen in all three FTIR spectra (see Figure 8a).

Table 3. Overview of the typical FTIR spectra of gelatin, PLA, PAN and MED610.

Polymers	FTIR Spectra	Reference
Gelatin	Gelatin spectra bands: 2700–3600 cm^{-1}—amides A and B 3400 cm^{-1}—amide A 1640, 1540 and 1240 cm^{-1}—C=O stretching of amide I, N-H bending of amide II and N-H bending of amide III 900–1900 cm^{-1}—amides I, II and III 900 cm^{-1}—amide IV 700 cm^{-1}—amide IV	[37,49,50]
PLA	1745 cm^{-1}—C=O group 1452 cm^{-1}—CH3 asymmetric group 1361 cm^{-1}—CH3 symmetric group 1080 cm^{-1}—C-O group 695–760 cm^{-1} and 1740–1750 cm^{-1}—C=O group	[51]
PAN	2240 cm^{-1}—C≡N nitrile stretching vibration mode 1732 cm^{-1}—carbonyl (C=O) stretching vibration mode 2938, 1452 and 1380 cm^{-1}—CH$_2$ stretching vibrations	[52]
MED610	2870 cm^{-1}—C-H stretching vibration mode 1724 cm^{-1}—C=O stretching vibration mode 1247 cm^{-1}—C-H stretching vibration mode 1193 cm^{-1}—C-O stretching vibration mode 2870 cm^{-1} 1724 cm^{-1} 1638 cm^{-1} 1247 cm^{-1} 1193 cm^{-1} 846 and 947 cm^{-1}	[53,54]

Looking at the Figure 8a, it becomes visible that the characteristic FTIR spectra of 1732, 1745, 1452, 1380 and 1361 cm^{-1} occur in PLA, PAN/gelatin, and PAN nanofiber mats (see Table 3). At this point, it is not possible to say precisely that these peaks belong only to a particular material.

Moreover, it is assumed that 3D-printed PLA materials shine through the layer of nanofiber mat (see Figure 8a) and even the characteristic peak for PLA at 1080 cm^{-1} is also slightly visible for PAN/gelatin and PAN nanofiber mats.

When looking at Figure 8b, some peaks, such as 1724 cm^{-1} of MED610 and 1732 cm^{-1} of PAN/gelatin nanofiber mats, are in a narrow range and at this point overlap slightly. The peaks at 1638 cm^{-1} of MED610 and at 1640 cm^{-1} of PAN/gelatin nanofiber mat are in a narrow range, and the peaks at 1451 cm^{-1} of MED610 and at 1452 cm^{-1} of PAN/gelatin nanofiber mat are also in a narrow range and can be detected in the Figure 8a by all three FTIR spectra.

3.2. CLSM Results

CLSM images of composites are depicted in Figures 9–14. Figure 9 shows the PAN nanofiber mats on PLA object. In addition, Figure 10 shows the surface of PLA object without nanofiber mats. Figure 11 presents the PAN/gelatin nanofiber mats on PLA object. In addition, Figure 12 shows the 3D printed side of PLA. Figure 13 presents CLSM images of the PAN/gelatin nanofiber mats MED610 object. In addition, Figure 14 shows the surface of MED610 object without nanofiber mat.

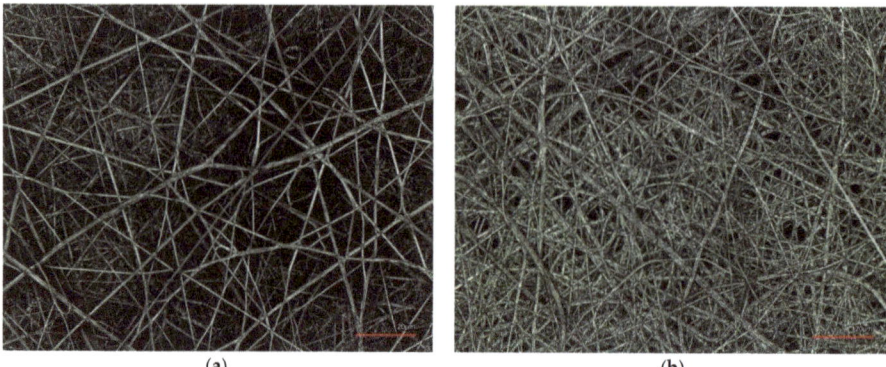

Figure 9. Confocal laser scanning microscope (CLSM) images of PAN nanofiber mats on PLA object. PLA object is covered with nanofibers and is not visible: (**a**) sample—1, (**b**) sample—2. Scale bars indicate 20 µm.

Figure 10. Confocal laser scanning microscope (CLSM) image of PLA object, sample—1. Scale bar indicates 200 µm.

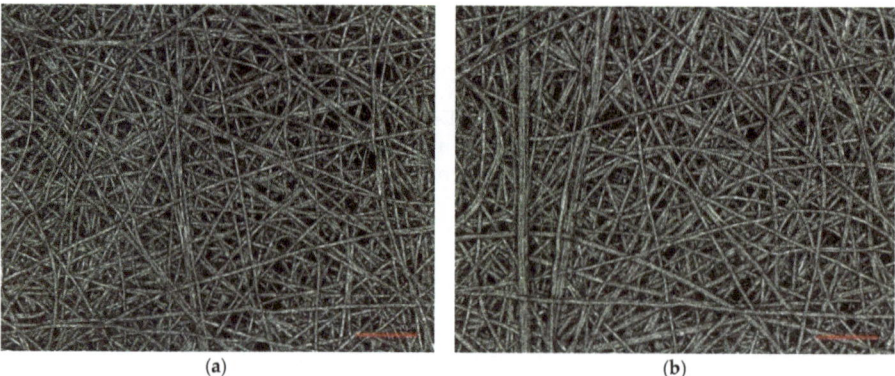

Figure 11. Confocal laser scanning microscope (CLSM) images of of PAN/gelatin nanofiber mats on PLA object. PLA object is covered with nanofibers and is not visible: (**a**) sample—7, (**b**) sample—8. Scale bars indicate 20 µm.

Figure 12. Confocal laser scanning microscope (CLSM) image of PLA object, sample—7. Scale bar indicates 200 μm.

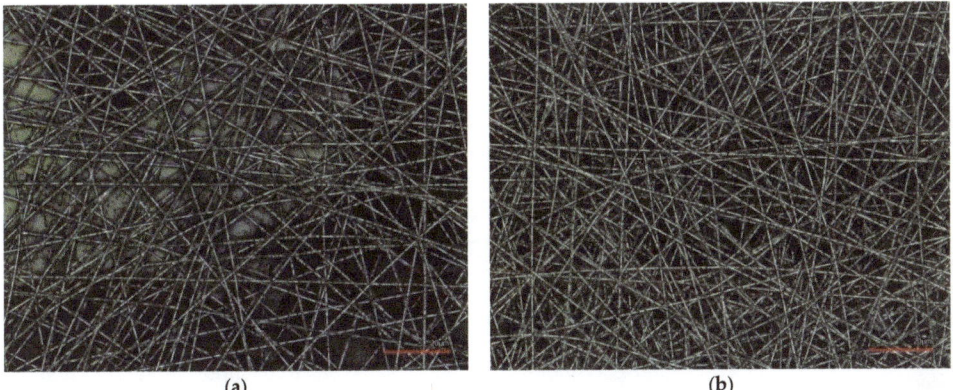

(a)   (b)

Figure 13. Confocal laser scanning microscope (CLSM) images of PAN/gelatin nanofiber mats on PJM object. PJM object is covered with nanofibers and is not visible: (**a**) sample—37, (**b**) sample—38. Scale bars indicate 20 μm.

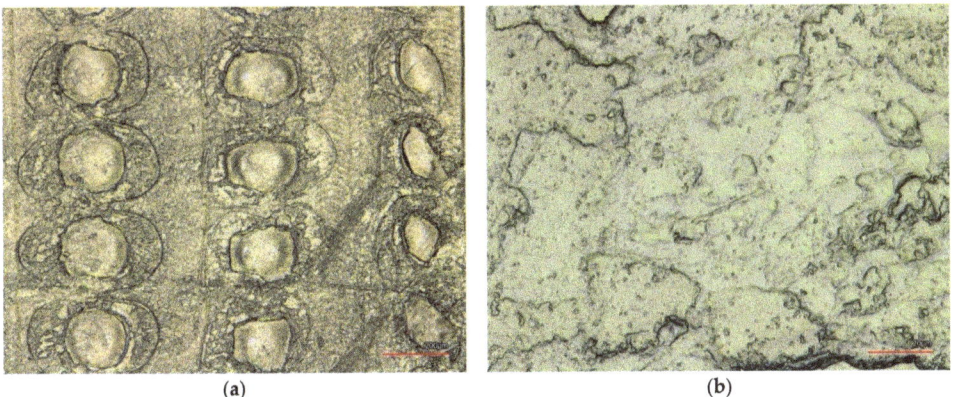

(a)   (b)

Figure 14. Confocal laser scanning microscope (CLSM) images of MED610 object, sample—29. Scale bars indicate: (**a**) 200 μm (**b**) 20 μm.

The surface morphology was analyzed using CLSM. Figure 9 shows CLSM images of the PAN nanofiber mats on PLA object. Here the nanofibers show mostly straight orientation with some tangled parts. Thicker nanofiber strands can also be observed. Figure 10 shows the 3D-printed side of PLA object. The PLA filament clearly shows distinct distances to other PLA filaments; the surface is not fused together and does not form a closed unit. In addition, the shape of the printed filaments is not linear and straight, but partly curved and not uniform. This could stem from discontinuous material release in the nozzle of the 3D printer, or from the temperature partially being too low in the upper layer of filament compared to the filament layer below.

Figure 11 shows CLSM images of the PAN/gelatin nanofiber mats on PLA object. In these samples, it is clearly visible that the fiber diameter has increased significantly compared to pure PAN nanofibers without gelatin (see Figures 9 and 15b). This parameter can be analyzed in terms of using the test results to build devices with filtering properties. That the addition of gelatin to PAN significantly increases the fiber diameter compared to pure PAN nanofibers, is an effect known from previous studies [37]. Figure 12 shows results for the 3D-printed side of PLA object. Compared to Figure 10, the filaments here form a closed unit with straight linear shapes and the layers of filament are close and fused to each other. Figure 13 shows CLSM images of the of PAN/gelatin nanofiber mats on MED610 object. Interestingly, at this point the nanofibers show much smaller diameters compared to the samples in Figure 11 (see Figure 15c), even though the nanofibers contain gelatin, which is known to increase the diameter of the nanofibers according to a previous study [37]. In this case, the nanofibers are on average much smaller and show strict straight alignment without tangles (see Figure 15c). It seems that the 3D material MED610 plays a major role here, in that the behavior of the nanofibers is completely different compared to that of the nanofibers with the same content of PAN and gelatin parts in Figure 11 that were electrospun on the PLA 3D object. This effect is unexpected and should be followed up. It should be noted that the material used for 3D printing has a clear influence on the filtering properties of the manufactured objects due to the filtering capabilities of nanofibers created during the electrospinning process. The production of fibers with a specific diameter affects the possibilities of filtration and materials separation. It should be stated that depending on the given 3D printing technology and materials, the control-process analysis of the electrospinning technology should be subjected to the appropriate selection of process parameters in order to obtain the composite structure and nanofiber diameters with a specific degree of filtration.

The diameter distributions of the PAN and PAN/gelatin nanofiber mats are presented in Figure 15.

The fiber diameter of PAN nanofibers on PLA resulted in a diameter distribution of (241 ± 84) nm. The PAN nanofibers with gelatin admixture resulted in a diameter distribution of (528 ± 157) nm on PLA and (388 ± 178) nm on MED610. Figure 14 shows the 3D printed side (MED610) with 200 and 20 μm enlargement respectively. It can be seen that the resin MED610 is completely melted without visible defects and voids. Figures 10, 12 and 14 show the surface of samples viewed from the side of the printed material. With the FFF technology, the vertically distributed lines of PLA material are clearly visible, whereas in Figure 10 two successive, crossing layers of the material are visible. A completely different structure is presented for samples made of MED610 material and PJM technology. In Figure 14 there is no clear view of the layers, which is undoubtedly an advantage affecting the mechanical properties and the quality of the composites produced. In Figure 14a, uniformly distributed cylindrical geometric features (with a diameter of less than 200 μm) are visible, which correspond to the characteristics of the object building process using the PJM technology, where the visible traces correspond to thin sprayed drops of material.

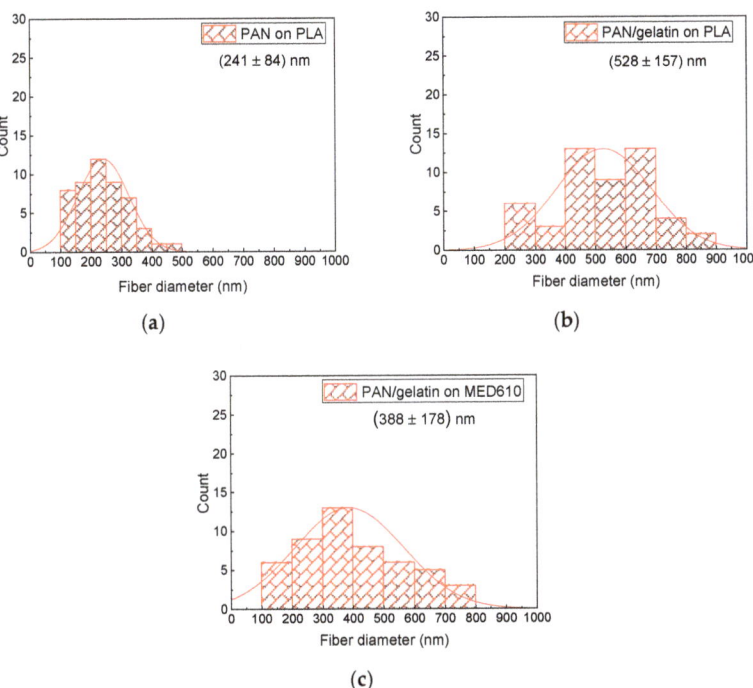

Figure 15. Diameter distribution of (**a**) PAN nanofiber mats on PLA object; (**b**) PAN/gelatin nanofiber on PLA object and (**c**) PAN/gelatin on MED610 object.

3.3. Contact Angle Results

The results of the contact angle measurements are presented in Tables 4 and 5 and in Figure 16. Tables 4 and 5 show the mean value of the contact angle measurement for 3D printed samples. The measurement of the contact angle of nanofiber mats was not possible because the droplet was immediately absorbed into the nanofiber mat due to the capillary effect.

Table 4. Contact angle measurement results for sample—18, PLA object.

Calculated Feature	Image in Print Direction	Image Perpendicular to Print Direction
Mean values and standard deviation of contact angle (°)	86.8 (±3.4)	62.7 (±2.8)

Table 5. Contact angle measurement results for sample—29—PJM, MED610 material.

Calculated Feature	Image in Print Direction	Image Perpendicular to Print Direction
Mean values and standard deviation of contact angle (°)	54.7 (±5.5)	45.2 (±4.0)

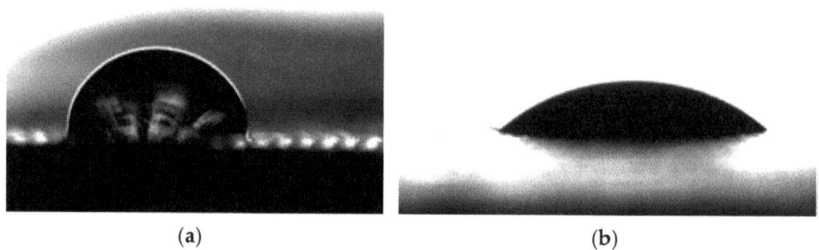

Figure 16. Contact angle measurement results (**a**) PLA object—sample 18, (**b**) MED610 object—sample 29.

The samples were examined in 'printing direction' and 'perpendicular to printing direction', because due to the 3D printing technology, the layers are built on top of each other, and the surface morphology as seen, for example, in Figure 11, has an impact on the contact angle measurement results (see Tables 4 and 5 and Figure 16). By analyzing the results of the contact angle measurement presented in Tables 3 and 4 for PLA and MED610 materials, it can be concluded that for two completely different production technologies there are clear differences in the values of the measured parameters. The highest values of 86.8° on average were obtained for FFF and image in print direction technologies. In the case of PJM technology and MED610 material, the highest contact angle values are 54.7° also for image in print direction. In the case of the PJM technology, a greater dispersion of the measurement results is clearly visible, amounting to SD—5.5 in the presented case image in 'print direction'. Based on the SD parameter, it can be concluded that the distribution of the results is relatively small and that the results are repeatable for both 3D printing technologies.

4. Discussion

The analysis of the results leads to the conclusion that 3D printed objects have an influence on the orientation of the nanofibers as well as on the nanofiber diameters, as shown by the results of this study. Generally, it is known that the addition of gelatin increases the nanofiber diameters compared to pure PAN nanofibers, which is observed for nanofiber mats electrospun on PLA. This conclusion has direct implications for the analysis of composites in terms of their filtration and medical applications. However, when the nanofibers are electrospun with gelatin on MED610, the morphology changes. In this case, the nanofibers are straighter and more oriented, and the fiber diameters decrease rapidly compared to PLA 3D printed objects. Moreover, with the increase in the diameter of nanofibers and their improper/unplanned arrangement, the filtration properties of the composite decrease due to the impossibility of filtration of small fractions of materials. When considering the engineering applications of the test results, it can be concluded that the MED610 material in combination with PAN/gelatin shows much better engineering properties (fibers' orientation), which may have influence into better quality control of the production of composite models with filtration properties. The analysis of the contact angle measurement results in real applications which allow us to conclude that the MED610 material will have a much better adhesion force. In addition, in the case of FFF technology and samples made on the basis of PLA, local densities and areas with lower fiber density are clearly visible, which can have a huge impact on the strength of the composite as can be seen in the Figures 9 and 10. It can be noticed, especially in Figure 9, Figure 11 (for PLA) and Figure 13 (for MED610), that the distribution of nanofibers in the areas of lower density generates gaps up to 1–2 µm in size, enabling the permeability of media such as liquids and gas/air in filtration applications. It seems that the next research effort should focus on an in-depth analysis of the sizes of gaps in order to determine the filtration capabilities of the composite and also on the distribution of the gaps in the composites that may also have an influence on the mechanical properties. In addition, the PLA object surface morphology clearly shows gaps in the layers, which in the case of the MED610 Object can hardly be discovered and the surface is not permeable.

Due to the structure of the samples, it is not possible to determine the adhesion force between the 3D printed part of the composite and the electrospinning technology. As the research results have shown [52,55], there is, however, a certain relationship between the contact angle and the adhesion force. As the results of previously conducted research work have shown, the contact angle has a significant impact on the adhesion force [52,55]. The results of the work showed that as the contact angle adhesion force increases, its value decreases. The contact angle results in the case of PLA material and FFF technology are characterized by the value for Image in print direction—86.8° (SD—3.4), and for Image perpendicular to print direction—62.7° (SD—2.8). Nevertheless, in the case of the PJM technology and the MED610 material, the mean value of the contact angle for the tested surface takes a much lower value, respectively: Image in print direction—54.7° (SD—5.5), and for *Image perpendicular to print direction*—45.2° (SD—4.0). Based on the results of the research presented in the literature, where the impact of the contact angle on adhesion force was analyzed in an indirect manner, it can be concluded that in the case of composites produced with the PJM and electrospinning adhesion force technology, it should show a higher value compared to the samples made with the 3D printing technology—FFF and electrospinning technology.

As numerous research studies show, the technological parameters of 3D printing affect the strength and quality of the surface; in the case of combining the printed objects with the electrospinning technology, however, there was no effect on the quality of the connection after the above-mentioned tests were carried out.

5. Conclusions

The article presents the results of research on composites manufactured with two 3D printing technologies: FFF from PLA material and PJM technology—MED610, and electrospinning technology from PAN and gelatin that also have medical uses. The analyzed parameters included print parameters such as the printing direction, layer thickness, and degree of filling (in FFF) with 3D printing material, and electrospinning process parameters such as electricity voltage, humidity, and chamber temperature. The analysis of the quality of the produced composite was based on the results of the surface morphology research using confocal laser scanning microscope (CLSM), fourier transform infrared spectroscopy (FTIR), and contact angle analysis.

It can also be concluded that a lower value of the contact angle will positively affect the adhesion force of the composite. In this study, it is shown that with variations of the applied 3D printing technology, such as FFF or PJM, the printing material, such as PLA and MED610, the polymers used for electrospinning for the production of nanofibers, such as PAN nanofibers or PAN combined with gelatin, versatile composites with different properties for defined medical applications can be produced. The influence of 3D printed materials on nanofibers has been observed here, allowing for the production of denser nanofibers with larger diameters or a denser matrix, etc. For example, depending on the defined filtration needs, smaller nanofibers with tighter interstitial spaces can filter out more particles, and a denser matrix of 3D printed layers can filter out coarser particles while protecting the sensitive nanofiber layer from mechanical influences. These variations of composites with denser or more flexible nanofibers are diverse. There is a need for further research in this area.

It seems that the use of both composite materials characterized by a high degree of biocompatibility, confirmed by ISO standards and preliminary analyzes of the structure of nanofibers and their orientation, dimensions, allows us to conclude that by selecting 3D printing technology and electrospinnning materials, we are able to control the production process of composites of a specific quality in order to ensure appropriate the desired degree of filtration and the biomedical application results from the fact that the analyzed biomedical materials allow for the creation of durable models with desirable properties, in this case filtration.

In conclusion, compare these two methodology of manufacturing composite it could be stated that the PLA and MED610 materials have an influence on the nanofiber diameter in different way. The CLSM images look different from the surface morphology of the 3D printed objects made of PLA and MED610. In the case of PLA, the spaces between the individual strands of the PLA material are clearly visible. There is a denser surface morphology of MED610 compared to PLA. This means that, depending on the defined purpose, with variations of materials, different surface morphologies can be produced, which is advantageous for defined filter applications.

Author Contributions: Conceptualization, T.K.; methodology, T.K.; software, M.T., L.S. and A.M.; validation, T.K.; formal analysis, T.K.; investigation, L.S., M.T., A.M. and T.K.; resources, T.K. and A.M.; data curation, L.S., M.T., A.M. and T.K.; writing—original draft preparation, T.K., A.M., M.T. and L.S.; writing—review and editing, T.K., A.M., M.T. and L.S.; visualization, T.K.; supervision, T.K. and L.S.; project administration, T.K.; funding acquisition, T.K. All authors have read and agreed to the published version of the manuscript.

Funding: This research was funded by National Science Center of Poland under the Miniatura 4, grant number 2020/04/X/ST5/00057 entitled: Analysis of polymer composites produced by 3D printing and electrospinning technologies in the applications of filtering devices.

Institutional Review Board Statement: Not applicable.

Informed Consent Statement: Not applicable.

Data Availability Statement: The data created in this study are fully depicted in the article.

Acknowledgments: The authors acknowledge personal funding from the internal PhD funds of Bielefeld University of Applied Sciences, Erasmus+ and the internal funds for female students in STEM programs at Bielefeld University of Applied Sciences.

Conflicts of Interest: The authors declare no conflict of interest.

References

1. Adamczak, S.; Zmarzly, P.; Kozior, T.; Gogolewski, D. Analysis of The Dimensional Accuracy of Casting Models Manufactured by Fused Deposition Modeling Technology. In Proceedings of the Engineering Mechanics 2017, Svratka, Czech Republic, 15–18 May 2017.
2. Kozior, T.; Bochnia, J.; Zmarzły, P.; Gogolewski, D.; Mathia, T.G. Waviness of freeform surface characterizations from austenitic stainless steel (316L) manufactured by 3D printing-selective laser melting (SLM) technology. *Materials* **2020**, *13*, 4372. [CrossRef]
3. Varela, L.; Araújo, A.; Ávila, P.; Castro, H.; Putnik, G. Evaluation of the relation between lean manufacturing, industry 4.0, and sustainability. *Sustainability* **2019**, *11*, 1439. [CrossRef]
4. Kozior, T. The influence of selected selective laser sintering technology process parameters on stress relaxation, Mass of models, and their surface texture quality. *3D Print. Addit. Manuf.* **2020**, *7*, 126–138. [CrossRef]
5. Madej, M.; Styp-Rekowski, M.; Wieciński, P.; Płociński, T.; Ozimina, D.; Kurzydłowski, K.; Matuszewski, M. Properties of diamond-like carbon coatings deposited on cocrmo alloys. *Trans. Famena* **2015**, *39*, 79–88.
6. Ozimina, D.; Madej, M.; Kałdoński, T. The wear resistance of HVOF sprayed composite coatings. *Tribol. Lett.* **2011**, *41*, 103–111. [CrossRef]
7. Madej, M.; Ozimina, D. Electroless Ni-P-Al$_2$O$_3$ composite coatings. *Kov. Mater.* **2006**, *44*, 291–296.
8. Okolo, C.; Rafique, R.; Iqbal, S.S.; Saharudin, M.S.; Inam, F. Carbon nanotube reinforced high density polyethylene materials for offshore sheathing applications. *Molecules* **2020**, *25*, 2960. [CrossRef] [PubMed]
9. Bochnia, J.; Blasiak, M.; Kozior, T. Tensile strength analysis of thin-walled polymer glass fiber reinforced samples manufactured by 3D printing technology. *Polymers* **2020**, *12*, 2783. [CrossRef]
10. Saharudin, M.S.; Hajnys, J.; Kozior, T.; Gogolewski, D.; Zmarzły, P. Quality of surface texture and mechanical properties of PLA and PA-based material reinforced with carbon fibers manufactured by FDM and CFF 3D printing technologies. *Polymers* **2021**, *13*, 1671. [CrossRef]
11. Mousapour, M.; Salmi, M.; Klemettinen, L.; Partanen, J. Feasibility study of producing multi-metal parts by Fused Filament Fabrication (FFF) technique. *J. Manuf. Process.* **2021**, *67*, 438–446. [CrossRef]
12. Dziegielewski, W.; Kowalczyk, J.; Kulczycki, A.; Madej, M.; Ozimina, D. Tribochemical interactions between carbon nanotubes and ZDDP antiwear additive during tribofilm formation on uncoated and DLC-Coated steel. *Materials* **2020**, *12*, 2409. [CrossRef]

13. Kozior, T.; Kundera, C. Viscoelastic properties of cell structures manufactured using a photo-curable additive technology—PJM. *Polymers* **2021**, *13*, 1895. [CrossRef]
14. Kundera, C.; Martsynkowskyy, V.; Gudkov, S.; Kozior, T. Effect of rheological parameters of elastomeric ring materials on dynamic of face seals. *Procedia Eng.* **2017**, *177*, 307–313. [CrossRef]
15. Kozior, T.; Mamun, A.; Trabelsi, M.; Wortmann, M.; Sabantina, L.; Ehrmann, A. Electrospinning on 3D printed polymers for mechanically stabilized filter composites. *Polymers* **2019**, *11*, 2034. [CrossRef]
16. Kozior, T.; Trabelsi, M.; Mamun, A.; Sabantina, L.; Ehrmann, A. Stabilization of electrospun nanofiber mats used for filters by 3D printing. *Polymers* **2019**, *11*, 1618. [CrossRef] [PubMed]
17. Martens, Y.; Ehrmann, A. Composites of 3D-Printed Polymers and Textile Fabrics. In Proceedings of the IOP Conference Series: Materials Science and Engineering, Xiamen, China, 20–22 October 2017.
18. Unger, L.; Scheideler, M.; Meyer, P.; Harland, J.; Görzen, A.; Wortmann, M.; Dreyer, A.; Ehrmann, A. Increasing adhesion of 3D printing on textile fabrics by polymer coating. *Tekstilec* **2018**, *61*, 265–271. [CrossRef]
19. Upadhyay, M.; Sivarupan, T.; El Mansori, M. 3D printing for rapid sand casting—A review. *J. Manuf. Process.* **2017**, *29*, 211–220. [CrossRef]
20. Hlinka, J.; Kraus, M.; Hajnys, J.; Pagac, M.; Petru, J.; Brytan, Z.; Tanski, T. Complex corrosion properties of aisi 316L steel prepared by 3D printing technology for possible implant applications. *Materials* **2020**, *13*, 1527. [CrossRef]
21. Budzik, G.; Przeszlowski, L.; Wieczorowski, M.; Rzucidlo, A.; Gapinski, B.; Krolczyk, G. Analysis of 3D printing parameters of gears for hybrid manufacturing. *AIP Conf. Proc.* **2018**, *1960*, 140005. [CrossRef]
22. Kozior, T.; Mamun, A.; Trabelsi, M.; Sabantina, L.; Ehrmann, A. Quality of the surface texture and mechanical properties of FDM printed samples after thermal and chemical treatment. *Stroj. Vestn./J. Mech. Eng.* **2020**, *66*, 105–113. [CrossRef]
23. Chalgham, A.; Ehrmann, A.; Wickenkamp, I. Mechanical properties of fdm printed pla parts before and after thermal treatment. *Polymers* **2021**, *13*, 1239. [CrossRef] [PubMed]
24. Mamun, A.; Blachowicz, T.; Sabantina, L. Electrospun Nanofiber Mats for Filtering Applications—Technology, Structure and Materials. *Polymers* **2021**, *13*, 1368. [CrossRef] [PubMed]
25. Kozior, T.; Bochnia, J. The influence of printing orientation on surface texture parameters in powder bed fusion technology with 316L steel. *Micromachines* **2020**, *11*, 639. [CrossRef] [PubMed]
26. Bochnia, J.; Blasiak, S. Fractional relaxation model of materials obtained with selective laser sintering technology. *Rapid Prototyp. J.* **2019**, *25*, 76–86. [CrossRef]
27. Hanon, M.M.; Marczis, R.; Zsidai, L. Influence of the 3D printing process settings on tensile strength of PLA and HT-PLA. *Period. Polytech. Mech. Eng.* **2021**, *65*, 38–46. [CrossRef]
28. Ehrmann, G.; Ehrmann, A. Pressure orientation-dependent recovery of 3D-printed PLA objects with varying infill degree. *Polymers* **2021**, *13*, 1275. [CrossRef]
29. Ehrmann, G.; Ehrmann, A. Investigation of the shape-memory properties of 3D printed pla structures with different infills. *Polymers* **2021**, *13*, 164. [CrossRef] [PubMed]
30. Kozior, T.; Kundera, C. Surface texture of models manufactured by FDM technology. In Proceedings of the AIP Conference Proceedings, Maharashtra, India, 5–6 July 2018.
31. Thompson, A.; Maskery, I.; Leach, R.K. X-ray computed tomography for additive manufacturing: A review. *Meas. Sci. Technol.* **2016**, *27*, 072001. [CrossRef]
32. Wortmann, M.; Frese, N.; Sabantina, L.; Petkau, R.; Kinzel, F.; Gölzhäuser, A.; Moritzer, E.; Hüsgen, B.; Ehrmann, A. New polymers for needleless electrospinning from low-toxic solvents. *Nanomaterials* **2019**, *9*, 52. [CrossRef] [PubMed]
33. Sabantina, L.; Klöcker, M.; Wortmann, M.; Mirasol, J.R.; Cordero, T.; Moritzer, E.; Finsterbusch, K.; Ehrmann, A. Stabilization of polyacrylonitrile nanofiber mats obtained by needleless electrospinning using dimethyl sulfoxide as solvent. *J. Ind. Text.* **2020**, *20*, 224–239. [CrossRef]
34. Wehlage, D.; Blattner, H.; Sabantina, L.; Böttjer, R.; Grothe, T.; Rattenholl, A.; Gudermann, F.; Lütkemeyer, D.; Ehrmann, A. Sterilization of pan/gelatine nanofibrous mats for cell growth. *Tekstilec* **2019**, *62*, 78–88. [CrossRef]
35. Wehlage, D.; Blattner, H.; Mamun, A.; Kutzli, I.; Diestelhorst, E.; Rattenholl, A.; Gudermann, F.; Lütkemeyer, D.; Ehrmann, A. Cell growth on electrospun nanofiber mats from polyacrylonitrile (PAN) blends. *AIMS Bioeng.* **2020**, *7*, 43–54. [CrossRef]
36. Trabelsi, M.; Mamun, A.; Klöcker, M.; Sabantina, L. Investigation of metallic nanoparticle distribution in PAN/magnetic nanocomposites fabricated with needleless electrospinning technique. *Commun. Dev. Assem. Text. Prod.* **2021**, *2*, 8–17. [CrossRef]
37. Sabantina, L.; Wehlage, D.; Klöcker, M.; Mamun, A.; Grothe, T.; García-Mateos, F.J.; Rodríguez-Mirasol, J.; Cordero, T.; Finsterbusch, K.; Ehrmann, A. Stabilization of electrospun PAN/gelatin nanofiber mats for carbonization. *J. Nanomater.* **2018**, *2018*, 61310. [CrossRef]
38. Grothe, T.; Sabantina, L.; Klöcker, M.; Junger, I.; Döpke, C.; Ehrmann, A. Wet relaxation of electrospun nanofiber mats. *Technologies* **2019**, *7*, 23. [CrossRef]
39. Fokin, N.; Grothe, T.; Mamun, A.; Trabelsi, M.; Klöcker, M.; Sabantina, L.; Döpke, C.; Blachowicz, T.; Hütten, A.; Ehrmann, A. Magnetic properties of electrospun magnetic nanofiber mats after stabilization and carbonization. *Materials* **2020**, *13*, 1552. [CrossRef]

40. Sabantina, L.; Böttjer, R.; Wehlage, D.; Grothe, T.; Klöcker, M.; García-Mateos, F.J.; Rodríguez-Mirasol, J.; Cordero, T.; Ehrmann, A. Morphological study of stabilization and carbonization of polyacrylonitrile/TiO_2 nanofiber mats. *J. Eng. Fiber. Fabr.* **2019**, *14*, 1–8. [CrossRef]
41. Yusof, M.R.; Shamsudin, R.; Zakaria, S.; Hamid, M.A.A.; Yalcinkaya, F.; Abdullah, Y.; Yacob, N. Electron-beam irradiation of the PLLA/CMS/β-TCP composite nanofibers obtained by electrospinning. *Polymers* **2020**, *12*, 1593. [CrossRef]
42. Kozior, T.; Blachowicz, T.; Ehrmann, A. Adhesion of three-dimensional printing on textile fabrics: Inspiration from and for other research areas. *J. Eng. Fiber. Fabr.* **2020**, *15*, 1–6. [CrossRef]
43. Greiner, A.; Wendorff, J.H. Electrospinning: A fascinating method for the preparation of ultrathin fibers. *Angew. Chemie Int. Ed.* **2007**, *46*, 5670–5703. [CrossRef] [PubMed]
44. Makerbot FDM Technology. Available online: https://www.makerbot.com/3d-printers/method/tech-specs/ (accessed on 27 November 2021).
45. Stratasys Biocompatible Clear MED610-Sheet Data. Available online: https://www.sys-uk.com/wp-content/uploads/2016/01/MSDS-Clear-Bio-Compatible-MED610-English-US-1.pdf (accessed on 27 November 2021).
46. ISO 10993. Biological evaluation of medical devices. In *Biological Evaluation of Medical Devices*; ISO: Geneva, Switzerland, 2009.
47. GTMDB05 Biological Reactivity Tests, In Vivo. In *Phyllanthin, United States Pharmacopeia*; USP: Rockville, MA, USA, 2018.
48. PLA Structural Formula. Available online: https://omnexus.specialchem.com/selection-guide/polylactide-pla-bioplastic (accessed on 27 November 2021).
49. Molnár, K.; Szolnoki, B.; Toldy, A.; Vas, L.M. Thermochemical stabilization and analysis of continuously electrospun nanofibers. *J. Therm. Anal. Calorim.* **2014**, *117*, 1123–1135. [CrossRef]
50. Al-Saidi, G.S.; Al-Alawi, A.; Rahman, M.S.; Guizani, N. Fourier transform infrared (FTIR) spectroscopic study of extracted gelatin from shaari (Lithrinus microdon) skin: Effects of extraction conditions. *Int. Food Res. J.* **2012**, *19*, 1167–1173.
51. Chieng, B.W.; Ibrahim, N.A.; Yunus, W.M.Z.W.; Hussein, M.Z.; Then, Y.Y.; Loo, Y.Y. Effects of graphene nanoplatelets and reduced graphene oxide on poly(lactic acid) and plasticized poly(lactic acid): A comparative study. *Polymers* **2014**, *6*, 2232–2246. [CrossRef]
52. Kozior, T.; Döpke, C.; Grimmelsmann, N.; Juhász Junger, I.; Ehrmann, A. Influence of fabric pretreatment on adhesion of three-dimensional printed material on textile substrates. *Adv. Mech. Eng.* **2018**, *10*, 1–8. [CrossRef]
53. Sato, S.; Swihart, M.T. Propionic-acid-terminated silicon nanoparticles: Synthesis and optical characterization. *Chem. Mater.* **2006**, *18*, 4083–4088. [CrossRef]
54. Stratasys. *PolyJet 3D Printers Systems and Materials Overview*; Stratasys: Rehovot, Israel, 2018.
55. Korger, M.; Bergschneider, J.; Lutz, M.; Mahltig, B.; Finsterbusch, K.; Rabe, M. Possible applications of 3D printing technology on textile substrates. *IOP Conf. Ser. Mater. Sci. Eng.* **2016**, *141*, 01201. [CrossRef]

Article

Waste Biopolymers for Eco-Friendly Agriculture and Safe Food Production

Elio Padoan [1], Enzo Montoneri [2,*], Giorgio Bordiglia [1], Valter Boero [1], Marco Ginepro [3], Philippe Evon [4,*], Carlos Vaca-Garcia [4], Giancarlo Fascella [5] and Michéle Negre [1]

[1] Dipartimento di Scienze Agrarie, Forestali e Alimentari, Università di Torino, Largo P. Braccini 2, 10095 Grugliasco, Italy; elio.padoan@unito.it (E.P.); giorgio.bordiglia@unito.it (G.B.); valter.boero@unito.it (V.B.); negre.michele@gmail.com (M.N.)
[2] Dipartimento di Scienze delle Produzioni Agrarie e Alimentari, Università di Catania, Via S. Sofia 98, 95123 Catania, Italy
[3] Dipartimento di Chimica, Università di Torino, Via P. Giuria 7, 10125 Torino, Italy; marco.ginepro@unito.it
[4] Laboratoire de Chimie Agro-Industrielle (LCA), Université de Toulouse, ENSIACET, INRAE, Toulouse INP, 4 Allée Emile Monso, 31030 Toulouse, France; carlos.vacagarcia@toulouse-inp.fr
[5] CREA Research Centre for Plant Protection and Certification, 90011 Bagheria, Italy; giancarlo.fascella@crea.gov.it
* Correspondence: enzo.montoneri@gmail.com (E.M.); philippe.evon@toulouse-inp.fr (P.E.)

Citation: Padoan, E.; Montoneri, E.; Bordiglia, G.; Boero, V.; Ginepro, M.; Evon, P.; Vaca-Garcia, C.; Fascella, G.; Negre, M. Waste Biopolymers for Eco-Friendly Agriculture and Safe Food Production. *Coatings* 2022, 12, 239. https://doi.org/10.3390/coatings12020239

Academic Editor: Fengwei (David) Xie

Received: 12 January 2022
Accepted: 10 February 2022
Published: 12 February 2022

Publisher's Note: MDPI stays neutral with regard to jurisdictional claims in published maps and institutional affiliations.

Copyright: © 2022 by the authors. Licensee MDPI, Basel, Switzerland. This article is an open access article distributed under the terms and conditions of the Creative Commons Attribution (CC BY) license (https://creativecommons.org/licenses/by/4.0/).

Abstract: This work addresses environmental problems connected with biowaste management, the chemical industry, and agriculture. These sectors of human activity cause greenhouse gas (GHG) emissions in the air, climate change, leaching of excess mineral fertilizers applied to soil into ground water, and eutrophication. To mitigate this problem in agriculture, controlled release fertilizers (CRFs) are made by coating mineral fertilizers granules with synthetic polymers produced from the fossil-based chemical industry. This strategy aggravates GHG emission. In the present work, six formulations containing sunflower protein concentrate (SPC) and a new biopolymer (BP) obtained from sunflower oil cake and by hydrolysis of municipal biowaste, respectively, and commercial urea were tested as CRFs for spinach cultivation against the control growing substrate Evergreen TS and commercial Osmocote®. The results show large differences in plants' nitrate concentration due to the different treatments, although the same nitrogen amount is added to the substrate in all trials. BP is the key component mitigating nitrate accumulation in plants. The plants grown in the substrates containing BP together with SPC and/or urea, although exhibiting relatively high total N uptake (47–52 g kg^{-1}), have significantly lower nitric to total N ratio (9.6–12.0) than that (15.3–16.5) shown by the plants grown in the substrates containing SPC and/or urea, but no BP. The data confirm that all composites containing BP yield the safest crop coupled with high biomass production. Replication of BP effects for the cultivation of different plants will contribute to the development of a biobased chemical industry exploiting biowastes as feedstock.

Keywords: biopolymers; municipal bio-waste; spinach; agriculture; nitrates

1. Introduction

The augmentation of human population, along with its concentration in cities and increasing consumption habits, is causing several problems connected with biowaste management, the chemical industry, and agriculture. For example, municipal biowaste (MBW) production in Europe is 100 Mt yr^{-1} [1], with more than half still landfilled releasing 25,000 Mm^3 yr^{-1} GHG [2]. To satisfy the food demand, crop production is boosted by applying mineral fertilizer doses higher than those adsorbed by soil and plants. Excess nutrients accumulate in soil, leach into ground water, and cause eutrophication [3]. In this context, for a long time the Universities of Torino and Toulouse focused their research on the valorization of biowaste from urban (MBW) and agro-industrial sources as feedstock for the production of value added biobased chemical specialties and materials to use in place of commercial products obtained from fossil sources [4].

With specific reference to the agriculture sector, very recently the authors of the present work have reported the manufacturing and thermomechanical properties of a composite material made by twin-screw extrusion followed by injection-molding for use as controlled release fertilizer (CRF) [5]. This material contains urea and two polymers, herein after named sunflower proteins concentrate (SPC) and biopolymer (BP). These are obtained from sunflower oil cake and by hydrolysis of the anaerobic fermentation digestate of municipal unsorted food wastes, respectively. All three components contain nitrogen, have soil fertilizing power, have different solubility in water and, therefore, can release nitrogen in soil. Their fertilizing properties depend on the nitrogen release rate. In addition to the fertilizing property, SPC is well known for its good processability and is suitable to be used as a thermoplastic matrix for manufacturing the CRF composite pellets by twin-screw extrusion followed by injection-molding [5].

To test and assess the performance of the SPC and BP biopolymers, the experimental plant reported in the present work included three key elements, i.e., urea as widely used fertilizer, spinach as a sensitive probe to measure SPC/BP effects, and Osmocote® as a reference of widely used commercial CRF, representing a typical material made by conventional coating technology. Urea is one of the most important N-fertilizers. Its world consumption is 51 Mt yr^{-1} [6]. Urea has environmental drawbacks deriving from the release of excess nitrogen over the plant uptake rate. Urea is highly soluble in water. In soil, it is hydrolyzed to ammonia [7,8] and then, transformed into nitrates. These are adsorbed by the plant roots and transferred to the leaves. Nitrates in leaves are reduced to ammonia and then converted to proteins. In the presence of excess ammonia, proteins' production is slowed down. Nitrates accumulate in leaves and soil. They may impact the environment negatively, leaching from soil into ground water and causing eutrophication [9]. Nitrates in food plants may also have carcinogenic effects for humans [10]. In the human body, nitrate can be reduced to nitrite, which may cause methemoglobinemia, and the possibility of gastric cancer and other diseases [11,12].

Leaves of spinach are a typical food that can exhibit the nitrates' accumulation effects [13]. Spinach is a high value crop that requires sufficient N fertilizer to ensure optimal growth and to meet high quality criteria [14]. It absorbs NO_3^- from the soil efficiently but is known to be relatively inefficient in NO_3^- reduction [15]. The European Commission issued the regulation No. 1258/2011, stating the maximum acceptable concentration of nitrates in spinach [16]. These are 3.5 g kg^{-1} in fresh spinach, 2 g kg^{-1} in preserved, frozen spinach, and 0.2 g kg^{-1} in baby food. Spinach leaves should have high nutritional value from protein content and no toxicity from nitrates.

Commercial Osmocote® is a typical material belonging to the family of CRFs containing urea granules coated with synthetic polymers [17,18]. These are used to mitigate nitrate accumulation and effects in the environment [19]. The drawbacks of these CRFs are those typical of the fossil-based chemical industry. To substitute synthetic organic materials derived from fossil sources with natural or biobased materials [20], other CRFs have been investigated, based on natural polymers derived from plants [21]. No CRF is known to contain biopolymers derived from biowastes as SPC and BP.

In the previous work [5], the mechanical properties' advantages of the injection-molding technology, compared to conventional coating processes, were discussed. The inclusion of Osmocote® in the experimental plan of the present work has allowed the vis-à-vis direct comparison of the twin-screw extruded/injection-molded SPC composite and the coated Osmocote® composite for their performance as CRF for the cultivation of spinach.

2. Materials and Methods

2.1. Materials

The components to fabricate the SPC composites were available from previous work [5]. SPC was obtained by sieving a sunflower oil cake (SOC) using a Ritec (Signes, France) 600 vibrating sieve shaker fitted with a 1 mm grid. The undersize (SPC) was enriched with the smaller particles coming from the kernel of the seed, thus having a high protein content (51%, in proportion to the SPC dry weight). The rest (oversize) was constituted mainly by particles from the seed hull, and it was therefore rich in lignocellulosic fibers and minerals. BP was obtained by hydrolysis at pH 13 and 60 °C of the anaerobic digestate of MBW [4]. The anaerobic digestate was collected from the MBW waste treatment plant owned by the Acea Pinerolese Industriale SpA (ACEA) company in Pinerolo (TO). This plant processes, by fermentation, urban food wastes from separate source collection to produce biogas and solid anaerobic digestate containing the recalcitrant lignocellulosic fraction of the pristine MBW. To prepare the BP, the ACEA anaerobic digestate was taken up in water to yield a slurry at 4 water/digestate (w/w) ratio. Potassium hydroxide was added to the slurry until it reached pH 13. The alkaline slurry was heated up to 60 °C for 4 h. Afterwards, the slurry was allowed to settle until the liquid hydrolysate separated from the insoluble phase. The liquid hydrolysate was filtered through a 5 kDa polysulphone membrane to separate the retentate containing BP and the permeate containing the excess unreacted alkali reagent. The retentate was finally dried at 60 °C to yield the solid BP, which was used in the present work. Major components of BP were 45% protein, 13% lignin, and 15% minerals. Urea was a common commercial product.

2.2. Fabrication and Composition of SPC Composites

The SPC composites were fabricated and characterized for their elemental composition (Table 1) in the previous work [5]. Granules containing urea and/or BP dispersed into the SPC matrix were made by twin-screw extrusion. These were converted into dense dark pellets, all having the same (10 mm × 10 mm × 4 mm) dimension, by injection-molding. The tested SPC composites had the following compositions: neat SPC, SPC-U containing 10% urea, SPC-BP containing 10% BP, and SPC-BP-U containing 5% BP and 5% urea. For the neat components and the composite materials, Table 1 reports the C and soil/plant N, P, and K nutrient contents.

Table 1. Total C, N, P, and K concentrations (%, $w\,w^{-1}$) in neat SPC, BP, urea (U), and in the SPC pellets' samples [1].

Formulation	C	N	P_2O_5	K_2O
BP	39.6	6.6	1.1	5.5
U	20.0	46.6	-	-
SPC	38.7 ± 5.2 [a]	6.9 ± 0.9 [a]	2.5 ± 0.1 [a]	1.5 ± 0.1 [a]
SPC-U	37.7 ± 0.2 [a]	10.5 ± 0.1 [c]	2.3 ± 0.1 [a]	1.3 ± 0.1 [a]
SPC-BP	39.5 ± 2.6 [a]	7.0 ± 0.6 [a]	2.6 ± 0.3 [a]	1.8 ± 0.1 [b]
SPC-BP-U	38.6 ± 0.4 [a]	8.5 ± 0.0 [b]	2.1 ± 0.1 [a]	1.5 ± 0.1 [a]

[1] Values in the same column followed by different letters are significantly different ($p < 0.05$).

2.3. Plant Growth Trials

Spinacia oleracea L. "Gigante d'inverno" seedlings, purchased from Beltrame Roberto nursery, strada Sanda, Moncalieri, Italy, were transplanted on the commercial growing substrate Evergreen TS (Turco S.a.s, Moncalieri, Italy) in 2 L pots with 13 cm × 13 cm × 13 cm size (5 plants per pot, and 5 pots per trial). The plants were grown during November and December 2019 in an unheated greenhouse at an average temperature of 18 °C and 90% relative humidity under natural lighting conditions. Surface irrigation was performed manually every three days. After three days, the neat components and the composites were added to the substrate in the following amounts:

- Trial 0: control (growing substrate) containing 25 mg organic N.
- Trial 1: growing substrate + SPC (4.1 g per pot, 283 mg N).
- Trial 2: growing substrate + SPC-U (2.7 g per pot, 284 mg N).
- Trial 3: growing substrate + SPC-BP (4.0 g per pot, 280 mg N).
- Trial 4: growing substrate + SPC-BP-U (3.3 g per pot, 281 mg N).
- Trial 5: growing substrate + urea (0.6 g per pot, 280 mg N).
- Trial 6: growing substrate + BP (4.3 g per pot, 284 mg N).
- Trial 7: growing substrate + urea (0.3 g per pot, 140 mg N) + BP (2.2 g per pot, 145 mg N).
- Trial 8: Osmocote® (1.8 g per pot, 282 mg N).

The trials' plan included the control (trial 0) without added N fertilization, trials 1–7 in the presence of added SPC, BP, and urea products, and trial 8 in the presence of commercial slow-release N, P, K fertilizer Osmocote® [22], which was used as reference material. The amount of SPC materials (trial 1–4) and of urea and BP (trial 5–7) added in each pot was calculated to provide, in each pot, nearly the same total N amount (280–285 mg N) as in trial 8. After 55 days, the shoots and roots were collected, and their respective fresh weights were measured. Two plants per pot were dried at 60 °C for 3 days for the determination of the dry weight of leaves and roots.

2.4. Analyses

Total N was determined by the flash combustion method (i.e., "Dumas method") with thermal conductivity detection using the UNICUBE elementar analyzer (Elementar, Langenselbold, Germany) according to the CSN EN 13654-2 standard method. Nitrate concentrations in leaves and roots at the end of the trial have been analyzed on the fresh material. For each plant, 100 mg of fresh tissue was ground in liquid nitrogen and suspended in 10 mL of deionized water. Suspensions were incubated for 1 h at 45 °C and then centrifuged at 5000 rpm for 15 min. The extract was filtered, and the nitrate N concentration was determined spectrophotometrically by the Griess reaction [23]. The determination of chlorophyll a and b and of carotenoids was performed on each plant by extraction of 300 mg fresh foliar tissue ground in liquid nitrogen with 10 mL 96% (v/v) ethanol. The samples were kept in the dark for 2 days at 4 °C, and the extracts were filtered and then analyzed by spectrophotometry using a Hitachi (Tokyo, Japan) U-2000 spectrophotometer. The absorbance readings were performed at 665 nm for chlorophyll a, at 649 nm for chlorophyll b, and at 470 nm for total carotene. Chlorophyll a, b, and total carotenoid concentrations were calculated according to the literature [24].

2.5. Statistical Treatment of Data

The data were evaluated by one-way ANOVA ($p < 0.05$ or 0.01) followed by the Tukey's test for multiple comparison procedures.

3. Results

For the spinach cultivated in the different trials, Table 2 reports the fresh and dry weights of leaves and roots. At the end of the experiment, the plants were healthy and reached the marketable size without apparent differences between the trials, except for the fresh weight of the leaves as higher values were recorded in trials 2 and 3, compared to trial 6. Overall, the experimental values suggest that the control substrate provides a sufficient amount of nutrients to allow a regular and constant growth of the plants.

Table 3 reports the chlorophyll a, chlorophyll b, and carotenoids concentrations in the fresh spinach leaves. It shows that the chlorophyll a content in the plants fertilized with SPC and urea (trials 1 and 5) is significantly higher than that of the control (trial 0). The plants fertilized with SPC and SPC-U (trials 1 and 2) exhibit significantly higher concentrations of chlorophyll b than the control (trial 0). The treatments appear not to significantly affect the carotenoid concentration compared to the control (trial 0).

Table 2. Fresh and dry weight of spinach leaves and roots (mean value ± standard deviation) [1] cultivated in trials 0–8.

Trial	Leaves		Roots	
	Fresh Weight (g)	Dry Weight (g)	Fresh Weight (g)	Dry Weight (g)
0	18.7 ± 3.2 [ab]	1.6 ± 0.55 [a]	0.86 ± 0.29 [a]	0.11 ± 0.04 [a]
1	25.6 ± 5.3 [ab]	2.3 ± 0.56 [a]	1.16 ± 0.22 [a]	0.13 ± 0.02 [a]
2	27.3 ± 4.5 [a]	2.4 ± 0.39 [a]	1.08 ± 0.23 [a]	0.12 ± 0.03 [a]
3	27.7 ± 6.1 [a]	2.3 ± 0.61 [a]	0.99 ± 0.26 [a]	0.10 ± 0.03 [a]
4	23.5 ± 5.1 [ab]	2.2 ± 0.54 [a]	0.99 ± 0.18 [a]	0.11 ± 0.03 [a]
5	24.5 ± 5.0 [ab]	2.3 ± 0.60 [a]	0.98 ± 0.16 [a]	0.12 ± 0.03 [a]
6	18.1 ± 1.9 [b]	1.8 ± 0.20 [a]	0.89 ± 0.10 [a]	0.12 ± 0.02 [a]
7	23.2 ± 3.4 [ab]	2.3 ± 0.54 [a]	1.08 ± 0.14 [a]	0.13 ± 0.03 [a]
8	22.8 ± 2.7 [ab]	2.2 ± 0.27 [a]	1.04 ± 0.08 [a]	0.13 ± 0.01 [a]

[1] Within columns, mean values followed by the same letter are not significantly different ($p < 0.01$).

Table 3. Chlorophyll and carotenoids content in spinach leaves expressed as mg kg^{-1} fresh weight (mean value ± standard deviation) [1] cultivated in trials 0–8.

Trial	Chlorophyll a	Chlorophyll b	Carotenoids
0	242 ± 49.0 [a]	123 ± 28.9 [a]	19 ± 0.9 [a]
1	330 ± 32.3 [b]	176 ± 11.0 [b]	8 ± 0.3 [a]
2	323 ± 46.6 [ab]	172 ± 15.0 [b]	8 ± 0.4 [a]
3	258 ± 41.0 [ab]	125 ± 26.9 [a]	16 ± 0.8 [a]
4	321 ± 29.4 [ab]	158 ± 11.0 [ab]	7 ± 0.3 [a]
5	341 ± 29.9 [b]	172 ± 34.1 [ab]	18 ± 0.6 [a]
6	241 ± 55.4 [a]	138 ± 23.3 [ab]	24 ± 0.7 [a]
7	309 ± 22.1 [ab]	158 ± 19.4 [ab]	11 ± 0.4 [a]
8	289 ± 42.2 [ab]	168 ± 34.5 [ab]	17 ± 0.6 [a]

[1] Within columns, mean values followed by the same letter are not significantly different ($p < 0.01$).

Table 4 reports the total N concentration in the plants. The experimental data in Table 4 show that the nitrogen uptake per plant is much lower in roots than in leaves. This is expected since most of the nitrate absorbed by the roots is transported in the xylem to the leaves for further reduction to ammonia, which is then used to synthesize the leaf proteins. The results also point out some significant differences between the trials. In trials 1–5, and 7 and 8, the total N concentration in both leaves and roots is higher than that for the control (trial 0). The highest total N concentration values are shown in the leaves of the plants grown on SPC-based fertilizers (trials 1–4) and urea (trial 5).

The data in Table 5 point out large differences in nitrate concentration among the trials, although the same nitrogen amount was added to the substrate in all trials. In all trials, the nitrate concentration in fresh spinach is well below the 3.5 g kg^{-1} safe limit recommended by the European Commission [18]. The highest concentration is found in the spinach fertilized with urea (2816 mg kg^{-1} in trial 5) and in the spinach fertilized with the sunflower proteins-based fertilizer (1890–2559 mg kg^{-1} in trials 1, 2, and 4). The lowest nitrate concentration is found in the plants grown on the control substrate (101 mg kg^{-1} in trial 0) and in the plants fertilized with BP (247 mg kg^{-1} in trial 6). However, these plants exhibit a rather low total nitrogen concentration (Table 4), probably due to the low level of nitrogen mineralization in the substrate. This is confirmed by the low nitric to total N ratio (1.0 and 2.2 in trial 0 and trial 6, respectively), which suggests that most of the absorbed inorganic nitrogen is promptly transformed into amino acids and proteinaceous matter. In all other cases, the nitric to total N ratio ranges from 9.6 to 16.5.

Table 4. Total nitrogen concentration and N uptake for spinach plants (mean value ± standard deviation) [1] cultivated in trials 0–8.

Trial	Total N (g kg^{-1} Dry Matter)		N Uptake (mg Plant^{-1})	
	Leaves	Roots	Leaves	Roots
0	34.0 ± 1.82 [a]	20.9 ± 2.15 [a]	54.4 ± 21.6 [a]	2.3 ± 1.01 [a]
1	51.9 ± 1.32 [bcd]	29.8 ± 1.72 [bc]	117.6 ± 32.0 [b]	3.7 ± 0.88 [a]
2	52.7 ± 1.31 [bd]	30.4 ± 2.09 [bc]	128.3 ± 23.6 [bc]	3.6 ± 1.09 [a]
3	48.5 ± 2.04 [bce]	28.2 ± 0.81 [bcd]	113.0 ± 34.3 [ab]	2.9 ± 1.01 [a]
4	52.1 ± 2.99 [bcd]	29.1 ± 1.77 [bc]	114.5 ± 34.5 [b]	3.3 ± 1.12 [a]
5	55.5 ± 1.71 [d]	30.7 ± 2.36 [b]	129.7 ± 37.4 [bc]	3.6 ± 1.20 [a]
6	34.3 ± 3.44 [a]	20.7 ± 3.44 [a]	63.1 ± 13.3 [ab]	2.6 ± 0.87 [a]
7	47.2 ± 2.55 [ce]	26.9 ± 2.14 [bc]	106.2 ± 31.1 [ab]	3.5 ± 0.96 [a]
8	45.1 ± 3.47 [e]	26.5 ± 2.11 [c]	99.8 ± 19.8 [ab]	3.3 ± 0.57 [a]

[1] Within columns, mean values with the same letter are not significantly different ($p < 0.01$).

Table 5. Nitric nitrogen (mean value ± standard deviation) [1] in spinach leaves cultivated in trials 0–8.

Trial	NO^{3-} N (g kg^{-1} Dry Matter)	NO^{3-}/Total N ($w\ w^{-1}$)	NO^{3-} N (mg kg^{-1} Fresh Matter)
0	1.1 ± 0.60 [a]	1.0 ± 0.58 [a]	101 ± 53.7 [a]
1	26.0 ± 4.9 [bcd]	15.3 ± 3.3 [c]	2281 ± 410 [bcd]
2	28.6 ± 2.8 [bc]	16.5 ± 2.0 [c]	2559 ± 297 [bc]
3	15.4 ± 7.5 [e]	9.6 ± 5.1 [b]	1373 ± 575 [e]
4	20.5 ± 7.0 [bde]	12.0 ± 4.8 [bc]	1890 ± 593 [bde]
5	29.8 ± 1.9 [c]	16.3 ± 1.5 [c]	2816 ± 125 [c]
6	2.5 ± 1.4 [a]	2.2 ± 1.5 [a]	247 ± 122 [a]
7	16.4 ± 5.1 [de]	10.5 ± 3.9 [b]	1541 ± 398 [de]
8	14.9 ± 4.0 [e]	10.1 ± 3.5 [b]	1438 ± 323 [e]

[1] Within columns, mean values followed the same letter are not significantly different ($p < 0.01$).

4. Discussion

4.1. Effect of BP on Plant Performances

The data obtained from the present experiment show significant differences only between trials 2, 3, and 6 for the leaf fresh biomass, but not the dry one (Table 2). The average dry weight of the leaves (2.2 g per plant) is consistent with literature data for cultivated spinach under different conditions: 0.6–2.0 g per plant for hydroponic cultures [15,25], 0.2–1.2 g per plant in pot experiments, depending on the amount of added nitrogen [26], and 4.3–5.4 g per plant in trials where different composts were used as fertilizers [27].

The chlorophyll content of spinach leaves was also affected by BP application (Table 3). The concentration values reported in Table 3 fall in the range of values reported by other workers for spinach cultivated under various operational conditions. For example, for chlorophyll a, the following values are reported: 300–400 mg kg^{-1} in spinach cultivated in field experiments at different N and water applied levels [28], and 860 mg kg^{-1} in spinach cultivated under hydroponic conditions at 105 mg L^{-1} N applied doses [29]. The data for the group of trials 1–7, in which SPC, BP, and/or urea were used, show that trials 3 and 6 exhibit lower contents of chlorophyll and higher contents of carotenoids than the other 5 trials. The statistical analysis of these groups of data does not prove significant differences between most of the trials. Yet, specifically for chlorophyll, the apparent lower chlorophyll concentration recorded in leaves from trials 3 and 6 may be correlated to the lower total N content and uptake (Table 4) measured in leaves. Indeed, it is well known that leaf chlorophyll concentration is linked to leaf N content due to the presence of N atoms in chlorophyll molecules [30]. Some authors observed that BP application to plants enhanced the availability of N requested for chlorophyll formation [31], with positive effects on photosynthetic activity. The apparent higher carotenoid content and lower chlorophyll content in trials 3 and 6 can be related to a trend that plants show in some growing conditions (particularly under stress), i.e., when chlorophyll values decrease, carotenoids tend to increase, and vice versa [32,33].

The BP application also influenced the total nitrogen content in spinach leaves and roots (Table 4). To be assimilated by plants, nitrogen must be in nitric or ammonia forms. Therefore, nitrogen present in organic forms in soil or substrate must be made available through mineralization by microorganisms. The prevailing form is the nitric one because ammonium ion is promptly oxidized and made less available by adsorption on the soil surfaces. Once absorbed by the roots, nitrates are transferred to the shoots where they are reduced to ammonium and used to first synthesize glutamate and glutamine through the enzymatic nitroreductase and GOGAT systems, and then other amino acids and N-containing compounds. The total N concentration values are consistent with those reported in literature for spinach leaves (22.7–51.5 g kg^{-1}) [15,27]. The low nitrogen concentration found in the plants grown on the BP-added substrate (trial 6) suggests that BP is recalcitrant to the biochemical attack by microorganisms, in accordance with the high lignin-like chemical moieties present in its macromolecular structure [4]. On the other hand, the low nitrogen uptake of the plants grown in trial 6 does not seem to negatively affect the leaf biomass yield as shown in Table 2. This is consistent with previous results reported by other authors, who used BPs in floriculture trials [30,31,34]. These authors agreed that the positive effects exhibited by the investigated biopolymers were due to the biopolymers' chemical structure interacting with the microorganisms and stimulating the plant metabolism, more than to their fertilizing power stemming from their contribution of organic nitrogen as soil fertilizer. On the other hand, although a high N content in the leaves, as shown in Table 4, can be considered a positive result because it is correlated to a high protein concentration, it should not be accompanied with a high nitrate concentration. Accumulation of nitrates in plants generally occurs when the plant uptakes more NO^{3-} than the NO^{3-} assimilable in protein form [35]. Most absorbed nitrate is stored in the vacuole until release for reduction in the cytosol [36]. In the present work, this issue is addressed by the collected data shown in Table 5.

Actually, BP application also affected nitric nitrogen content in spinach leaves (Table 5). Accumulation of nitrates in the vacuoles of the cells occurs when the enzymatic systems leading to the reduction of nitrates and further synthesis of protein matter are inhibited by excessive nitrate uptake. In this scenario, it is noteworthy that the plants grown in trials 3, 4, and 7 substrates, all containing BP together with SPC and/or urea, although exhibiting a high total N uptake (Table 4), have significantly lower nitric to total N ratio (9.6–12.0) than that (15.3–16.5) shown in the other trials (1, 2, and 5) containing SPC and/or urea but no BP. The best fertilizers in terms of high total N content and low nitrates accumulation were the SPC-BP (trial 3) and SPC-BP-U (trial 4) composites, the mix of urea and BP in trial 7, and Osmocote® (trial 8). In all these trials, the nitrate concentration in the spinach leaves is even below the limit of 2 g kg^{-1} recommended for preserved frozen spinach by the European Commission [16]. The data confirm that all composites containing BPs yield the safest crop coupled with high biomass production.

4.2. Plausible Explanation of the BP Effects

The experimental data point out that BP can mitigate nitrates' accumulation in spinach plants. The fertilizer used in trial 3 was SPC loaded with 10% BP. About 90% of the total nitrogen present in this specimen comes from SPC. Therefore, most of the N uptake of the plants in trial 3 is to be attributed to SPC nitrogen. Indeed, trial 6 has demonstrated that BP nitrogen is relatively less available for the plant to take up. In trial 7, in which urea was the nitrogen source and the same amount of BP was added, high nitrogen uptake is also accompanied by a relatively low nitrate content. The results suggest that BP, although not supplying to the plant as much nitrogen as SPC and urea, strongly affects the pathways responsible for the mineralization of organic nitrogen. A similar effect was observed in the previous work [5] reporting the kinetics of ammonia and organic N release rates of the SPC composites in water. BP was shown to retard the formation of ammonia from urea hydrolysis and enhance the release of organic nitrogen from SPC. In this case, the effect might be due to a plausible interaction of BP functional groups with urea and SPC.

BP belongs to a family of biopolymers obtained by the same hydrolysis process from fermented lignocellulose biowastes of different sources [4]. All these biopolymers keep the memory of the sourcing lignocellulose materials, are constituted by a mix of macromolecules containing the same carbon types and various acid and basic functional groups capable of interacting with other molecules by protonation and donor-acceptor complexing reactions. The relative ratios of the chemical functionalities in these families of biopolymers depend on the sourcing materials. As a consequence, they have the same multiple properties and performances, although at different levels depending on the sourcing materials. Other BPs, obtained from composted green wastes or composted mixes of green wastes and MBW anaerobic digestates, have been investigated as auxiliaries in the anaerobic fermentation of MBW [37,38] carried out in bioreactors dedicated to the production of biogas. These biopolymers can reduce the ammonia and/or the nitrate level in the process digestate. Other workers have used BPs as animal diet supplements [39]. They have shown that BPs reduce the proteolysis occurring in the caecum intestine of pigs with consequent reduction of ammonia formations. Biagini and coworkers [40] tested the biopolymer obtained from composted mixes of green wastes and MBW anaerobic digestates as a supplement for rabbits' protein diet. They demonstrated that the rabbits fed with the biopolymer-supplemented diet produced manure with significantly lower ammonia and GHG emissions compared to the animals fed with the control diet.

While the confirmation and replicability of the effects of the above biopolymers under different operational conditions and of the BP used in the present work are unquestionable, the role of these biopolymers is not yet clear. So far, it cannot be established definitely whether the biopolymer effects involve pure chemical reactions or biochemical processes with participation of a microorganism. The second hypothesis is the most likely, according to Baglieri and coworkers [41]. These researchers cultivated bean plants using a biopolymer obtained from the hydrolysis of exhausted tomato plants as fertilizers. The biopolymer sourced from the agriculture biowastes bears strong chemical similarities with BP sourced from MBW. The former was found [41] to significantly enhance nitrate reductase, glutamine synthetase, and glutamate synthase activities, and to increase soluble proteins' concentration in shoots and roots, compared to the control. Based on the lack of differences between the concentrations of mineral nitrogen in the control and treated cultivation substrate, as opposed to the significant differences observed for enzymatic activity and soluble proteins' concentration in the plants, Baglieri and coworkers [41] concluded that the biopolymer acts as plant biostimulant with a possible auxin-like effect, more than as soil fertilizer.

On the other hand, a possible action of BP as chemical catalyst must be considered. This is also in view of previous work reporting the property of BPs to catalyze oxidation reactions in the absence of any microorganism [42]. According to the current view of the behavior of urea in soil and the fate of the produced ammonia [7,8,35] for the system investigated in the present work, the following reaction scheme may help clarify the BPs effects:

$$CO(NH_2)_2 + H_2O \leftrightarrows 2\,NH_3 + CO_2 \qquad (1)$$

$$NH_3 + 2\,O_2 \leftrightarrows HNO_3 + H_2O \qquad (2)$$

$$R\text{-}CH(OH)\text{-}COOH + CH_3\text{-}COOH + NH_3 \leftrightarrows R\text{-}CH[NH\text{-}(C=O)\text{-}CH_3]\text{-}COOH + 2\,H_2O \qquad (3)$$

The scheme shows that urea is hydrolyzed to ammonia (reaction (1)) and then, transformed into nitrates (reaction (2) forward). These are adsorbed by the plant roots and transferred to the leaves. Nitrates in leaves are reduced to ammonia (reaction (2) backward) and then converted to proteins (reaction (3)). A recent work [38] investigated the BPs' assisted anaerobic fermentation of MBW in 150 mL shake flasks. It demonstrated that BPs, in the investigated MBW fermentation system comprising organic, nitrate and ammonia N, catalyze the following chemical redox reaction:

$$HNO_3 + NH_3 \leftrightarrows N_2 + 2\,H_2O + 1/2\,O_2 \qquad (4)$$

The calculated Gibbs free energy from literature data [43] for this reaction, i.e., −360 kJ/N_2 mole, shows that reaction (4) is thermodynamically favored. Occurrence of the chemical catalysis by BP in the system investigated in the present work, and the consequent reduction of nitrates by reaction (4), may be a plausible explanation for the lower nitric to total N ratio reported in Table 4 for the composite materials containing BPs (trials 3, 4 and 7), compared to the other materials in trials 1, 2, and 5 that contain SPC and/or urea but no BP.

4.3. Perspectives for a New Biowaste-Based Chemical Industry

The spinach case study reported in the present work shows that, for use in the agriculture sector, materials obtained from urban and agro-industrial biowastes are competitive with materials obtained from fossil sources. Particularly, the BP material obtained from MBW exhibits unique properties that allow modulating the N release rate and fate in soil and in the plant crop. These properties are very important to safeguard the quality of soil, water, and crops. More ambitiously, the present article has relevance also for the sectors of waste management, pollution, and the chemical industry.

Montoneri [4] and Tabasso and coworkers [44] have reviewed the sustainability of the hydrolysis process to produce BPs, as well as the BPs' multipurpose performance and related economic, environmental, and social benefits for several sectors of the chemical industry and agriculture. The engineered composite materials tested in the present work disclose a further benefit offered by the tested BP for developing safe agriculture and food. They prove a new property of BP capable of reducing fertilizer nitrate leaching through soil and eutrophication effects, and at the same time diminish nitrate accumulation in crops. They add further important incentives for valorizing MBW as feedstock for the production of BPs and their use at commercial scale. They also prospect the feasibility of substituting products from fossil sources with products from biowastes. Particularly, because of their origin and special properties, BPs have high potential for developing a sustainable, waste-based industry integrating chemical and biochemical processes.

MBW treatment plants are the ideal settings to this end. At present, they are service providers for citizens, as they perform the collection, disposal, and recycling of urban biowastes. To reduce landfill disposal, the most advanced plants process MBW by anaerobic fermentation yielding biogas and digestate, and by aerobic fermentation producing compost. The value of these products is not enough to cover the plant operational costs [4]. The missing revenue is covered by citizens' taxes. No MBW plant applies chemical processes. Yet, MBWs are a potential source of valuable renewable organic C, which could be recycled in the form of valued-added, biobased products for consumer use.

5. Conclusions

It has been demonstrated that composite controlled release fertilizers made by twin-screw extrusion followed by injection-molding can achieve the same performance in terms of spinach growth, nitrogen uptake, and nitrate accumulation as the commercial Osmocote® controlled release fertilizer, which contains urea coated with synthetic polymers. This result is achieved thanks to the municipal biowaste derived biopolymer. This component can control the nitrogen release rate of urea and from the sunflower protein concentrate, and reduce nitrogen accumulation by the plant while maintaining the same biomass growth and nitrogen uptake of Osmocote®. This finding poses a worthwhile scope for testing the twin-screw extruded/injection-molded biopolymer composites in the cultivation of other plant species in order to contribute to the replacement of current commercial materials coated with synthetic polymers from fossil sources. However, the ambition of the authors of the present work is far beyond developing controlled release fertilizers.

Previous work [4] has demonstrated that the hydrolysis of municipal biowaste from different sources is a cost-effective, feasible process that yields a family of biopolymers for sustainable use in different sectors of agriculture and the chemical industry. The demonstration of the BP biopolymer properties in the present work is a new finding. It widens the fields of application and the potential benefits of BPs. It adds a further argument for developing a waste-based chemical industry, which exploits biowaste as feedstock for the production of value-added chemical specialties and materials rather than plants specifically cultivated for this purpose. This approach, extended to the exploitation of biowastes from urban, agriculture, and agro-industrial sources, would allow not only the replacement of chemicals and materials from fossil sources, but would also dismiss environmentally unfriendly waste disposal practices and keeping agriculture soil for food production rather than for the production of chemicals.

6. Patent

The French patent application submitted 16 October 2020 under number FR2010597 to Institut National de la Propriété Industrielle (INPI) in France and entitled *"Produit pour l'agriculture, et procédé de préparation"* results from both [5] and the work reported in the present manuscript. The inventors of this patent are Philippe Evon, Carlos Vaca-Garcia, Laurent Labonne, and Antoine Rouilly. The owners of this patent are Institut National Polytechnique de Toulouse (INPT) and Institut National pour l'Agriculture, l'Alimentation et l'Environnement (INRAE).

Author Contributions: E.P., E.M., G.B., V.B., M.G., P.E., C.V.-G., G.F. and M.N. contributed equally to this research article. All authors have read and agreed to the published version of the manuscript.

Funding: This research was supported partly by endowed funds of the authors' institutions, and partly funded by the European Commission in order to support the implementation of the actions pursued in the LIFE16 ENV/IT/000179-LIFECAB and the LIFE19 ENV/IT/000004-LIFEEBP projects.

Institutional Review Board Statement: Not applicable.

Informed Consent Statement: Not applicable.

Data Availability Statement: Data is contained within the article.

Conflicts of Interest: The authors declare no conflict of interest.

References

1. European Commission. Biodegradable Waste. Available online: https://ec.europa.eu/environment/topics/waste-and-recycling/biodegradable-waste_en (accessed on 30 December 2021).
2. Eurostat. Municipal Waste Statistics. Available online: https://ec.europa.eu/eurostat/statistics-explained/index.php?title=Municipal_waste_statistics (accessed on 30 December 2021).
3. IFA. Fertilizers, for Productive and Sustainable Agriculture Systems. Available online: https://www.fertilizer.org/ (accessed on 30 December 2021).
4. Montoneri, E. Municipal waste treatment, technological scale up and commercial exploitation: The case of bio-waste lignin to soluble lignin-like polymers. In *Food Waste Reduction and Valorisation*; Morone, P., Papendiek, F., Tartiu, V.E., Eds.; Springer: Berlin/Heidelberg, Germany, 2017; Chapter 6.
5. Evon, P.; Labonne, L.; Padoan, E.; Vaca-Garcia, C.; Montoneri, E.; Boero, V.; Negre, M. A new composite biomaterial made from sunflower proteins, urea, and soluble polymers obtained from industrial and municipal biowastes to perform as slow release fertiliser. *Coatings* **2021**, *11*, 43. [CrossRef]
6. IFA. Consumption Urea World. Available online: https://www.ifastat.org/databases/plant-nutrition (accessed on 30 December 2021).
7. Sigurdarson, J.J.; Svane, S.; Karring, H. The molecular processes of urea hydrolysis in relation to ammonia emissions from agriculture. *Agric. Rev. Environ. Sci. Biotechnol.* **2018**, *17*, 241–258. [CrossRef]
8. Beig, B.; Niazi, M.B.K.; Jahan, Z.; Kakar, S.J.; Shah, G.A.; Shahi, M.; Zia, M.; Haq, M.U.; Rashid, M.I. Biodegradable polymer coated granular urea slows down N release kinetics and improves spinach productivity. *Polymers* **2020**, *12*, 2623. [CrossRef] [PubMed]

9. Leip, A.; Billen, G.; Garnier, J.; Grizzetti, B.; Lassaletta, L.; Reis, S.; Simpson, D.; Sutton, M.A.; Vries, W.; Weiss, F.; et al. Impacts of European live-stock production: Nitrogen, sulphur, phosphorus and greenhouse gas emissions, land-use, water eutrophication and biodiversity. *Environ. Res. Lett.* **2015**, *10*, 115004. [CrossRef]
10. Anjana, A.; Umar, S.; Iqbal, M. Nitrate accumulation in plants, factors affecting the process, and human health implications. A review. *Agron. Sustain. Dev.* **2007**, *27*, 45–77. [CrossRef]
11. Breimer, T. Environmental Factors and Cultural Measures Affecting the Nitrate Content in Spinachs. Doctoral Dissertation, Wageningen University & Research (WUR), Wageningen, The Netherlands, 1982. Available online: https://library.wur.nl/WebQuery/wurpubs/fulltext/201755 (accessed on 30 December 2021).
12. Chen, B.; Wang, Z.; Li, S.; Wang, G.; Song, H.; Wang, X. Effects of nitrate supply on plant growth, nitrate accumulation, metabolic nitrate concentration and nitrate reductase activity in three leafy vegetables. *Plant Sci.* **2004**, *167*, 635–643. [CrossRef]
13. Shoji, S. Innovative use of controlled availability fertilizers with high performance for intensive agriculture and environmental conservation. *Sci. China Life Sci.* **2005**, *48*, 912–920.
14. Heinrich, A.; Smith, R.; Cahn, M. Nutrient and water use of fresh market spinach. *Horttechnology* **2013**, *23*, 325–333. [CrossRef]
15. Chan-Navarrete, R.; Kawai, A.; Dolstra, O.; Lammerts van Bueren, E.T.; van der Linden, C.G. Genetic diversity for nitrogen use efficiency in spinach (*Spinacia oleracea* L.) cultivars using the ingestad model on hydroponics. *Euphytica* **2014**, *199*, 155–166. [CrossRef]
16. European Commission. Regulation n° 1258/2011 of 2 December 2011 Amending Regulation (EC) n° 1881/2006 as Regards Maximum Levels for Nitrates in Foodstuffs. Available online: https://eur-lex.europa.eu/LexUriServ/LexUriServ.do?uri=OJ:L:2011:320:0015:0017:EN:PDF (accessed on 30 December 2021).
17. Trenkel, M.E. *Slow- and Controlled-Release and Stabilized Fertilizers: An Option for Enhancing Nutrient Use Efficiency in Agriculture*; International Fertilizer Industry Association (IFA): Paris, France, 2010. Available online: https://www.fertilizer.org/images/Library_Downloads/2010_Trenkel_slow%20release%20book.pdf (accessed on 30 December 2021).
18. Incrocci, L.; Maggini, R.; Cei, T.; Carmassi, G.; Botrini, L.; Filippi, F.; Clemens, R.; Terrones, C.; Pardossi, A. Innovative controlled-release polyurethane-coated urea could reduce n leaching in tomato crop in comparison to conventional and stabilized fertilizers. *Agronomy* **2020**, *10*, 1827. [CrossRef]
19. Naz, M.Y.; Sulaiman, S.A. Slow release coating remedy for nitrogen loss from conventional urea: A review. *J. Control. Release* **2016**, *225*, 109–120. [CrossRef]
20. European Commission. Bio-Based Products. Available online: http://ec.europa.eu/growth/sectors/biotechnology/bio-based-products_it (accessed on 30 December 2021).
21. Khalid, N.N.A.; Ashaari, Z.; Mohd, A.H.; Mohamed, H.A.; Lee, S.H. Nitrogen deposition and release pattern of slow release fertiliser made from urea-impregnated oil palm frond and rubberwood chips. *J. For. Res.* **2019**, *30*, 208762094.
22. Osmocote®. Product Information. Available online: https://icl-sf.com/uploads/ITALY/Ita_PI/Ita_PI_OH/8756_PI%20OsmocotePro_12-14M%20new.pdf (accessed on 30 December 2021).
23. Miranda, K.M.; Espey, M.G.; Wink, D.A. A rapid, simple spectrophotometric method for simultaneous detection of nitrate and nitrite. *Nitric Oxide* **2001**, *5*, 61–71. [CrossRef]
24. Wellburn, A.R.; Lichtenthaler, H. Formulae and program determine carotenoids and chlorophyll *a* and *b* of leaf extracts inferent solvents. *Adv. Photosyn. Res.* **1984**, *2*, 272–284.
25. Ikeura, H.; Tsukad, K.; Tamaki, M. Effect of microbubbles in deep flow hydroponic culture on spinach growth. *J. Plant Nutr.* **2017**, *40*, 2358–2364. [CrossRef]
26. Liu, I.J.; Tong, Y.; Zhu, Y.; Ding, H.; Smith, F.A. Leaf chlorophyll readings as an indicator for spinach yield and nutritional quality with different nitrogen fertilizer applications. *J. Plant Nutr.* **2006**, *29*, 1207–1217. [CrossRef]
27. Ebid, E.A.; Ueno, H.; Ghoneim, A.; Asagi, N. Nitrogen uptake by radish, spinach and "chingensai" from composted tea leaves, coffee waste sediment and kitchen garbage. *Compost. Sci. Util.* **2008**, *16*, 152–158. [CrossRef]
28. Zhang, J.; Liang, Z.; Jiao, D.; Tian, X.; Wang, C. Different water and nitrogen fertilizer rates effects on growth and development of spinach. *Comm. Soil Sci. Plant Anal.* **2018**, *49*, 1922–1933. [CrossRef]
29. Lefsrud, M.; Kopsell, D.; Sams, C.; Wills, J.; Both, A.J. Dry matter content and stability of carotenoids in kale and spinach during drying. *Hort. Sci.* **2008**, *43*, 1731–1736. [CrossRef]
30. Fascella, G.; Montoneri, E.; Francavilla, M. Biowaste versus fossil sourced auxiliaries for plant cultivation: The lantana case study. *J. Clean. Prod.* **2018**, *185*, 322–330. [CrossRef]
31. Fascella, G.; Montoneri, E.; Rouphael, Y. Biowaste-derived humic-like substances improve growth and quality of Orange Jasmine (*Murraya paniculata* L. Jacq.) plants in soilless potted culture. *Resources* **2021**, *10*, 80. [CrossRef]
32. Fascella, G.; Mammano, M.M.; D'Angiolillo, F.; Rouphael, Y. Effects of conifers wood biochar as substrate component on ornamental performance, photosynthetic activity and mineral composition of potted *Rosa rugosa*. *J. Hort. Sci. Biotech.* **2017**, *93*, 519–528. [CrossRef]
33. Fascella, G.; Mammano, M.; D'Angiolillo, F.; Pannico, A.; Rouphael, Y. Coniferous wood biochar as substrate component of two containerized Lavender species: Effects on morpho- physiological traits and nutrients partitioning. *Sci. Hort.* **2020**, *267*, 109356. [CrossRef]
34. Massa, D.; Lenzi, A.; Montoneri, E.; Ginepro, M.; Prisa, D.; Burchi, G. Plant response to biowaste soluble hydrolysates in hibiscus grown under limiting nutrient availability. *J. Plant Nutr.* **2018**, *41*, 396–409. [CrossRef]

35. Gülser, F. Effects of ammonium sulphate and urea on NO^{3-} and NO^{2-} accumulation, nutrient contents and yield criteria in spinach. *Sci. Hortic.* **2005**, *106*, 330–340. [CrossRef]
36. Stagnari, F.; Di Bitetto, V.; Pisante, M. Effects of N fertilizers and rates on yield, safety and nutrients in processing spinach genotypes. *Sci. Hortic.* **2007**, *114*, 225–233. [CrossRef]
37. Francavilla, M.; Beneduce, L.; Gatta, G.; Montoneri, E.; Monteleone, M.; Mainero, D. Biochemical and chemical technology for a virtuous bio-waste cycle to produce biogas without ammonia and speciality bio-based chemicals with reduced entrepreneurial risk. *J. Chem. Technol. Biotechnol.* **2016**, *91*, 2679–2687. [CrossRef]
38. Photiou, P.; Kallis, M.; Samanides, C.; Vyrides, I.; Padoan, E.; Montoneri, E.; Koutinas, M. Integrated chemical biochemical technology to reduce ammonia emission from fermented municipal biowaste. *ACS Sustain. Chem. Eng.* **2021**, *9*, 8402–8413. [CrossRef]
39. Montoneri, C.; Montoneri, E.; Tomasso, L.; Piva, A. Compost derived substances decrease feed protein N mineralization in swine cecal fermentation. *J. Agric. Sci.* **2013**, *13*, 31–44. [CrossRef]
40. Biagini, D.; Montoneri, E.; Rosato, R.; Lazzaroni, C.; Dinuccio, E. Reducing ammonia and GHG emissions from rabbit rearing through a feed additive produced from green urban residues. *Sustain. Prod. Consum.* **2021**, *27*, 1–9. [CrossRef]
41. Baglieri, A.; Cadilia, V.; Mozzetti Monterumici, C.; Gennari, M.; Tabasso, S.; Montoneri, E.; Nardi, S.; Negre, M. Fertilization of bean plants with tomato plants hydrolysates. Effect on biomass production, chlorophyll content and N assimilation. *Sci. Hortic.* **2016**, *176*, 194–199.
42. Gomis, J.; Bianco Prevot, A.; Montoneri, E.; González, M.C.; Amat, A.M.; Mártire, D.O.; Arque, A.; Carlos, L. Waste sourced biobased substances for solar-driven wastewater remediation: Photodegradation of emerging pollutants. *Chem. Eng. J.* **2014**, *235*, 236–243. [CrossRef]
43. Rossini, F.D.; Wagman, D.D.; Evans, W.H.; Levine, S.; Jaffe, I. *Circular of the Bureau of Standards n° 500: Selected Values of Chemical Thermodynamic Properties*; Nat. Bureau of Standards, Circ.; U.S. Government, Printing Office: Washington, DC, USA, 1952.
44. Tabasso, S.; Ginepro, M.; Tomasso, L.; Nisticò, R.; Francavilla, M. Integrated biochemical and chemical processing of municipal bio-waste to obtain bio based products for multiple uses. The case of soil remediation. *J. Clean. Prod.* **2020**, *245*, 119191. [CrossRef]

Article

Performance of Polyvinyl Alcohol/Bagasse Fibre Foamed Composites as Cushion Packaging Materials

Baodong Liu [1], Xinjie Huang [1], Shuo Wang [1], Dongmei Wang [2,*] and Hongge Guo [1,*]

[1] Department of Packaging Engineering, School of Light Industry Science and Engineering, Qilu University of Technology, Jinan 250353, China; lbd980918@163.com (B.L.); 17854117642@163.com (X.H.); 17854115944@163.com (S.W.)
[2] Packaging Planning and Design, School of Communication, Shenzhen Polytechnic, Shenzhen 518000, China
* Correspondence: sxxawdm@szpt.edu.cn (D.W.); ghg@qlu.edu.cn (H.G.)

Abstract: This work was designed to determine the mechanical properties and static cushioning performance of polyvinyl alcohol (PVA)/bagasse fibre foam composites with a multiple-factor experiment. Scanning electron microscopy (SEM) analysis and static cushioning tests were performed on the foamed composites and the results were compared with those of commonly used expanded polystyrene (EPS). The results were as follows: the materials had a mainly open cell structure, and bagasse fibre had good compatibility with PVA foam. With increasing PVA content, the mechanical properties of the system improved. The mechanical properties and static cushioning properties of the foam composite almost approached those of EPS. In addition, a small amount of sodium tetraborate obviously regulated the foaming ratio of foamed composites. With increasing sodium tetraborate content, the mechanical properties of foamed composites were enhanced. The yield strength and Young's modulus of the material prepared by reducing the water content to 80.19 wt% were too high and not suitable for cushioned packaging of light and fragile products.

Keywords: polyvinyl alcohol; bagasse fibre; foamed composites; cushion packaging material

Citation: Liu, B.; Huang, X.; Wang, S.; Wang, D.; Guo, H. Performance of Polyvinyl Alcohol/Bagasse Fibre Foamed Composites as Cushion Packaging Materials. *Coatings* **2021**, *11*, 1094. https://doi.org/10.3390/coatings11091094

Academic Editor: Philippe Evon

Received: 8 August 2021
Accepted: 7 September 2021
Published: 10 September 2021

Publisher's Note: MDPI stays neutral with regard to jurisdictional claims in published maps and institutional affiliations.

Copyright: © 2021 by the authors. Licensee MDPI, Basel, Switzerland. This article is an open access article distributed under the terms and conditions of the Creative Commons Attribution (CC BY) license (https://creativecommons.org/licenses/by/4.0/).

1. Introduction

Packaging is an indispensable part of commodity circulation. In the logistics and transportation processes, not only is external packaging needed, but internal packaging is also needed to pack objects or fill gaps [1]. The foaming materials often used in the cushion packaging area can be divided into four categories: organic foam material, plant fiber foam material [2,3], metal foam material [4–7] and ceramic foam material [8,9]. Organic foam material is a closed cell foam made of organic materials with emulsifiers, foaming agents, curing agents and other auxiliaries such as expanded polystyrene (EPS), polyethylene foam (EPE), etc. These are widely used due to their low density, excellent buffer performance and low cost. However, they are hard to biodegrade, thus causing "white pollution". EPS also precipitates toxic substances at high temperatures and harms human health [10,11]. Plant fiber foam is a polymer thermoplastic polymer made from plant fibers (such as orange sticks, seaweed, etc.) and starch (such as rice, corn, potato, etc.), using principles of biological recombination and molecular recombination technology, and special production technology. Research on environmentally friendly cushion packaging materials has therefore become an inevitable trend in the 21st century [12]. In recent years, metal foam material and ceramic foam material are seldom used in packaging.

The main biodegradable packaging materials currently on the market are corrugated cardboard, honeycomb cardboard and paper pulp moulding. Corrugated board and honeycomb paperboard are expensive and exhibit unsatisfactory buffer performance [13,14]. The production process for paper pulp moulding is complex and is only used to make small cushioned packaging material [15]. In summary, all of these buffer packaging materials have some limitations.

Due to their environmental friendliness, low cost, and unlimited market potential, the development of plant fibre cushioning materials has attracted increasing interest. Numerous studies have made great breakthroughs in the development of plant fibre cushioning materials [16–18], and research on plant fibre foaming materials has become a new area of interest. These materials use plant fibre as the main raw material and contain various additives, including foaming agents, nucleating agents, adhesives, foam stabilizers, thickeners, and plasticizers; when mixed in a slurry, the foam is then prepared by moulding foaming, baking foaming, extrusion foaming or microwave foaming, after which the material is finally dried and shaped [15]. The buffer packaging material prepared by this method has the advantages of a simple preparation process, a widely available source of raw materials, and complete susceptibility to degradation, and its mechanical performance is essentially equal to that of EPS. The plant fibre foam material expands spontaneously in response to temperature and pressure. Although it is difficult to obtain materials featuring low density and substantial foaming [19], the density of these fibres is less than 0.1 g/cm^3, and the mechanical strength of the material is roughly equal to that of EPS. At present, the density of the most commonly used EPS is between 0.005 and 0.015 g/cm^3, which means that there is a large gap between the densities of the two materials [20,21]. To overcome this difficulty, researchers have used various methods to reduce the density of the material, and low-density plant fibre cushion packaging material has become popular in the packaging industry [22,23]. Bagasse fibre is the residue remaining from sugarcane after squeezing [24]. As a natural plant fibre material it can be naturally degraded [25], exhibits a low cost and its output is seven times the total output of jute [26–28], kenaf, and hemp [29]. Polyvinyl alcohol is a polyhydroxy polymer produced by the hydrolysis of polyvinyl acetate, and it exhibits good mechanical properties [30,31]. It is nontoxic, biocompatible, water soluble, semicrystalline, and fully biodegradable, and it has a relatively lower cost than the currently dominant foam precursor, polyurethane [32]. Bagasse fibre and polyvinyl alcohol meet the requirements of environmental protection.

In this work, the components of foamed composites were PVA and bagasse fibre, and the production process utilized a nucleating agent and a cross-linking agent, with water as a plasticizer. The foamed composites used as cushion packaging were prepared by the mechanical foaming method. The advantage of composite materials is that the components can complement each other [33] and produce synergies that avoid the high-density defects of plant fibre foaming material while providing good mechanical performance. Through multifactorial experiments the mechanical properties, static cushioning properties, and variation laws of the microstructures of these materials were studied via single factor changes (content of PVA, sodium tetraborate and water).

2. Experiments

2.1. Experimental Materials

Polyvinyl alcohol was purchased from Shanghai Aladdin Biological Technology Co., Ltd. (Shanghai, China), specifically chemically pure PVA 1788; the degree of polymerization was 1700 and the degree of alcoholysis 88%, while the average Mw was 85,000–124,000; it had good water solubility and dissolved quickly in cold or hot water. Bagasse fibre (chemical pulp) was supplied by Nanning Sugar Industry Co., Ltd. (Nanning, China). The nucleating agent and sodium tetraborate (analytically pure) were obtained from Tianjin Bodi Chemical Co., Ltd. (Tianjin, China). Deionized water was prepared with filtering equipment in our laboratory.

2.2. Experimental Apparatus

Experiments involved the use of an electronic balance (Accuracy: 0.001 g, Mettler Toledo Co., Ltd., Greifensee, Switzerland), a mechanical stirrer (stepless speed change, IKA RW20 Digital, Shanghai Yikong Mechanical and Electrical Co., Ltd., Shanghai, China), a thermostat water bath (HH-2, Changzhou Putian Instrument Manufacturing Co., Ltd., Changzhou, China), a vacuum freeze dryer (74200-30, Labconco Co., Ltd., Kansas, MO,

USA), an electric blast drying oven (DHG-9140A, Shanghai Jinghong macro laboratory equipment Co., Ltd., Shanghai, China), a universal testing machine (M-3050, Shenzhen Reger Instrument Co., Ltd., Shenzhen, China), and a scanning electron microscope (JCM6000, Japanese Electronics Maker Hitachi Ltd., Tokyo, Japan).

2.3. Specimen Preparation

First, bagasse cellulose was blended with polyvinyl alcohol solution at room temperature. Then, calcium carbonate and sodium tetraborate were added in turn. After mixing for 5 min at a speed of 1600 r/min, the mixture was cast in the cube mould (length × width × height = 5 cm × 5 cm × 4 cm) and put into the refrigerator at low temperature to freeze for 12 h. Finally, it was dried and shaped in a vacuum freeze dryer. Formulations are listed in Tables 1–3.

Table 1. Formulations of composites with different PVA contents.

Specimens	PVA (g)	Fibre (g)	Water (g)	Sodium Tetraborate (g)	Nucleating Agent (g)
1: PVA of 7.11 wt%	5	5	60	0.06	0.3
2: PVA of 8.41 wt%	6	5	60	0.06	0.3
3: PVA of 9.67 wt%	7	5	60	0.06	0.3
4: PVA of 10.91 wt%	8	5	60	0.06	0.3
5: PVA of 12.10 wt%	9	5	60	0.06	0.3

Table 2. Formulations of composites with different sodium tetraborate content.

Specimens	Sodium Tetraborate (g)	PVA (g)	Fibre (g)	Water (g)	Nucleating Agent (g)
a: sodium tetraborate of 0 wt%	0	8	5	60	0.3
b: sodium tetraborate of 0.027 wt%	0.02	8	5	60	0.3
c: sodium tetraborate of 0.055 wt%	0.04	8	5	60	0.3
d: sodium tetraborate of 0.082 wt%	0.06	8	5	60	0.3
e: sodium tetraborate of 0.109 wt%	0.08	8	5	60	0.3

Table 3. Formulations of the composites with different water contents.

Specimens	Water (g)	PVA (g)	Fibre (g)	Sodium Tetraborate (g)	Nucleating Agent (g)
I: Water of 80.19 wt%	54	8	5	0.04	0.3
II: Water of 80.76 wt%	56	8	5	0.04	0.3
III: Water of 81.30 wt%	58	8	5	0.04	0.3
IV: Water of 81.81 wt%	60	8	5	0.04	0.3
V: Water of 82.29 wt%	62	8	5	0.04	0.3
VI: Water of 82.75 wt%	64	8	5	0.04	0.3
VII: Water of 83.19 wt%	66	8	5	0.04	0.3

2.4. Testing and Characterization

2.4.1. Static Compression Properties

Static compression experiments were performed with a microcomputer-controlled electronic universal testing machine. After the material was formed, the temperature and humidity of the samples were established according to ISO 2233-2000; the compression performance was tested according to the GB/T 18942.1-2003: static compression test method of cushioning materials for packaging at a compression speed of 12 mm/min. Each sample was subjected to three repeated experiments. The results presented are the average values of the three experiments. The material load-displacement curve was obtained, and the stress-strain curve of the material was calculated according to the sample size. The samples had almost the same length × width (5 cm × 5 cm); the thickness was around 3 cm.

2.4.2. Yield Strength and Young's Modulus

The yield stress value was regarded as the compression strength (δ_*) when the corresponding strain was 10%. The Yield modulus is the compression strength (δ_*)/10%.

The formula for Young's modulus is as follows:

$$E_{sj} = \frac{1}{2}\left(\frac{\delta_{j+1} - \delta_j}{\varepsilon_{j+1} - \varepsilon_j} + \frac{\delta_j - \delta_{j-1}}{\varepsilon_j - \varepsilon_{j-1}}\right) \tag{1}$$

where δ is the stress of the material and ε is the strain of the material.

The modulus variance of the material with strains between 0.05–5% is the smallest, and the corresponding modulus is the Young's modulus of the material.

2.4.3. Cushion Coefficient

The energy absorption formula for the material is

$$e = \int_0^\varepsilon \delta d\varepsilon \tag{2}$$

where e is the energy absorbed by the material, that is, the area between the stress-strain curve and the x-axis. The cushion coefficient formula is

$$C = \frac{\delta}{e} \tag{3}$$

where C is the cushion coefficient.

2.4.4. SEM Analysis

The external and internal structures of the material were observed by scanning electron microscopy (JCM6000). The material was first cut into slices approximately 2 mm thick and then processed by spraying gold. Finally, the samples were studied with an accelerating voltage of 15 kV. Samples for which the PVA content was 10.91 wt% were used for surface SEM analysis at magnifications of 30, 90, 150 and 300×. Samples with a PVA content of 8.41%, 9.67% and 10.91% were used for SEM analysis of internal sections at magnifications of 30, 50, 90, 300 and 500×.

3. Result and Analysis
3.1. Effect of PVA Content on Material Properties
3.1.1. SEM Analysis

The surface morphology of the materials when the PVA content was 10.91 wt% is shown in Figure 1, with magnifications of 30, 90, 150 and 300×. It is obvious that the fibres were staggered and evenly distributed in the mixture, forming a support system for the skeleton structure. In Figure 1d, the inside of the cell can be clearly observed, so a preliminary estimate suggests that the material has an open cell structure. Bagasse fibres were distributed uniformly in PVA without agglomeration on the surface.

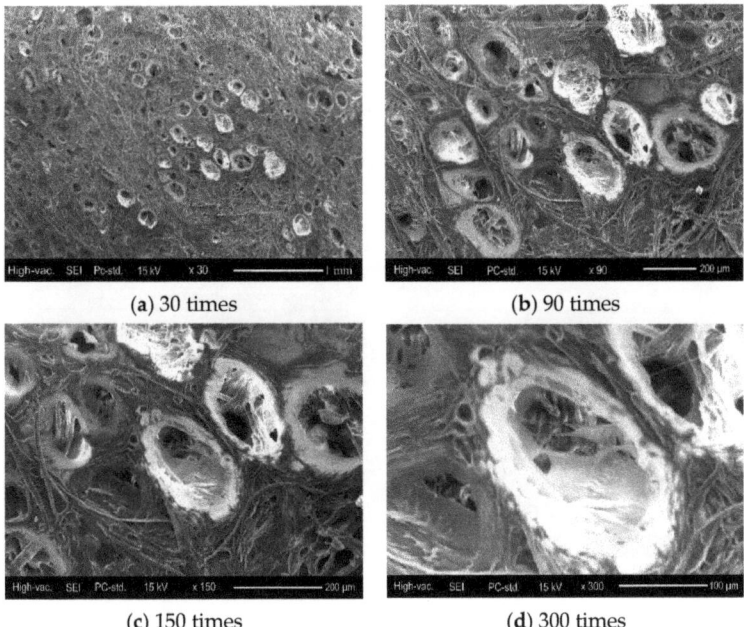

(a) 30 times (b) 90 times (c) 150 times (d) 300 times

Figure 1. Morphology of the surface of a sample (PVA content of 10.91%) at different magnifications. (a) 30 times; (b) 90 times; (c) 150 times; (d) 300 times.

Figure 2 shows that with continuous increases in PVA content, the quantity of cells gradually increased, cell diameters gradually decreased, cell walls gradually thickened, and the macroscopic observation was that the strength of the material increased. The cells in the material consisted of two parts. The main part consisted of the voids formed by air bubbles in the process of mechanical stirring, and the relatively smaller part formed because of the evaporation of water in the drying process.

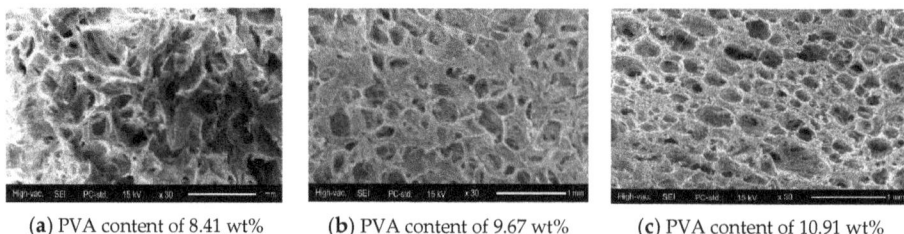

(a) PVA content of 8.41 wt% (b) PVA content of 9.67 wt% (c) PVA content of 10.91 wt%

Figure 2. Morphologies of internal sections of the samples at 30× magnification. (a) PVA content of 8.41 wt%; (b) PVA content of 9.67 wt%; (c) PVA content of 10.91 wt%.

As shown in Figure 3a, the distribution of cells on the surface of the material was relatively sparse. Figure 3d shows a cell interior through a channel with surface bubble cracks, indicating that the material has an open structure. The vesicles are round or oval in shape. The interface between bagasse fibre and polyvinyl alcohol cannot be clearly distinguished from an interrupted surface in the figure. Only the fibre extends to the outside of the composite from the materials inside the cell, indicating that bagasse fibre and polyvinyl alcohol have good compatibility and are closely combined with each other.

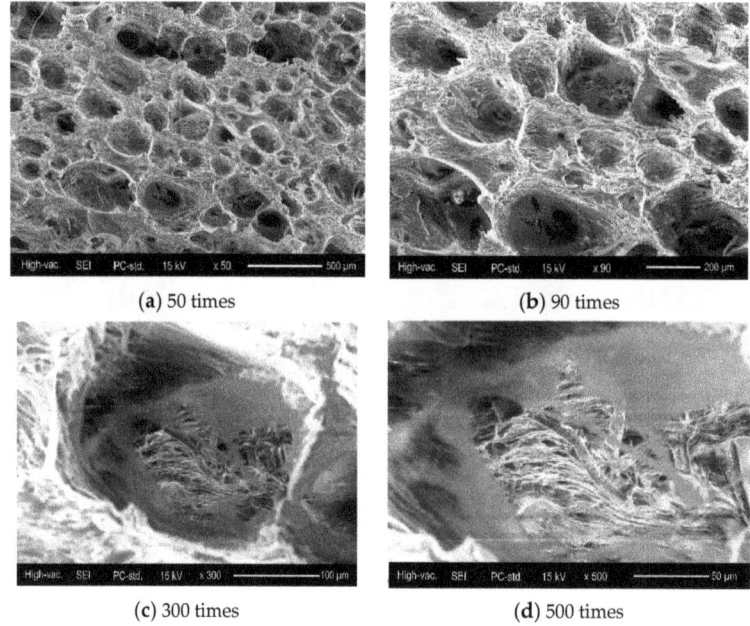

(a) 50 times (b) 90 times

(c) 300 times (d) 500 times

Figure 3. Morphology of an internal section of a sample (PVA content of 10.91%) with different magnifications. (a) 50 times; (b) 90 times; (c) 300 times; (d) 500 times.

3.1.2. Stress and Strain Curve Analysis

Figure 4 shows the stress and strain curves of different samples. Samples 1–5 contained PVA contents ranging from 7.11 wt% to 12.10 wt%, as listed in Table 1. Sample 6 comprised high-density polystyrene (HDEPS, with a density of 0.015 g/cm^3) and sample 7 comprised low-density polystyrene (LDEPS, with a density of 0.005 g/cm^3). HDEPS and LDEPS are high foaming materials.

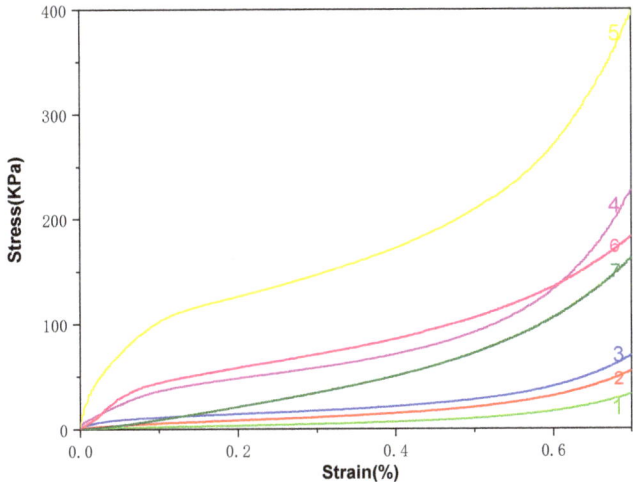

Figure 4. Stress and strain curves for materials with different PVA contents.

Sample 5 showed a higher stress and strain curve than HDEPS and LDEPS. The stress for sample 4 was higher than that for LDEPS and similar to that for HDEPS. After entering the densification stage (strain exceeding 0.6), the stress of the composite gradually surpassed that of the HDEPS because the density of the composite is three times larger than that of HDEPS. Therefore, composites first entered the compaction stage, and the stress-strain curve of the material was determined by the characteristics of the material itself rather than by elastic compression of the cell. When the PVA content was less than that in sample 4, the overall strength of the material was gradually lower than that of LDEPS, which indicates that more than 9.67 wt% PVA should be added in this formula to produce a relatively higher stress and strain curve.

3.1.3. Analysis of Yield Strength and Young's Modulus

Figure 5 shows that with increasing PVA content, the yield strength and Young's modulus of the material gradually increased, indicating that when PVA was at a low level in these formulas the viscosity of the material had reached the requirements for maintaining cell stability.

The increase in PVA content increased the viscosity of the material, increased the resistance to cell growth, and finally increased the thicknesses of cell walls. In this way, the bearing capacity of the material was stronger. On the other hand (as indicated by the SEM analysis of the material), the higher the PVA content was, the finer the cell of the foamed composite. When stress concentration occurred during the compression process the propagation of the crack tip was effectively hindered with higher PVA content. As greater stress was required to destroy the cell structure, the mechanical strength of the material was better. As can be seen from Figure 6, the crosslinking of PVA with sodium tetraborate increased the molecular weight of PVA and the spatial network structure of PVA was formed, thus the strength of the matrix was greatly improved. This indicates that such crosslink can prevent propagation of the crack tip [34].

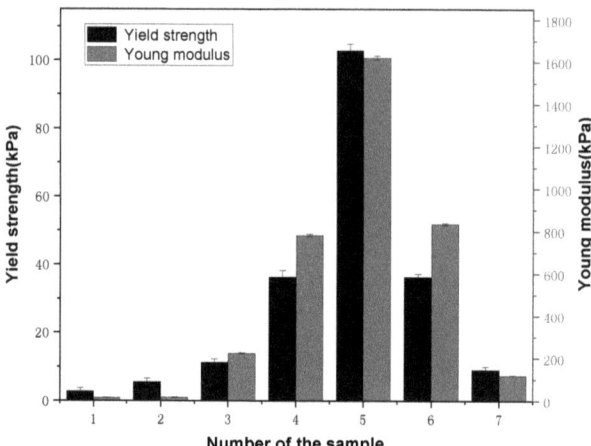

Figure 5. Yield strength and Young's modulus of materials with different PVA contents.

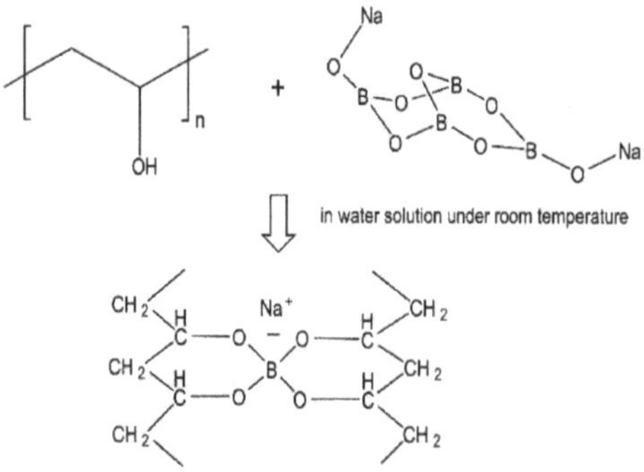

Figure 6. Crosslinking reaction equation of PVA.

When the PVA content was 12.10 wt%, the Young's modulus and yield strength of the composite were much higher than those of HDEPS and LDEPS. When PVA content was 9.67%, they remained higher than those of LDEPS, and similar to those of HDEPS. The yield strength and the Young's modulus of the material exhibited the same change trend because the larger the Young's modulus was, the faster the material stress rose in the online elastic stage, and the higher the yield point of the material was. This trend fully explains the wired elastic stage and yield platform stage in the static compression curve of the material.

3.1.4. Cushion Coefficient Analysis

Figure 7 shows that when the PVA content was at a low level, the cushion coefficient of the material quickly reached the lowest point with increasing strain, after which the cushion coefficient of the material increased rapidly. At a high stress level, the cushion coefficient was high and the energy absorption level was low, indicating that although the foaming ratio of the material was high, the mechanical strength was poor; thus the material was only suitable for use as a cushioning packaging filler for special lightweight materials.

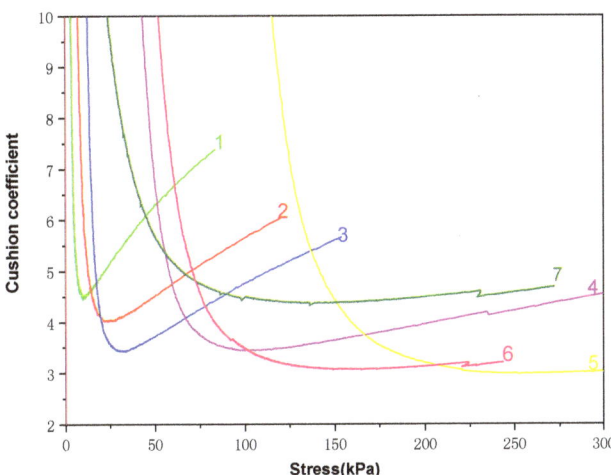

Figure 7. Cushion coefficient of materials as a function of stress, with different PVA contents.

When the PVA contents were 10.91 wt% and 12.10 wt%, the minimum cushion coefficients of the material were 3.43 and 2.96, respectively. After the cushion coefficient reached the lowest point, the values did not rise rapidly. The material still had a low cushion coefficient at a higher stress level and exhibited a good energy absorption level.

3.2. Effect of Sodium Tetraborate Content on the Mechanical Properties of Materials

Samples a–e contain sodium tetraborate contents ranging from low to high, as listed in Table 2. Sample f is high-density polystyrene (HDEPS, with a density of 0.015 g/cm^3) and Sample g is low-density polystyrene (LDEPS, with a density of 0.005 g/cm^3). As shown in Figure 8, with continuous increases in sodium tetraborate content the stress-strain level of the material continued to improve, indicating that the added sodium tetraborate served as a crosslinker and increased the degree of PVA crosslinking and improved the viscosity of the system, increasing cell growth resistance, reducing the foaming ratio of the material, and enhancing the relevant mechanical properties. When the sodium tetraborate content exceeded 0.082 wt% (curve d), the stress-strain curve of the material was higher than that of HDEPS.

Figure 9 shows that with increasing sodium tetraborate content the yield strength and Young's modulus of the material gradually increased. During processing, the addition of sodium tetraborate increased the viscosity of the material and regulated the foaming ratio of the foaming material. When the amount of sodium tetraborate reached 0.109 wt% (sample e), stirring of the material became very difficult. If sodium tetraborate was added continuously, the foaming rate suddenly decreased. As the viscosity of the system increased, stirring aggravated the reaction of PVA and sodium tetraborate, further gel reactions occurred, and the walls of the cells disappeared. This resulted in high density and a sudden decrease in the foaming ratio.

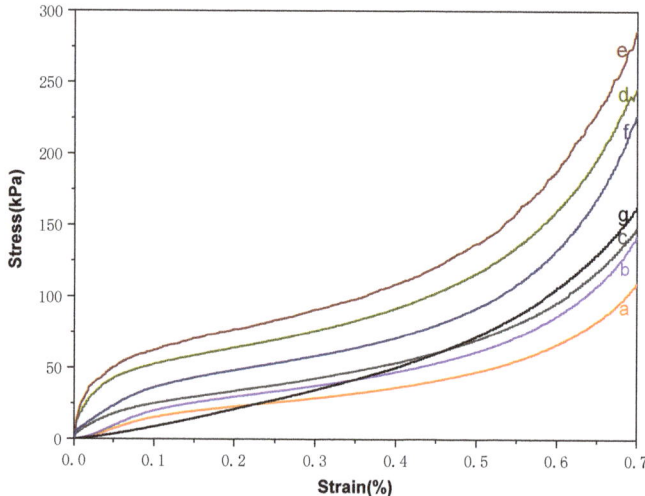

Figure 8. Stress and strain curves of materials with different sodium tetraborate contents.

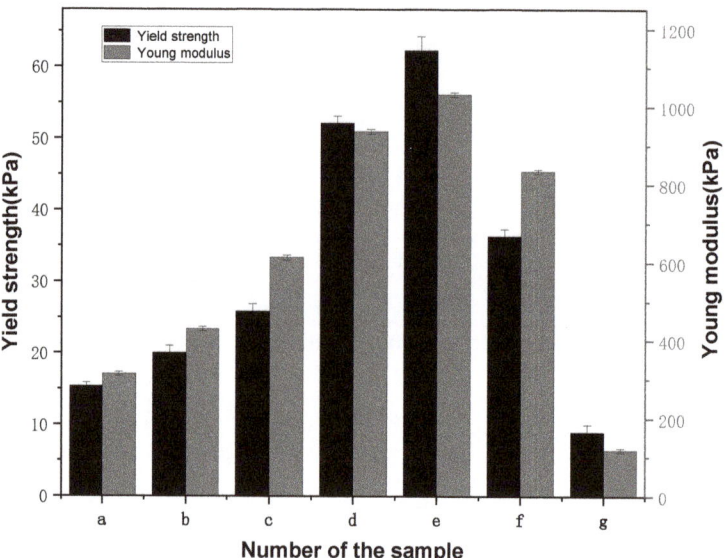

Figure 9. Yield strength and Young's modulus for materials with different sodium tetraborate contents.

In Figure 10, the curves for samples a–c exhibited trends relatively similar to that of LDEPS, which indicated that the lower cushion coefficient was obtained when the stress was low, and the minimum cushion coefficient was obtained when the stress was between 50–100 kPa. The material was therefore suitable for cushion packaging of light materials, and the energy absorption capacity was higher than that of LDEPS. When the stress increased, the cushion coefficient rose more rapidly than that of LDEPS, which means that the energy absorption capacity was relatively poor. At this point the composite entered the compaction stage and the buffer capacity was weakened.

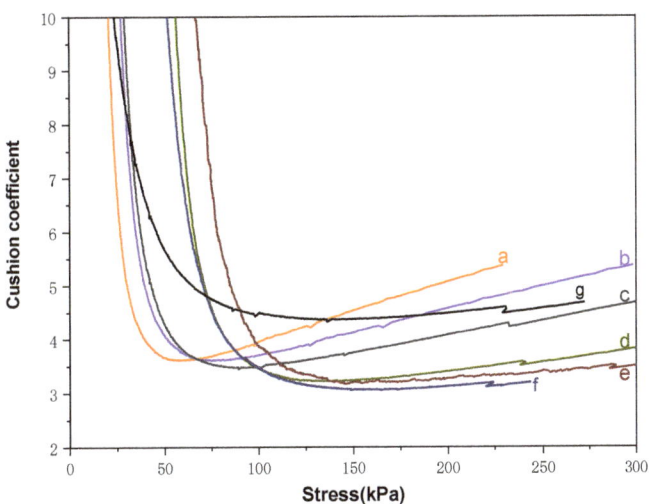

Figure 10. Cushion coefficient of materials with different sodium tetraborate contents.

The curves for samples d and e were very close to that of HDEPS, and the cushion coefficients of the materials did not increase rapidly after reaching the minimum point, indicating that the composite was in the yield platform stage like HDEPS, and had good energy absorption capacity.

3.3. Effect of Water Content on the Mechanical Properties of Materials

Samples I–IX contained water contents ranging from low to high, as listed in Table 3. Sample VIII was HDEPS and Sample IX was LDEPS. Figure 11 shows that increasing water content meant that the Young's Modulus from the stress-strain curve dropped down and the viscosity of the mixture decreased continuously; moreover, the foams were softer, absorbed compressive energy, and deformed.

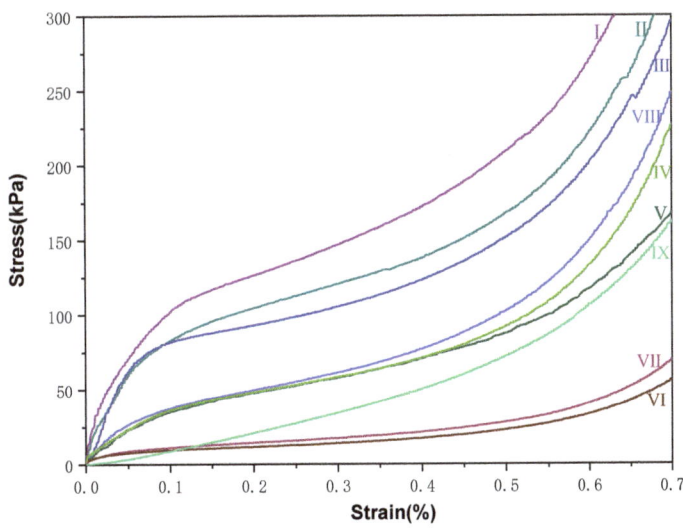

Figure 11. Stress and strain curves of materials with different water contents.

Figure 12 shows that with increasing water content, the yield strengths and Young's moduli of the materials decreased. The Young's modulus of sample I was 1920.37 kPa, 130.21% higher than that of sample VIII (HDEPS), indicating that the foamed material made with this water content was relatively hard, and that increasing the water content was conducive to reducing the hardness.

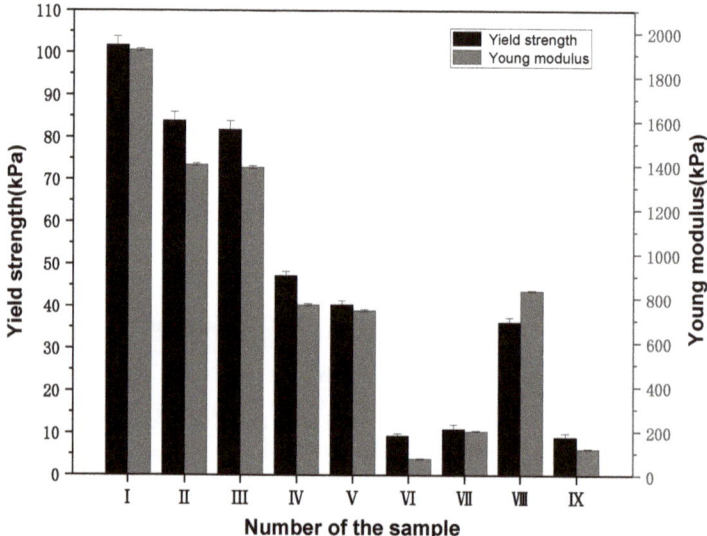

Figure 12. Yield strengths and Young's moduli of materials with different water contents.

As shown in Figure 13, with continuous increases in water content, the stress at which materials reached a lower cushion coefficient gradually decreased. Sample I had a low water content, and the stress at the low point for C was too high and not suitable for cushioning protection of lightweight items.

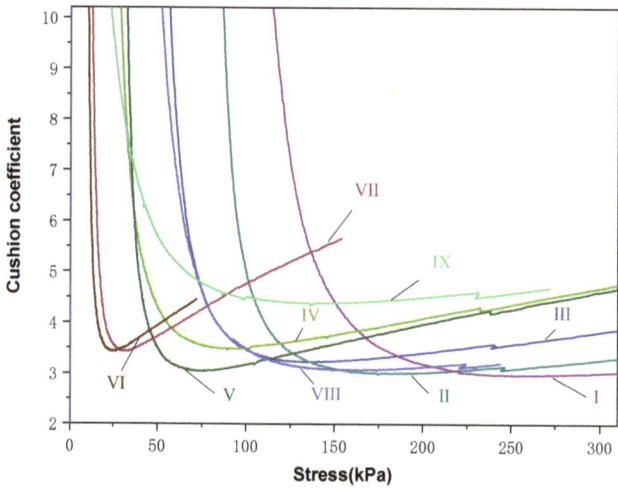

Figure 13. Cushion coefficient of materials with different water contents.

4. Conclusions

Foams have an open cell structure, and bagasse fibre has good compatibility with PVA. With increasing PVA content, the number of cells in the material gradually increases, the diameters gradually decrease, the cell walls gradually thicken, and the macroscopic observation is that the mechanical properties of the material are enhanced. When the PVA content is 10.91 wt%, the comprehensive mechanical properties and static cushioning properties of the material are closest to those of HDEPS.

When the content of sodium tetraborate constitutes 0.027 wt% or 0.055 wt% of the basic formulation, the mechanical properties of the material are similar to those of LDEPS, and when the content of sodium tetraborate is 0.082 wt% or 0.109 wt%, the mechanical properties of the material are similar to those of HDEPS. The added amount cannot exceed 0.109 wt%, or the degree of gelation is high and not conducive to cell stability.

As a plasticizer, water plays a role in adjusting the hardness and cushioning of foaming materials. If the amount of water added is too small, the material is too hard, and if the amount of water added is too large, the cell strength of the material is low and the deformation is large.

With an appropriate formula, we can obtain PVA/bagasse fibre cushioning materials suitable for replacement of LDEPS or HDEPS. The new foamed material exhibits the characteristics of environmental protection, degradability and low price.

In the future, the dynamic cushioning properties of these foams should be analysed to complete performance simulations under actual conditions. Both matrix materials contain hydroxyl groups. Under high humidity conditions, hydroxyl groups absorb water, resulting in a decline in cushioning performance. In addition, a suitable waterproof coating is needed to protect the cushioning material.

Author Contributions: B.L.: Writing-original draft, Investigation, Data curation; X.H.: Software, Visualization; S.W.: Validation, Data curation; D.W.: Conceptualization, Formal analysis; H.G.: Resources, Supervision, Data curation. All authors have read and agreed to the published version of the manuscript.

Funding: This research was funded by Qilu University of Technology (Shandong Academy of Sciences) International Cooperation Research Special Fund Project (QLUTGJHZ2018028). Projects funded by Shenzhen Science and Technology Plan International Cooperation Fund (GJHZ20180928161004981).

Institutional Review Board Statement: Not applicable.

Informed Consent Statement: Not applicable.

Data Availability Statement: Not applicable.

Conflicts of Interest: The authors declare no conflict of interest.

References

1. Gao, Z.; Dong, W.; Xiaomin, J. Research on the reduction design of fragile product cushion packaging under the green development concept. *IOP Conf. Ser. Earth Environ. Sci.* **2021**, *657*, 012057. [CrossRef]
2. Chen, Q.; Du, X.; Chen, G. Preparation of Modified Montmorillonite—Plant Fiber Composite Foam Materials. *Materials* **2019**, *12*, 420. [CrossRef]
3. Li, J.; Yang, X.; Xiu, H.; Dong, H.; Song, T.; Ma, F.; Feng, P.; Zhang, X.; Kozliak, E.; Ji, Y. Structure and performance control of plant fiber based foam material by fibrillation via refining treatment. *Ind. Crop. Prod.* **2019**, *128*, 186–193. [CrossRef]
4. Duarte, I.; Banhart, J. A study of aluminium foam formation—kinetics and microstructure. *Acta Mater.* **2000**, *48*, 2349–2362. [CrossRef]
5. Movahedi, N.; Linul, E.; Marsavina, L. The Temperature effect on the compressive behavior of closed-cell aluminum-alloy foams. *J. Mater. Eng. Perform.* **2018**, *27*, 99–108. [CrossRef]
6. Movahedi, N.; Murch, G.E.; Belova, I.V.; Fiedler, T. Effect of heat treatment on the compressive behavior of zinc alloy ZA27 syntactic foam. *Materials* **2019**, *12*, 792. [CrossRef]
7. Orbulov, I.N. Metal matrix composite syntactic foams for light-weight structural materials. In *Encyclopedia of Materials: Composites*; Brabazon, D., Ed.; Elsevier: Oxford, UK, 2021; pp. 781–797.

8. Liu, J.; Ren, B.; Zhang, S.; Lu, Y.; Chen, Y.; Wang, L.; Yang, J.; Huang, Y. Hierarchical ceramic foams with 3D interconnected network architecture for superior high-temperature particulate matter capture. *ACS Appl. Mater. Interfaces* **2019**, *11*, 40585–40591. [CrossRef]
9. Semerci, T.; Innocentini, M.D.D.M.; Marsola, G.A.; Lasso, P.R.O.; Soraru, G.D.; Vakifahmetoglu, C. Hot air permeable preceramic polymer derived reticulated ceramic foams. *ACS Appl. Polym. Mater.* **2020**, *2*, 4118–4126. [CrossRef]
10. Li, F.; Guan, K.; Liu, P.; Li, G.; Li, J. Ingredient of Biomass packaging material and compare study on cushion properties. *Int. J. Polym. Sci.* **2014**, *2014*, 146509. [CrossRef]
11. Kaisangsri, N.; Kerdchoechuen, O.; Laohakunjit, N. Characterization of cassava starch based foam blended with plant proteins, kraft fiber, and palm oil. *Carbohydr. Polym.* **2014**, *110*, 70–77. [CrossRef]
12. Xingyun, Z.; Lingqing, Z.; Yuping, E.; Jiajun, W. Preparation of laminaria japonica foamed cushioning material and the research on its properties. *J. Zhejiang Sci.-Technol. Univ.* **2015**, *33*, 781–787.
13. Aboura, Z.; Talbi, N.; Allaoui, S.; Benzeggagh, M. Elastic behavior of corrugated cardboard: Experiments and modeling. *Compos. Struct.* **2004**, *63*, 53–62. [CrossRef]
14. Allaoui, S.; Aboura, Z.; Benzeggagh, M. Effects of the environmental conditions on the mechanical behaviour of the corrugated cardboard. *Compos. Sci. Technol.* **2009**, *69*, 104–110. [CrossRef]
15. Shenghua, Z. Discussion on the background, present condition, and craft of foaming vegetable fiber material. *Packag. Eng.* **2007**, *11*, 239–242.
16. Hongwei, J.; Xin, G.; Yuhua, G.; Dandan, W. Preparation and properties of bagasse cushioning materials. *Funct. Mater.* **2015**, *46*, 19101–19105.
17. Guangsheng, Z.; Ruizhen, L.; Liangjie, Z.; Lei, C.; Cong, M. Performance and preparation of foamed waste paper pulp reinforced starch-based composites. *Funct. Mater.* **2012**, *43*, 3054–3057.
18. Gang, L.; Fangyi, L.; Kaikai, G.; Peng, L.; Yu, L.; Xiujie, J.; Jianfeng, L. Preparation and properties of biomass cushion packaging material. *Funct. Mater.* **2013**, *44*, 1969–1973.
19. Allen, M.Y. Development of MLFB made from Micro-wood fiber. *Wood Process. Mach.* **2006**, *5*, 36–37.
20. Yan, M. Exploring the processing technology for low density boards with micro wood fiber. *Wood Ind.* **2006**, *4*, 19–21.
21. Lisheng, X.; Yingjun, L. Research on the manufacturing method and process of low-density fiber formed body. *China For. Prod. Ind.* **2005**, *32*, 20–23. [CrossRef]
22. Lawton, J.W.; Shogren, R.L.; Tiefenbacher, K.F. Effect of batter solids and starch type on the structure of baked starch foams. *Cereal Chem. J.* **1999**, *76*, 682–687. [CrossRef]
23. Shogren, R.; Lawton, J.; Tiefenbacher, K. Baked starch foams: Starch modifications and additives improve process parameters, structure and properties. *Ind. Crop. Prod.* **2002**, *16*, 69–79. [CrossRef]
24. Shibata, S.; Cao, Y.; Fukumoto, I. Effect of bagasse fiber on the flexural properties of biodegradable composites. *Polym. Compos.* **2005**, *26*, 689–694. [CrossRef]
25. Verma, D.; Gope, P.; Maheshwari, M.K.; Sharma, R. Bagasse-fiber composites—A review. *J. Mater. Environ. Sci.* **2012**, *3*, 1079–1092.
26. Gowda, T.M.; Naidu, A.; Chhaya, R. Some mechanical properties of untreated jute fabric-reinforced polyester composites. *Compos. Part A Appl. Sci. Manuf.* **1999**, *30*, 277–284. [CrossRef]
27. Ray, D.; Sarkar, B.K.; Rana, A.K.; Bose, N.R. Effect of alkali treated jute fibres on composite properties. *Bull. Mater. Sci.* **2001**, *24*, 129–135. [CrossRef]
28. Gassan, J. A study of fibre and interface parameters affecting the fatigue behaviour of natural fibre composites. *Compos. Part A Appl. Sci. Manuf.* **2002**, *33*, 369–374. [CrossRef]
29. Keller, A. Compounding and mechanical properties of biodegradable hemp fibre composites. *Compos. Sci. Technol.* **2003**, *63*, 1307–1316. [CrossRef]
30. Bai, J.; Li, Y.; Yang, S.; Du, J.; Wang, S.; Zheng, J.; Wang, Y.; Yang, Q.; Chen, X.; Jing, X. A simple and effective route for the preparation of poly(vinylalcohol) (PVA) nanofibers containing gold nanoparticles by electrospinning method. *Solid State Commun.* **2007**, *141*, 292–295. [CrossRef]
31. Rachipudi, P.; Kariduraganavar, M.; Kittur, A.; Sajjan, A. Synthesis and characterization of sulfonated-poly(vinyl alcohol) membranes for the pervaporation dehydration of isopropanol. *J. Membr. Sci.* **2011**, *383*, 224–234. [CrossRef]
32. Song, T.; Tanpichai, S.; Oksman, K. Cross-linked polyvinyl alcohol (PVA) foams reinforced with cellulose nanocrystals (CNCs). *Cellulose* **2016**, *23*, 1925–1938. [CrossRef]
33. Ding, F.F.; Zhang, M.; Dong-Cai, Y.U. Research Progress in Influencing Factors on Comprehensive Performance of Plant Fiber Composite Material. *Packag. Eng.* **2009**, *11*. [CrossRef]
34. Liguo, W.; Yueting, Z.; Shaohua, H. Recent Research and Development of Poly(vinyl alcohol) Hydrogels. *J. Donghua Univ. Nat. Sci.* **2001**, *6*, 114–118.

Article

Analysis of the Coloring and Antibacterial Effects of Natural Dye: Pomegranate Peel

Aicha Bouaziz [1,2,†], Dorra Dridi [3,†], Sondes Gargoubi [4,*], Souad Chelbi [2], Chedly Boudokhane [5], Abderraouf Kenani [6] and Sonia Aroui [6]

1. Bio-Resources, Integrative Biology & Valorization (BIOLIVAL, LR14ES06), Higher Institute of Biotechnology of Monastir, University of Monastir, Monastir 5000, Tunisia; bouaziz.aicha@gmail.com
2. Higher School of Health Sciences and Techniques of Sousse, University of Sousse, Sousse 4054, Tunisia; souad_chelbi@yahoo.fr
3. Unit of Analysis and Process Applied to the Environment (UR17ES32) Issat Mahdia, University of Monastir, Monastir 5000, Tunisia; dorra.dridi.jeddi@gmail.com
4. Textile Engineering Laboratory—LGTex, University of Monastir, Monastir 5000, Tunisia
5. Research Unity of Applied Chemistry and Environment, Faculty of Science of Monastir, University of Monastir, Monastir 5000, Tunisia; chimi.tex@planet.tn
6. Research Laboratory "Environment, Inflammation, Signaling and Pathologies" (LR18ES40), Faculty of Medicine of Monastir, University of Monastir, Monastir 5019, Tunisia; raouf.kenani@fmm.rnu.tn (A.K.); sonia_aroui2002@yahoo.fr (S.A.)
* Correspondence: gargoubisondes@yahoo.fr
† A. Bouaziz and D. Dridi contributed equally to this work as first authors.

Citation: Bouaziz, A.; Dridi, D.; Gargoubi, S.; Chelbi, S.; Boudokhane, C.; Kenani, A.; Aroui, S. Analysis of the Coloring and Antibacterial Effects of Natural Dye: Pomegranate Peel. *Coatings* **2021**, *11*, 1277. https:// doi.org/10.3390/coatings11111277

Academic Editor: Philippe Evon

Received: 19 August 2021
Accepted: 19 September 2021
Published: 21 October 2021

Publisher's Note: MDPI stays neutral with regard to jurisdictional claims in published maps and institutional affiliations.

Copyright: © 2021 by the authors. Licensee MDPI, Basel, Switzerland. This article is an open access article distributed under the terms and conditions of the Creative Commons Attribution (CC BY) license (https:// creativecommons.org/licenses/by/ 4.0/).

Abstract: This work aims to conduct an eco-friendly textile finishing process by applying agricultural by-products as a dye for the finishing of polyamide fabrics. A natural dye was obtained from pomegranate peel extract. Polyamide fabrics were dyed at different conditions, and four mordanting agents were tested. The finished fabrics were analyzed in terms of CIE L, a, b and color yield (K/S) values, as well as washing fastness, rubbing fastness, light fastness and antibacterial activity. Results show that pomegranate peel extract could dye polyamide fabrics. The rubbing and washing fastness of the finished samples was good. The light fastness was fair, and its antibacterial efficiency against the tested bacteria was good.

Keywords: polyamide; pomegranate peel; natural dye; mordant; fastness; antibacterial

1. Introduction

Vegetables and fruits play a significant part in our daily life. The request for such imperative commodities has expanded with the increase in world population [1]. Mass consumption has led to a higher generation of by-products and has created a disposal problem [2].

Vegetable and fruit by-products could be an important and profitable source of natural compounds [3]. Numerous research has shown that the generated compounds are great sources of phenolics, natural acids, sugar, colors and minerals. Some of these natural compounds exhibit bioactive functions such as antibacterial, antitumor, antifungal, antiviral, antimutagenic and cardioprotective [4,5].

Although some of the generated by-products can be considered unavoidable, others can be utilized in different domains, including pharmaceutical, food, textile and cosmetic industries [6–8]. The valorization of by-products may be a promising way to establish sustainable development and reduce environmental problems [9]. Waste management procedures must be installed with the increasing valorization of by-products and industries must develop new ways of recycling wastes.

In line with green trends towards sustainability, textile researchers have found great potential in using plant extracts as natural dyes [10]. Much research has focused on studying the chemical composition and functional properties of these natural dyes [11]. In this context, pomegranate peel extracts were extensively used as a source of yellow dye [11].

The pomegranate is one of the oldest cultivated fruits in the world [12]. This fruit is broadly spread throughout numerous nations, and it is well adapted to dry zones and the Mediterranean climate [13]. In addition, pomegranate peel accounts for nearly half of the entire weight of the fruit [14].

A review of the literature shows that there have been relatively few studies investigating the dyeing of polyamide fibers with pomegranate peel [15] or the bio-functional activities of the dyed substrates. In addition, there has been far less research on dyeing using mordant concentration, which does not exceed the limits specified by environmental and health standards. Therefore, the target of the present work is to demonstrate the feasibility of applying pomegranate peel extract as a natural functional dye for polyamide substrates. This research evaluated the effects of four mordants on color yield (K/S): color fastness, rubbing, light and washing. To the best of our knowledge, this is an original study on polyamide dyeing, using pomegranate peel extract to assess ecofriendly mordant dyeing and antibacterial finishing.

2. Materials and Methods

2.1. Materials

Pomegranate fruits were purchased from local shops. The peels were collected, dried in the sun, and then powdered. Commercially available polyamides (purchased from local shops) were used, and for the entire set of studies, analytical grade chemicals were employed. The used chemicals are: Folin–Ciocalteu (LOBA Chemie, Mumbai, India), Gallic acid solution, sodium carbonate (75%), sodium nitrate $NaNO_2$ (5%, w/v), aluminum trichloride $AlCl_3$ (10%, w/v) Catechin (kindly provided by FSM university), $C_3H_6O_3$, Sodium hydroxide solution (1M NaOH) (Kindly provided by Chimitex company), Phosphate buffer solution (PBS) (Kindly provided by local hospital).

2.2. Aqueous Extraction

A mass of 20 g of the powdered pomegranate peels was taken in a 250 mL flask and covered with 50 mL of pure water in order to keep the plant material fully immersed in water. The flask was then kept at 90 °C for 120 min. Thus, natural dye from pomegranate peels was obtained by the aqueous extraction technique. After the complete extraction of dye, the mixture was filtered, and residual entities were extracted from the liquor. A reflux system was used during extraction to avoid solvent evaporation.

2.3. Content of Total Phenols

The Folin–Ciocalteu reagent was used to evaluate the total phenolic content of the aqueous extract. Quantification of total phenols is made by comparing the absorbance observed with that obtained by a Gallic acid solution with a known concentration [15].

A mixture of 0.4 mL of the pomegranate peel extract and 10 mL of diluted Folin–Ciocalteu reagent was subjected to stirring and then kept at ambient temperature for 5 min. After stirring, 8 mL of an aqueous solution of sodium carbonate (75%) was added. After 1 h, the absorbances were determined at the wavelength of 765 nm using a spectrophotometer (Specord 210 plus, Analytik Jena AG, Jena, Germany). The concentration of phenolic compounds was determined using a calibration curve of Gallic acid and reported in mg Gallic acid equivalent (GAE)/g of extract.

2.4. Total Flavonoids Content

Flavonoids were quantified in accordance with the method defined by Zhishen et al. using aluminum trichloride and sodium hydroxide [16]. Aluminum trichloride forms a yellow complex with flavonoids, and sodium hydroxide forms a pink complex that absorbs in the visible range of 510 nm [15].

A volume of 1mL of pomegranate peel extract was added to 4 mL of pure water and 0.3 mL of sodium nitrate $NaNO_2$ (5%, w/v). The system was mixed and kept in the dark at room temperature for 5 min, then 0.3 mL of aluminum trichloride $AlCl_3$ (10%, w/v) was added. After 10 min in the dark, 2 mL of sodium hydroxide solution (1 M NaOH) was added. The mixture was stirred, and the absorbance was determined using a spectrophotometer (Specord 210 plus, Analytik Jena AG, Jena, Germany) at the wavelength 510 nm. Flavonoids were quantified using a calibration curve performed using Catechin and expressed in milligrams (mg) equivalent to Catechin per gram of extract (mg CE/g).

2.5. Visible Absorption Spectra

The visible absorbance spectra of the pomegranate peel aqueous extract were recorded using a visible spectrophotometer (DR/3900, Hach, Colorado, CO, USA). Measurements were performed by wavelength scan over the visible range of 320 to 800 nm.

2.6. Dyeing Process

To dye the polyamide fabrics with pomegranate peel extract, the exhaust process was applied using a bath ratio of 40:1. The pH was adjusted to 3 for all dye baths. During dyeing, the temperature was raised from 25 °C to the dyeing temperature (40 °C, 60 °C, 80 °C and 100 °C) at a rate of 2 °C/min. The samples were treated for 45 min. After dyeing, polyamide fabrics were subjected to water rinsing cycles (1 min at ambient temperature for each cycle). Finally, the dyed samples were air-dried.

To improve dyeability, a mordanting technique was applied. An amount of 3% of mordant was used during dyeing. The released metal ions were quantified using inductively coupled plasma optical emission spectrometry (ICP OES, Perkin Elmer, Norwalk, CT, USA). An artificial acidic sweat solution made according to the ISO 3160/2 standard was used to extract metal ions from polyamide fabrics. A dilution of 20 gL^{-1} NaCl, 17.5 gL^{-1} NH_4Cl, 5 g L^{-1} CH_3COOH, and 15 g L^{-1} $C_3H_6O_3$ was used to prepare the acidic sweat solution. To adjust the pH value at 4.7, sodium hydroxide (0.1 mol L^{-1}) was used [17].

A mass of 1.5 g of each polyamide fabric was vortexed with 25 mL of artificial sweat solution for 2 h at 40 °C. ICP OES was used to examine the solutions after they had been filtered.

The concentrations of removed metal ions from dyed fabrics were found not to exceed the limits specified by the Öko Tex standard [18].

2.7. Colorimetric Data

A spectrophotometer (Datacolor 650®, Lawrenceville, NJ, USA) was used to record light reflectance measurements under illuminant D65 and a 10° standard observer. These measurements allow for the analysis of colorimetric data. The International Commission on Illumination's CIELab color coordinates (L, a, and b) were recorded. The letter L stands for lightness–darkness values, whereas the letter A stands for red-green share and ranges from negative (green) to positive (red).

The yellow-blue share is represented by b values, which range from negative (blue) to positive (yellow). The standard color strength value (K/S) was calculated based on the Kubelka–Munk equation:

$$(1 - R_{\lambda max})^2 / 2R_{\lambda max} = K/S \tag{1}$$

where S is the scattering coefficient, R is the reflectance of the dyed fabric and K is the absorption coefficient.

2.8. Fastness Evaluation

The ISO standard methods were applied to evaluate the fastness properties of dyed fabrics. The considered standards were as follows: ISO 105-C06 for washing fastness, ISO 105-X12 for dry and wet rubbing fastness and ISO 105-B02 for light fastness.

2.9. Antibacterial Testing

Escherichia coli (ATCC 8739) and Staphylococcus aureus (ATCC 6538) were used to evaluate the antibacterial activity of polyamide fabrics according to ASTM E2149. In this method, the polyamide samples (1 g) were placed in a flask containing a diluted suspension of bacteria using a phosphate buffer solution (PBS) to obtain a concentration of 3×10^5 UFC/mL.

Treated and control samples were placed in laboratory flasks and shaken with 50 mL of the bacteria suspension for 1 h at 200 rpm. Then, 0.1 mL of prepared suspension was taken from each flask and distributed over a Petri dish. All Petri dishes were incubated for 24 h at 37 °C. Finally, the formed bacteria colonies were counted. Antibacterial activity was reported as a percent reduction of the bacteria cells after contact with treated samples compared to the number of bacteria cells remaining after being in contact with the untreated fabric. The percentage reduction ($R\%$) was determined based on the following equation:

$$R\% = \frac{B - A}{B} \times 100 \qquad (2)$$

where A and B are the bacteria cells expressed in CFU/mL for the suspension in contact with treated and control polyamide fabrics, respectively.

3. Results and Discussion

3.1. Total Phenolic and Flavonoid Contents

Table 1 shows the concentrations of the phenolic and flavonoid compounds of the pomegranate peel extract. The results evidenced that the extract possessed phenolic and flavonoid components. The brown color of the extract is chemically related to these compounds. The extract is rich in phenolics, while the flavonoids constitute a small part of the total phenolic compounds (10%). This finding agrees with previous studies that have proven that pomegranate peels polyphenols essentially consist of tannins.

Table 1. Total phenolic compounds and flavonoids of pomegranate peels extract.

Total Phenolics	Flavonoids
176 ± 11 mg GAE/g	18 ± 3 mg CE/g

3.2. Visible Spectroscopy Characterization

Figure 1 shows the visible spectrum of pomegranate peel extract. Two maxima absorptions were detected at 325 nm and 367 nm. These peaks appear due to the presence of flavonoids.

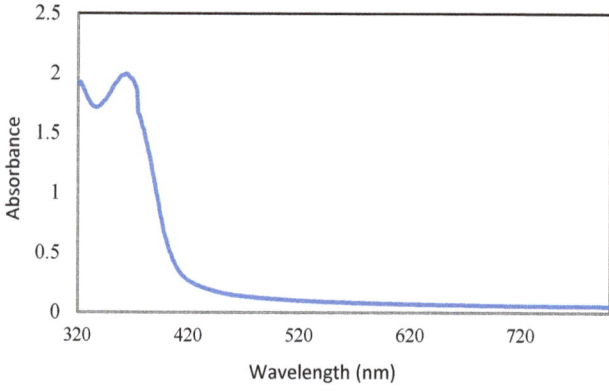

Figure 1. Visible spectrum of pomegranate peel extract.

Generally, the visible spectrum of flavonoids shows peaks in the 300–400 nm region. These peaks are associated with the cinnamoyl system (Figure 2) [19]. The absorbance shifts towards higher wavelengths by increasing conjugation [20].

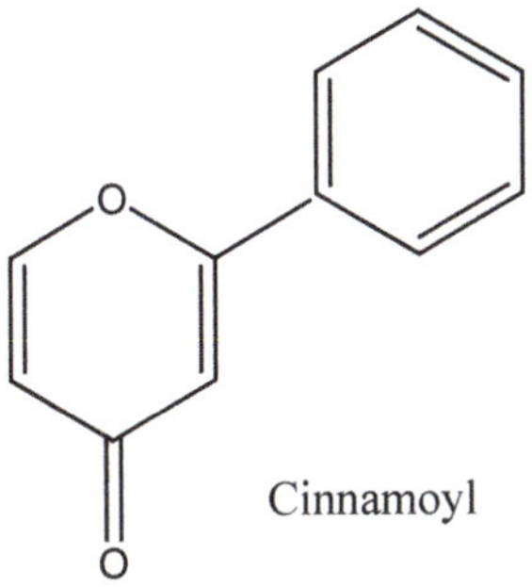

Figure 2. Structural base of the cinnamoyl system.

3.3. Effect of the Dye Concentration

Colorimetric data were considered to evaluate the influence of dye concentration during the dyeing of polyamide fabrics. Different dye concentrations (2%, 4%, 5%, 6% and 7%) were used. Colorimetric data are shown in Table 2.

Table 2. Shades and colorimetric data obtained for the dyed polyamide fabrics at different concentrations.

Extract (%)	Shade	K/S	L	a	b
2%		2.99	74.92	3.14	26.81
4%		3	75.34	3.47	27.49
5%		3.06	75.2	3.71	30.27
6%		3.99	72.57	4.27	30.35
7%		4.31	71.57	3.90	28.12

(Temperature: 100 °C).

The exploitation of the different values provides evidence that the increase in the dye concentration increases the shade intensity of dyed fabrics expressed by the color yield (K/S) values. The best value of (K/S) was obtained for a concentration of 7%. The colorimetric parameters L, a and b values are variable. The pomegranate peel extract gave yellow to brown shades.

The dye uptake on the polyamide fabric is primarily due to the electrostatic forces (Figure 3) created between the positively charged end amino groups of polyamide fibers and the dye molecules under acidic conditions [21]. At pH = 3, the protonation of amino groups is increased, and electrostatic interactions are enhanced. In addition to electrostatic interactions, hydrogen bonds occur between flavonoids and polyamide fibers (Figure 3). Moreover, hydrophobic interactions exist and Van der Waals attraction between the methylene groups of fibers and the aromatic moieties of flavonoids [22,23].

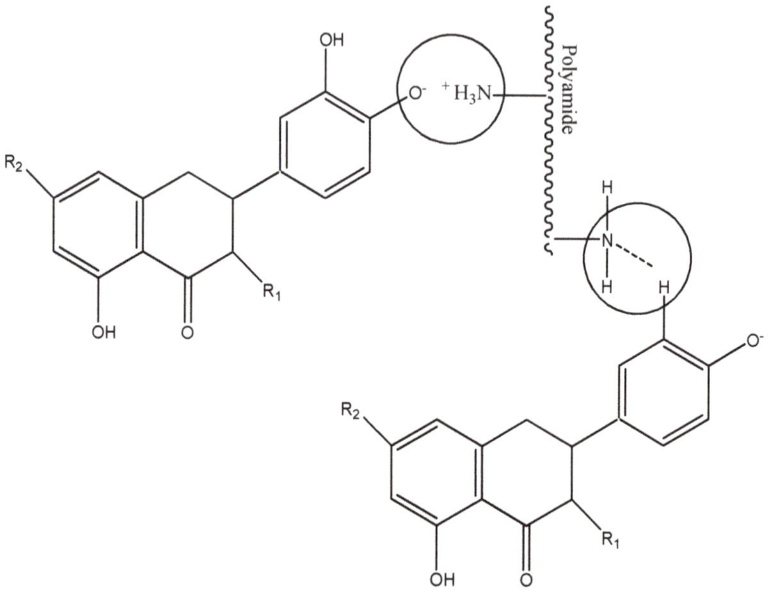

Figure 3. Electrostatic interactions and hydrogen bond between positively charged end amino groups of polyamide and flavonoids of the dye (pH = 3).

3.4. Effect of Dyeing Temperature

Dyeing was carried out at different temperatures to test the effect of heating on the dyeing properties. The obtained results are recapitulated in Table 3. They clearly show that increasing the dye bath temperature results in an increase in the color strength value (K/S). The higher color strength value was obtained at 100 °C. When the temperature exceeds the glass transition of polyamide, a good facility of penetration of dye takes place. Results show also that the luminosity, L, and the parameters, a and b, are influenced by dyeing temperature.

Table 3. Shades and colorimetric data obtained for the dyed polyamide fabrics at different dyeing temperatures.

Temperature (°C)	Shade	K/S	L	a	b
40		1.08	84.50	1.25	22.14
60		1.73	79.42	4.75	25.24
80		3.46	70.96	5.33	28.63
100		3.99	72.57	4.27	30.35

(Extract: 6%).

3.5. Mordanting Effect

During this work, simultaneous mordanting was used to dye polyamide fabrics with pomegranate peel extract. The mordants used were aluminum sulfate, iron sulfate, potassium dichromate and gallnut extract. Table 4 shows the influence of mordanting treatment on dyed samples in terms of their color shade and their colorimetric data. The difference in shades of the dyed fabrics can be noticed. This result can be attributed to the mordants effect. The use of iron sulfate and potassium dichromate has affected the shades of dyed fabrics. On the contrary, the use of aluminum sulfate and gallnut gives shades similar to that obtained without a mordant. In addition, the use of iron sulfate gave the highest K/S value.

Table 4. Shades and colorimetric data obtained for polyamide fabrics.

Mordant	Shade	K/S	L	a	B
Without mordant		2.99	74.92	3.14	26.81
Iron sulfate		3.27	53.99	2.52	9.04
Aluminum sulfate		2.07	72.93	7.52	24.47
Potassium dichromate		2.44	69.65	6.49	24.81
Gallnut		2.62	73.45	3.81	27.63

(Extract: 2%, temperature: 100 °C).

Metal mordants are known to bind with the natural dye by forming coordination complexes. The formed complex, via mordant metal, acts as a binding agent by attracting the fiber from one side and the natural dye from the other side.

3.6. Color Fastness

Table 5 shows the ratings of mordanted and un-mordanted dyed polyamide fabrics for washing, rubbing and light fastness. The rubbing fastness of the un-mordanted sample was found to be quite good. However, the washing fastness was also good, but the light fastness was poor.

Table 5. Fastness results.

Sample	Washing	Dry Rubbing	Wet Rubbing	Light
Without mordant	3	4	4	2
Iron sulfate	4	4–5	4–5	3
Aluminum Sulfate	3–4	4–5	4–5	2–3
Potassium dichromate	3–4	4–5	4–5	2–3
Gallnut	3–4	4–5	4–5	2–3

Natural dyes have long been known to have low light fastness. Using mordants, the wash fastness of mordanted samples was found to improve from a rating of three to a rating of four. These findings suggest that mordants are able to improve the wash fastness ratings of polyamide fabrics. Iron sulfate was the best mordant to improve light fastness.

3.7. Antibacterial Testing

Figure 4 shows the results of ASTM E2149 antibacterial testing against two strains: E. coli and S. aureus, after 1 h of contact. Results of testing against S. aureus, which is Gram positive, show low reduction rates. The finishing with natural dye does not significantly reduce the bacteria concentration. The best result was obtained for the dyed fabric after mordanting using iron sulfate with a reduction rate of 67%. By contrast, the efficacy against E. Coli shows better results, with reduction rates up to 91%. The differing efficacy responses may be related to the different Gram statuses. The antibacterial activity of dyed polyamide fabrics was ranked as iron sulfate > aluminum sulfate > potassium dichromate > gallnut >

un-mordant against S. aureus and E. coli. The use of mordants enhances antibacterial activity. It is well known that metallic salts exhibit toxic effects against microorganisms.

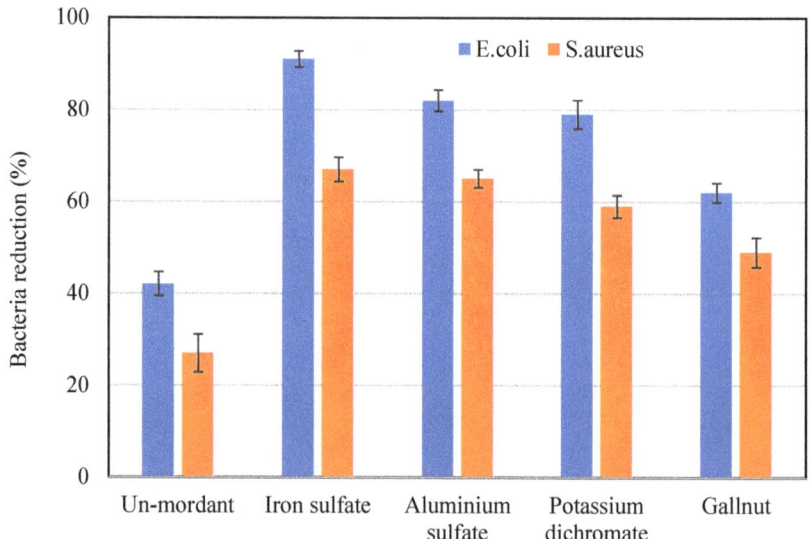

Figure 4. The antibacterial rates of dyed polyamide fabrics against bacteria strains.

4. Conclusions

The aim of this work was to use pomegranate by-products as natural dye for an environmentally friendly and sustainable textile dyeing process. During the polyamide dyeing process, yellow and brown hues were obtained. The effects of dye concentration and temperature on polyamide dyeability were evaluated, and it was discovered that the concentration and temperature had a significant impact on the colors obtained. High concentrations at 100 °C produced the best results. The rates of washing and rubbing fastness were good to excellent, and the rates of light fastness were fair to good. When mordants are used with extracts, the dyeability and fastness properties are improved. The antibacterial testing of the treated polyamide fabrics led to encouraging results. Further investigations are, however, needed to evaluate the durability of their antibacterial activity after washing and exposure to light.

Author Contributions: Conceptualization, A.B., D.D. and S.G.; methodology, A.B., D.D., S.C. and S.G.; experiments, A.B., D.D., S.G., S.A. and S.C.; validation, C.B., A.K. and S.A.; analysis, A.B., D.D., S.G. and S.A.; investigation, A.B., D.D. and S.G.; resources, C.B. and S.G.; writing—original draft preparation, A.B., D.D. and S.G.; writing—review and editing A.B., D.D., S.G., A.K. and S.A.; supervision, C.B., S.A., A.K. and S.C.; project administration, S.G. and C.B. All authors have read and agreed to the published version of the manuscript.

Funding: This research received no external funding.

Institutional Review Board Statement: Not applicable.

Informed Consent Statement: Not applicable.

Data Availability Statement: Data sharing is not applicable to this article.

Conflicts of Interest: The authors declare no conflict of interest.

References

1. Schieber, A.; Stintzing, F.C.; Carle, R. By-products of plant food processing as a source of functional compounds—Recent developments. *Trends Food Sci. Technol.* **2001**, *12*, 401–413. [CrossRef]
2. Sagar, N.A.; Pareek, S.; Sharma, S.; Yahia, E.M.; Lobo, M.G. Fruit and vegetable waste: Bioactive compounds, their extraction, and possible utilization. *Compr. Rev. Food Sci. Food Saf.* **2018**, *17*, 512–531. [CrossRef] [PubMed]
3. Trigo, J.P.; Alexandre, E.M.; Saraiva, J.A.; Pintado, M.E. High value-added compounds from fruit and vegetable by-products–Characterization, bioactivities, and application in the development of novel food products. *Crit. Rev. Food Sci. Nutr.* **2020**, *60*, 1388–1416. [CrossRef] [PubMed]
4. Đilas, S.; Čanadanović-Brunet, J.; Ćetković, G. By-products of fruits processing as a source of phytochemicals. *Chem. Ind. Chem. Eng. Q. CICEQ* **2009**, *15*, 191–202. [CrossRef]
5. Yahia, E.M.; García-Solís, P.; Celis, M.E.M. Contribution of fruits and vegetables to human nutrition and health. In *Postharvest Physiology and Biochemistry of Fruits and Vegetables*; Elsevier: Amsterdam, The Netherlands, 2019; pp. 19–45.
6. Haddar, W.; Baaka, N.; Meksi, N.; Ticha, M.B.; Guesmi, A.; Mhenni, M.F. Use of ultrasonic energy for enhancing the dyeing performances of polyamide fibers with olive vegetable water. *Fibers Polym.* **2015**, *16*, 1506–1511. [CrossRef]
7. Meksi, N.; Haddar, W.; Hammami, S.; Mhenni, M. Olive mill wastewater: A potential source of natural dyes for textile dyeing. *Ind. Crop. Prod.* **2012**, *40*, 103–109. [CrossRef]
8. Guesmi, A.; Dhahri, H.; Hamadi, N.B. A new approach for studying the dyeability of a multifibers fabric with date pits powders: A specific interest to proteinic fibers. *J. Clean. Prod.* **2016**, *133*, 1–4. [CrossRef]
9. Dos-Santos, M.J.P.L. Value Addition of Agricultural Production to Meet the Sustainable Development Goals. In *Zero Hunger. Encyclopedia of the UN Sustainable Development Goals*; Leal Filho, W., Azul, A.M., Brandli, L., Özuyar, P.G., Wall, T., Eds.; Springer International Publishing: Cham, Switzerland, 2020; pp. 1–8.
10. Gong, K.; Rather, L.J.; Zhou, Q.; Wang, W.; Li, Q. Natural dyeing of merino wool fibers with Cinnamomum camphora leaves extract with mordants of biological origin: A greener approach of textile coloration. *J. Text. Inst.* **2020**, *111*, 1038–1046. [CrossRef]
11. Peran, J.; Ercegović Ražić, S.; Sutlović, A.; Ivanković, T.; Glogar, M.I. Oxygen plasma pretreatment improves dyeing and antimicrobial properties of wool fabric dyed with natural extract from pomegranate peel. *Coloration Technol.* **2020**, *136*, 177–187. [CrossRef]
12. Çam, M.; Hışıl, Y.; Durmaz, G. Classification of eight pomegranate juices based on antioxidant capacity measured by four methods. *Food Chem.* **2009**, *112*, 721–726. [CrossRef]
13. Ozgen, M.; Durgaç, C.; Serçe, S.; Kaya, C. Chemical and antioxidant properties of pomegranate cultivars grown in the Mediterranean region of Turkey. *Food Chem.* **2008**, *111*, 703–706. [CrossRef]
14. Sreekumar, S.; Sithul, H.; Muraleedharan, P.; Azeez, J.M.; Sreeharshan, S. Pomegranate fruit as a rich source of biologically active compounds. *BioMed Res. Int.* **2014**, *2014*. [CrossRef] [PubMed]
15. Dif, M.; Benchiha, H.; Mehdadi, Z.; Benali-Toumi, F.; Benyahia, M.; Bouterfas, K. Étude quantitative des polyphénols dans les différents organes de l'espèce Papaver rhoeas L. *Phytothérapie* **2015**, *13*, 314–319. [CrossRef]
16. Zhishen, J.; Mengcheng, T.; Jianming, W. The determination of flavonoid contents in mulberry and their scavenging effects on superoxide radicals. *Food Chem.* **1999**, *64*, 555–559. [CrossRef]
17. Menezes, E.; Carapelli, R.; Bianchi, S.; Souza, S.; Matos, W.; Pereira-Filho, E.; Nogueira, A. Evaluation of the mineral profile of textile materials using inductively coupled plasma optical emission spectrometry and chemometrics. *J. Hazard. Mater.* **2010**, *182*, 325–330. [CrossRef]
18. Iva Rezić, I.S. ICP-OES determination of metals present in textile materials. *Microchem. J.* **2007**, *85*, 46–51. [CrossRef]
19. Mabry, T.; Markhan, K.; Thomas, M. *The Systemic Identification of Flavonoids*; Springer: New York, NY, USA, 1970; pp. 46–54.
20. Halbwirth, H. The creation and physiological relevance of divergent hydroxylation patterns in the flavonoid pathway. *Int. J. Mol. Sci.* **2010**, *11*, 595–621. [CrossRef]
21. Lokhande, H.; Dorugade, V.A. Dyeing nylon with natural dyes. *Am. Dyest. Report.* **1999**, *88*, 29–34.
22. Li, Y.-D.; Guan, J.-P.; Tang, R.-C.; Qiao, Y.-F. Application of natural flavonoids to impart antioxidant and antibacterial activities to polyamide fiber for health care applications. *Antioxidants* **2019**, *8*, 301. [CrossRef] [PubMed]
23. Tang, R.-C.; Tang, H.; Yang, C. Adsorption isotherms and mordant dyeing properties of tea polyphenols on wool, silk, and nylon. *Ind. Eng. Chem. Res.* **2010**, *49*, 8894–8901. [CrossRef]

Article

Synthesis and Characterization of Metal Complexes Based on Aniline Derivative Schiff Base for Antimicrobial Applications and UV Protection of a Modified Cotton Fabric

S. El-Sayed Saeed [1,*], Tahani M. Al-Harbi [1], Ahmed N. Alhakimi [1,2] and M. M. Abd El-Hady [3,4]

1. Department of Chemistry, College of Science, Qassim University, Buraidah 51452, Saudi Arabia
2. Department of Chemistry, College of Science, Ibb University, Ibb P.O. Box 70270, Yemen
3. National Research Centre, Institute of Textile Research and Technology, 33 El Bohouth Street, Dokki, Cairo P.O. Box 12622, Egypt
4. Department of Physics, College of Science and Arts in Al-Asyah, Qassim University, Buraidah 51452, Saudi Arabia

* Correspondence: s.saeed@qu.edu.sa

Citation: Saeed, S.E.-S.; Al-Harbi, T.M.; Alhakimi, A.N.; Abd El-Hady, M.M. Synthesis and Characterization of Metal Complexes Based on Aniline Derivative Schiff Base for Antimicrobial Applications and UV Protection of a Modified Cotton Fabric. *Coatings* **2022**, *12*, 1181. https://doi.org/10.3390/coatings12081181

Academic Editor: Jiri Militky

Received: 18 July 2022
Accepted: 9 August 2022
Published: 15 August 2022

Publisher's Note: MDPI stays neutral with regard to jurisdictional claims in published maps and institutional affiliations.

Copyright: © 2022 by the authors. Licensee MDPI, Basel, Switzerland. This article is an open access article distributed under the terms and conditions of the Creative Commons Attribution (CC BY) license (https://creativecommons.org/licenses/by/4.0/).

Abstract: Antimicrobial textiles have played an increasingly important protection role in the medical field. With this aim, Schiff bases and nanometal complexes on the cotton fabric were in situ synthesized for achieving the conventional cotton fabric's highly efficient and durable UV protection and antibacterial properties. Herein, a new Schiff base derived from the condensation reaction of 2,4-dihyroxybenzaldehyde with p-amino aniline was synthesized. Co, Ni, Cu, and Zn complexes of the Schiff base were also prepared and characterized by UV-Vis, Fourier-transform infrared spectroscopy, ^{1}HNMR, ^{13}CNMR, elemental analysis, and thermal analysis. The modified cotton fabric was also characterized via X-ray diffraction, Fourier-transform infrared spectroscopy (FTIR), scanning electron microscope (SEM), transition electron microscope (TEM), and Energy Dispersive X-Ray Analysis (EDX). Moreover, the microbial, UV protection, and tensile strength of the samples were investigated. The antimicrobial was studied against Gram-positive bacteria, Gram-negative bacteria, and fungal strains. Modified cotton fabric exhibited highly antibacterial activity in contrast with fungal activity. These results depended on the Schiff base and the type of metal complex. The results also show that the cotton fabric modified by in situ nanometal complexes provides excellent UV protection.

Keywords: cotton fabric; aniline Schiff base; 2,4-dihyroxybenzaldehyde; p-phenylenediamine; antimicrobial activity; UV protection properties

1. Introduction

Cotton is a plentiful natural fiber made up almost entirely of cellulose with hydroxyl functional groups (about 88%–96%). It is one of the most extensively used natural fibers in everyday life due to its high hygroscopicity, soft comfort, and biodegradability. In textile and biomedical engineering, the biopolymeric cotton fabric material offers a variety of benefits. On the other hand, cotton textiles promote the growth of microorganisms, such as bacteria and fungus, which spread diseases, significantly raise the risk of cross-infection, and even damage human health [1]. Hence, the search for the ways of imparting cotton fabrics with antibacterial properties is crucial. Different types of antibacterial agents, such as N-halamine [2], quaternary ammonium salts [3], nanomaterials [4–6], chitosan [7], reduced graphene oxide/silver nanocomplexs [8], and curcumin/titanium dioxide nanocomposites [9], have been developed to add antibacterial activity.

Schiff bases are the condensation products of primary amines with carbonyl compounds (aldehydes and ketones) [10]. Hugo Schiff, a German chemist, first reported Schiff bases in 1864; hence, the name Schiff bases. Schiff bases are sometimes known as azomethines or imines [11]. The common structural feature of Schiff bases is azomethine group linked with substituents (R-C=N-R′), these substituents may be cycloalkyl, alkyl, heterocyclic, or aryl groups [11]. The Schiff base metal complexes' importance has been recognized in the fields of material sciences, bioinorganic chemistry, biomedical applications, and supramolecular chemistry.

Schiff base metal complexes make available compounds naturally and synthetic oxygen carriers [12] and also offer compounds that operate as active stereospecific catalysts in redox, hydrolysis, and conversion reactions in organic and inorganic processes [13]. Metals, in addition to their complexes, have had important applications in medical applications for over 5000 years [14]. Schiff base metal complexes also function as pigments and dyes in polymerization and are used in the pharmaceutical sector. Biologically active Schiff bases moieties have various pharmacological activities [15].

Noticeable antimicrobial activity of Schiff bases containing aniline and phenolic fragments was reported [16]. Schiff base complexes based on benzimidazole derivative and furfural (Fur) or salicylaldehyde have a strong antibacterial and antitumor effect [17]. N-(Salicylidene)-2-hydroxyaniline is a salicylaldehyde Schiff base derivative, that was shown to be a powerful antibacterial agent against Mycobacterium tuberculosis [18]. Recently, reported work proved that Iron (III) and zinc (II) monodentate Schiff base metal complexes have antibacterial activity [19,20]. The phenols and amines play a large effect in medicinal chemistry due to their strong biological action. They are used in different cellulosic applications [21–23] and still attract chemists' and researchers' attention in several areas.

Studies indicate that the antifungal and antibacterial activities of metal complexes are better than their actual Schiff base ligands [24,25]. However, the Schiff base complex was used in the majority of publications, and only a few studies used the Schiff base in fabric finishing. Furthermore, employing a simple approach to insert highly effective functional molecules onto cotton fabric surfaces remains an essential topic for researchers. The previous works reported that azobenzene ring Schiff base treated cellulosic fabric for UV protection properties [26]. Benzyl vanillin Schiff base was used as finishing of polyester fabric and showed highly UPF (ultraviolet protection factor) [27]. Recently, cotton fabric was treated with piperazinyl Schiff for highly antibacterial efficacy [28].

The lack of published research performed on the application of Schiff base complexes for cellulose treatment has motivated us to explore its utilization of it as antimicrobial, and UV protection. In our study, we herein report the synthesis and characterization of Schiff base ligand derived from the condensation reaction of 2,4-dihyroxybenzaldehyde with p-amino aniline. Co, Ni, Cu, and Zn complexes of the ligand were prepared. The characterization of prepared ligand and complexes was analyzed by spectroscopic methods, such as UV-Vis, IR, ^{1}HNMR, ^{13}CNMR, elemental analysis, and thermal analysis.

Two techniques were used for modified cotton fabric with Schiff base and its metal complexes. The first is the modification of cotton fabric by the synthesized Co, Ni, Cu, or Zn (d-block metal) complexes. The second technique is the in situ formation of nanometal Schiff base complexes on the surface of cotton fabric. Furthermore, Schiff bases and their metal complexes were synthesized on cotton fabric for antibacterial finishing and UV protection properties. The tensile strength of the treated samples was also studied.

2. Experimental

2.1. Materials

Misr Company for spinning and weaving provided mill bleached pure 100 percent cotton fabric (138 g/m^2) at Mehalla El-Kobra, Egypt.

2.2. Chemicals

All solvents were purchased from Fisher Scientific, Loughborough, UK. $CuCl_2 \cdot 2H_2O$, $CoCl_2 \cdot 6H_2O$, $NiCl_2 \cdot 6H_2O$, and $ZnCl_2$ dry were purchased from Loba Chemie. 2,4-Dihydroxybenzaldehyde and p-phenylenediamine were purchased from Sigma-Aldrich, St. Louis, MO, USA.

2.3. Preparation of (E)-4-(((4-Aminophenyl)imino)methyl)benzene-1,3-diol (HL) Ligand

Scheme 1 represent the Schiff base ligand (E)-4-(((4-aminophenyl) imino) benzene-1,3-diol (HL). It was prepared by slowly adding of 20 mL ethanolic solution of 0.54 g (5 mmol) of p-phenylenediamine to 20 mL ethanolic solution of 0.69 g (5 mmol) of 2,4 dihydroxy benzaldehyde, and the mixture was stirred until complete dissolution. The mixture was left to stir under reflux for 4 h, during which an orange precipitate was formed. Glacial acetic acid few drops were added at the start of the reflux. The orange precipitate was then filtered and washed several times with distilled water followed by absolute ethanol. The obtained Schiff base was (E)-4-(((4-aminophenyl) imino) methyl) benzene-1,3-diol (HL).

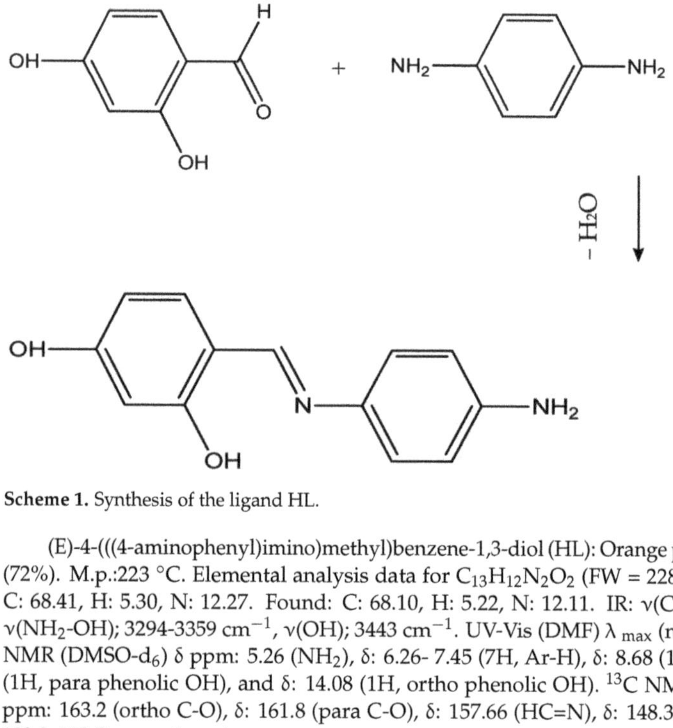

Scheme 1. Synthesis of the ligand HL.

(E)-4-(((4-aminophenyl)imino)methyl)benzene-1,3-diol (HL): Orange precipitate. Yield: (72%). M.p.:223 °C. Elemental analysis data for $C_{13}H_{12}N_2O_2$ (FW = 228.25); Calculated: C: 68.41, H: 5.30, N: 12.27. Found: C: 68.10, H: 5.22, N: 12.11. IR: ν(C=N); 1626 cm^{-1}, ν(NH$_2$-OH); 3294-3359 cm^{-1}, ν(OH); 3443 cm^{-1}. UV-Vis (DMF) λ max (nm): 311, 370. ^{1}H NMR (DMSO-d$_6$) δ ppm: 5.26 (NH$_2$), δ: 6.26- 7.45 (7H, Ar-H), δ: 8.68 (1H, CH), δ: 10.10 (1H, para phenolic OH), and δ: 14.08 (1H, ortho phenolic OH). ^{13}C NMR (DMSO-d$_6$) δ ppm: 163.2 (ortho C-O), δ: 161.8 (para C-O), δ: 157.66 (HC=N), δ: 148.3 (C-NH$_2$), and δ: 102.8–133.7 aromatic carbons.

2.4. Synthesis of Transition Metal Complexes (1–4)

We added 0.457 g (2 mmol) of HL in 30 mL ethanol to metal salt (2 mmol)—namely, CuCl$_2$·2H$_2$O, CoCl$_2$·6H$_2$O, NiCl$_2$·6H$_2$O, or ZnCl$_2$ dry; dissolved in the least amount of bi-distilled water. The mixture was left to stir under reflux for 10 h to ensure complete formation. The precipitate was filtered and washed several times with an ethanol-water mixture of 50% (v/v) to remove any unreacted reactants. Then, the precipitate was dried in anhydrous CaCl$_2$. The obtained complexes are [Co L Cl (H$_2$O)$_2$] (**1**), [Ni L Cl (H$_2$O)$_3$]·H$_2$O (**2**), [Cu L Cl (H$_2$O)$_3$]·H$_2$O (**3**), and [Zn L Cl (H$_2$O)$_2$]·H$_2$O (**4**). The chemical structure of the synthesized complexes is represented in Scheme 2.

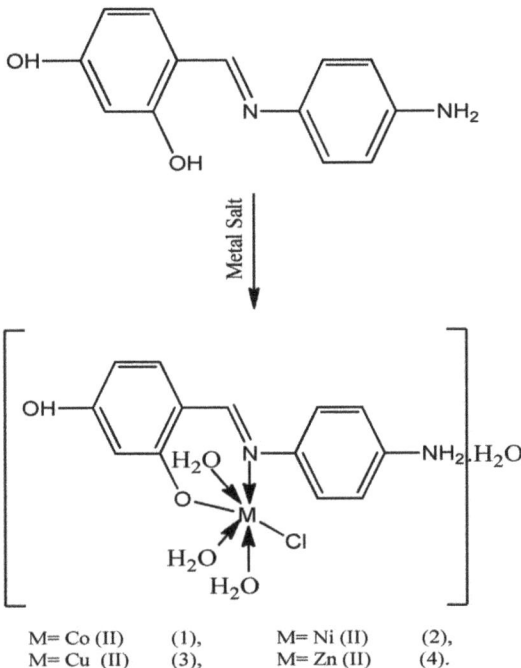

M= Co (II) (1),   M= Ni (II) (2),
M= Cu (II) (3),   M= Zn (II) (4).

Scheme 2. Synthesis of the metal complexes (1–4).

Co (II) Complex (**1**): Dark Brown Precipitate. Yield: (83%). M.p.:270 °C. Elemental analysis data for $C_{13}H_{19}$ Cl Co N_2O_6 (FW = 393.7); Calculated: C: 39.66, H: 4.86, N: 7.12. Found: C: 39.38, H: 4.24, N: 6.98. IR: ν(C=N); 1611 cm^{-1}, ν(NH$_2$-OH); 3142-3223cm^{-1}, ν(M-O); 589 cm^{-1}, ν(M-N); 534 cm^{-1}. UV-Vis (DMF) λ $_{max}$ (nm): 224, 307, 481.

Ni (II) Complex (**2**): Red Brown Precipitate. Yield: (90%). M.p.:254°C. Elemental analysis data for $C_{13}H_{19}$ Cl Ni N_2O_6 (FW = 393.45); Calculated: C: 39.68, H: 4.87, N: 7.12. Found: C: 39.50, H: 4.62, N: 7.09. IR: ν(C=N); 1619 cm^{-1}, ν(NH$_2$-OH); 3223-3251 cm^{-1}, ν(M-O); 592 cm^{-1}, ν(M-N); 541 cm^{-1}. UV-Vis (DMF) λ $_{max}$ (nm): 229, 317, 452.

Cu (II) Complex (**3**): Black Precipitate. Yield: (76%). M.p.: 298 °C. Elemental analysis data for $C_{13}H_{19}$ Cl Cu N_2O_6 (FW= 398.3); Calculated: C:39.20, H:4.81, N:7.03. Found: C:38.99, H:3.24 N:7.01. IR: ν(C=N); 1622 cm^{-1}, ν(NH$_2$- OH); 3166-3250 cm^{-1}, ν(M-O); 591cm^{-1}, ν(M-N); 543 cm^{-1}. UV-Vis (DMF) λ $_{max}$ (nm): 219, 306, 464.

Zn (II) Complex (**4**): Pale Brown Precipitate. Yield: (73%). M.p.: 290 °C. Elemental analysis data for $C_{13}H_{19}$ Cl Zn N_2O_6 (FW = 400.1); Calculated: C: 39.02, H: 4.79, N: 7.00. Found: C: 38.87, H: 4.54, N: 6.82. IR: ν(C=N), 1620 cm^{-1}, ν(NH$_2$- OH); 3263-3300 cm^{-1}, ν(M-O); 585 cm^{-1}, ν(M-N); 523 cm^{-1}. UV-Vis (DMF) λ $_{max}$ (nm):233, 320, 380. ^{1}HNMR (DMSO-d$_6$) δ: 5.29 (NH$_2$), δ: 6.25- 7.44 (7H, Ar-H), δ: 8.66 (1H, CH), δ: 10.07 (1H, para phenolic OH). ^{13}CNMR (DMSO-d$_6$) δ ppm: 164.34 (ortho C-O), δ: 162 (para C-O), δ: 157.63 (HC=N), δ: 148.3 (C-NH$_2$), and δ: 102.8-137 aromatic carbons.

2.5. Coating Techniques

Different two coating techniques were used for modification of the cotton fabric, and these techniques are represented in Figure 1a,b.

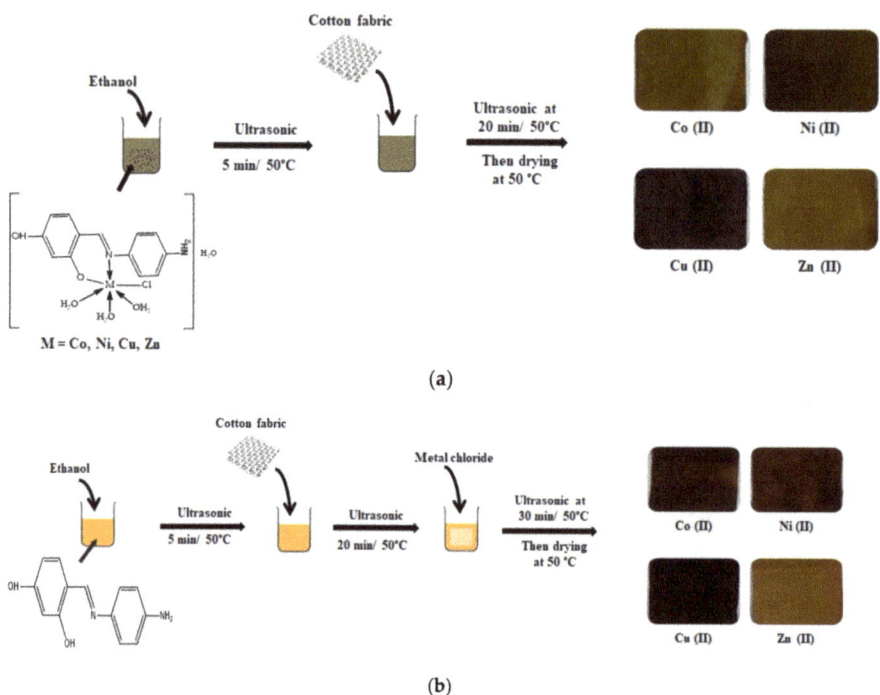

Figure 1. Photographic images of cotton fabric treated with Schiff base metal under two different techniques (a) Technique 1 and (b) Technique 2 (in situ formation).

2.5.1. Coating of Cotton Fabric by the Previously Synthesized Ligand or the Metal Complexes (Technique 1-T1)

As presented in Figure 1a, the mixture of 0.1 g of the synthesized ligand (or the complex) was dissolved in 30 mL of ethanol and sonicated for 5 min at 50 °C. We immersed 1 gm of cotton fabric in the prepared mixture and sonicated for 20 min. Then, we dried the samples at 50 °C for 10 min. Finally, the samples were washed with deionized water several times and dried.

2.5.2. In Situ Formation of Nanometal Complexes Schiff Base Coated Cotton Fabric (Technique 2-T2)

The in situ formation of Schiff base, Co complex is illustrated in Figure 1b; 0.1 g of the synthesized ligand was dissolved in 30 mL of ethanol under sonication for 5 min at 50 °C. Then, 1 gm of cotton fabric was immersed in the previous mixture with continuous sonication for 20 min. After that, 0.1 g of the $CoCl_2 \cdot 6H_2O$ was added to the solution mixture and sonicated for 30 min. Finally, the sample was washed with deionized water several times and dried. The in situ Ni, Co, and Zn complexes were synthesized with the same method.

2.6. Instruments

^{1}H and ^{13}C nuclear magnetic resonance (NMR) spectroscopies for the ligand and its diamagnetic complexes were performed using a Bruker spectrometer (Billerica, MA, USA) at 850 MHz; the used solvent was DMSO, the standard reference was tetramethylsilane, and the temperature of the probe was 25 °C. The FTIR spectra of the ligand, complexes, the cotton fabric samples were measured using an Agilent spectrometer (Cary 600 FTIR, Santa Clara, CA, USA), which was operated in the wavenumber range of 4000–400 cm^{-1}.

A Shimadzu UV-Vis spectrophotometer (UV-1650PC, Kyoto, Japan) was used to measure dimethylformamide (DMF) solutions (1×10^{-3} M) of the ligand and their metal complexes. A Vario EL M, Hanau, Germany was used to measure the CHN contents of the ligand and its complexes. A Shimadzu simultaneous TG apparatus (DTG-60AH, Kyoto, Japan) was used in the air with a heating rate of 10 °C/min; from (room temperature −700 °C) range was used. A Rigaku XRD diffractometer was used to measure the samples' XRD patterns (Ultima IV, Tokyo, Japan; using Cu Kα radiation (λ = 1.54180 Å).

The molar conductivity of (1×10^{-3} M) of samples dissolved in DMSO of the metal complexes at room temperature was measured using an Oakton (CON 700, Singapore) Conductivity meter. The Magnetic Susceptibility Balance—Auto (Sherwood Scientific, Cambridge, UK) was used to measure the magnetic susceptibility of the prepared solid metal complexes at room temperature. The used ultrasonic water base is a Wise Clean ultrasonic bath (WUC-D22H, Wertheim, Germany, frequency of 40 kHz, power input of 300 W).

The studies of electron microscopy were undertaken using JEOL -JEM-1230 transmission electron microscopy (TEM) (with a 40–120 kV accelerating voltage, Tokyo, Japan) and scanning electron microscopy (SEM) (Tescan Vega3) with an attached energy dispersive X-ray spectrometer (EDX) Model vega3 (Brno, Czech Republic). SEM samples were prepared on an appropriate disc and coated with gold to make the samples conductive to electrons.

2.7. Antimicrobial Activity

The biological activities of treated cotton samples were studied for antibacterial and antifungal properties using the disc diffusion method. Different types of bacteria *Staphylococcus aureus* (*S. aureus*) as Gram-positive and *Escherichia coil* (*E. coli*) as Gram-negative were used. *Candida albicans* (*C. albicans*) and *Aspergillus flavus* for fungus. *S.aureus*, *E. coil*, *C. albicans*, and *Aspergillus flavus* originated from ATCC12600, ATCC11775, ATCC 7102, and ATCC 9643, respectively. The antibacterial and antifungal properties were studied by the disc diffusion method.

2.8. Tensile Strength

The ASTM Test Method (D-1682-94, 1994) was used to determine the tensile strength of the cotton samples.

2.9. The Add-On (%) Loading

The add-on (%) loading was calculated as follows:

$$\text{Add} - \text{on}(\%) = \frac{W_2 - W_1}{W_1} \times 100 \tag{1}$$

where W_1 and W_2 are the weights of the fabric specimens before and after treatment, respectively.

2.10. UV Protection Factor

Ultraviolet protection factor (UPF) was measured using UV Shimadzu 3101 Spectrophotometer. UV protection and classification according to AS/NZS 4399:1996 were evaluated with a scan range was of 200–600 nm.

2.11. Durability Test

To evaluate the UPF protection values' durability to washing, the treated cotton samples were subjected to ten laundry cycles according to the ASTM standard test procedure (D 737-109 96).

2.12. Statistical Analysis

The results of add-on% and tensile strength values were expressed as follows: the mean of the repeating of each sample three times (n = 3) with its standard deviation (the mean ± S.D.).

3. Result and Discussion

3.1. Ligand and Its Metal Complexes Characterization

The Supplementary Materials describe the detailed characterization of the ligand metal complexes and the energy dispersive X-ray analysis of the modified cotton. This section will discuss the main characteristics of the complexes and their interactions with the cellulosic fibers.

UV-Visible Spectroscopy and Magnetic Susceptibility

Table 1 represent UV-visible spectroscopy and magnetic susceptibility of Schiff base ligand and its complexes. The electronic spectra of the Schiff base ligand and its complexes were recorded in dimethylformamide (DMF) solvent within the wavelength range of 200–700 nm. The electronic spectrum of the ligand is characterized by its absorption bands at 311 and 370 nm. The highest energy band (lower wavelength) is assigned to π–π^* transitions, while the lowest energy bands can be assigned to n–π^* transitions [29,30].

Table 1. UV–Vis spectral data of the ligand and its metal complexes and magnetic moment.

Comp. No.	λ_{max} (nm)	Wavenumber (cm^{-1})	Assignment	µeff (BM)
HL	311	32,154	$\pi \to \pi^*$	-
	370	27,027	n $\to \pi^*$	
Co (II) complex	224	44,642	$\pi \to \pi^*$	3.99
	307	32,573	n $\to \pi^*$	
	481	20,790	$^4T_{1g} \to {}^4T_{2g}$ (P)	
Ni (II) Complex	229	43,668	$\pi \to \pi^*$	2.7
	317	31,545	n $\to \pi^*$	
	452	22,124	$^3A_{2g} \to {}^3T_{2g}$	
Cu (II) Complex	219	45,662	$\pi \to \pi^*$	1.60
	306	32,679	n $\to \pi^*$	
	464	21,552	$^2B_1 \to {}^2Eg$	
Zn (II) Complex	233	42,918	$\pi \to \pi^*$	Dia
	320	31,250	n $\to \pi^*$	
	380	26,315	MLCT	

As illustrated in Figure 2, Co (II) complex (**1**) shows an absorption spectrum at 481 nm. as a result of d-d transition for $^4T_{1g} \to {}^4T_{2g}$ (P) [31]. The magnetic moment value of Co (II) complex equals 3.99 B.M. the low value may be due antiferromagnetic spin–spin interaction between cobalt (II) ions (d^7) through molecular association, which is in agreement with the values of Co (II) octahedral geometry [30–32].

The absorption spectrum of Ni (II) complex (**2**) shows a band at 452 nm. as a result of d-d transition for $^3A_{2g} \to {}^3T_{2g}$ [33]. The magnetic moment value of Ni (II) complex 2.7 BM confirms the presence of two unpaired electrons in the octahedral geometry [34].

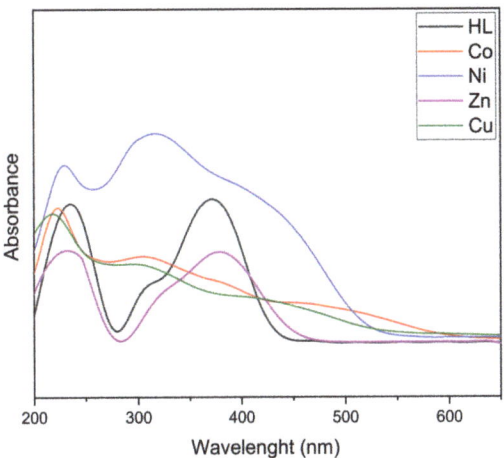

Figure 2. UV-Vis spectrum of the ligand and its metal complexes.

The electronic absorption spectrum of Cu (II) complex (3) shows two bands at 464 and 620 nm. The first is assigned to the $^2B_1 \rightarrow {}^2Eg$ transition, while the second is very weak and broad centered at 620 nm. The position and the broadness of this band indicated to tetragonally distorted octahedral geometry around the copper (II) ion. This broad band may consist of three superimposed transitions $^2B_{1g} \rightarrow {}^2A_{1g}$, $^2B_{1g} \rightarrow {}^2A_{1g}$, and $^2B_{1g} \rightarrow {}^2B_{2g}$ [16,20]. The magnetic moment value of Cu (II) complex is 1.60 BM, which suggests an octahedral geometry around the Cu (II) [35].

The electronic absorption spectrum of Zn (II) complex (4) shows a band at 380 nm, assignable to the metal to ligand charge transfer (MLCT) transition. Finally, the Zn (II) complex showed a diamagnetic value [35].

3.2. Characterizations of Cotton Fabric

3.2.1. Fourier Transform Infrared Spectroscopy (FTIR) Spectra

The FTIR spectrum gives an idea about the functional groups present in the macromolecules before and after the structural modification reaction. Figure 3 indicates the FTIR spectrum of modified cotton fabric by ligand and its metal complexes via technique 1 and technique 2. Figure 3 (T2, T1) indicates the FTIR spectrum of blank cotton fabric. A broad band appears around 3320 cm^{-1}, which can be attributed to (O–H) stretching. The existence of (C–H), (C–O), (O–H), and (C–O–C) vibrations produced the characteristic bands in the range of 1500–800 cm^{-1}. The C- H symmetric and antisymmetric stretching was observed at 2843 and 2904 cm^{-1}, respectively [36,37].

Furthermore, the spectra of all treated cotton fabrics showed distinctive peaks related to cellulose structure, as well as the addition of absorption peaks. According to Figure 3 (T2, T1), the stretching vibrations (C=N), and (N-M) (Nitrogen-Metal) were obtained for the structural conformation of ligands and their respective metal complexes modified cotton fabric. The FTIR bands were found to be slightly shifted for modified fabrics as indicated by the stretching data of functional groups of the complexes mentioned in Table S2 (Supplementary Materials).

In Figure 3 (T2), the band of (C=N) for ligand appeared at 1645 cm^{-1}, which is a distinctive characteristic feature of Schiff bases. After complexation, the peak of (C=N) was shifted towards the lower frequency, i.e., 1621 cm^{-1} for Co (II) and Ni(II), 1625 cm^{-1} for Zn(II) and Co (II) [38]. It was noticed from Figure 3 (T2), that the broader O-H band for cotton-modified ligand and their nano metal complexes.

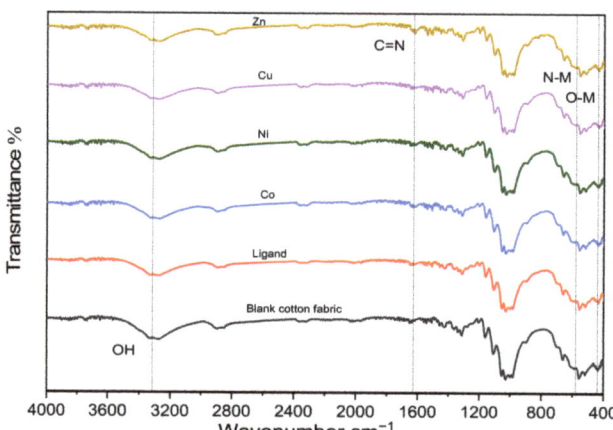

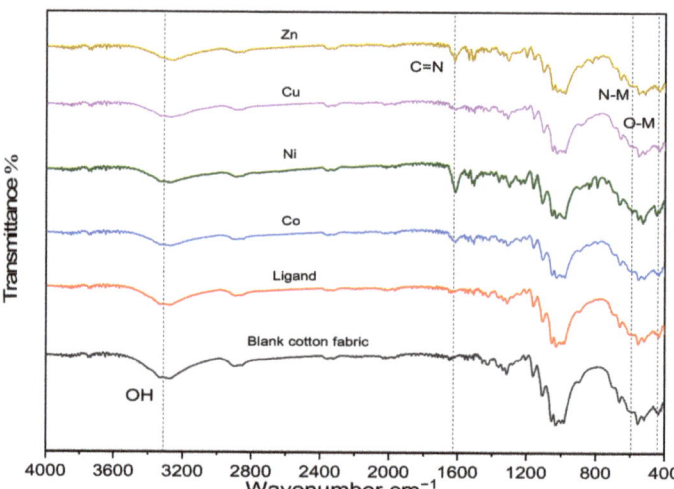

Figure 3. FTIR spectrum of unmodified and modified cotton fabric with ligand and their metal complexes via, (T2) technique 2 (in situ) and (T1) technique 1.

This may be due to the intermolecular hydrogen bond between terminal NH_2 and OH groups of Schiff base and primary (OH) groups of cellulosic cotton fabric [39] as shown in Figure 4. The weak peaks at 590 and 455 cm^{-1} are slightly shifted comparing with the blank cotton fabric are attributed to interactions between N-M (Nitrogen-Metal) and O-M (Oxygen-Metal) suggesting that coordination has occurred during the modification of cotton fabric [40]. In a similar way to technique 2, the same absorption peaks were observed in technique 1 but with weaker intensity than technique 1.

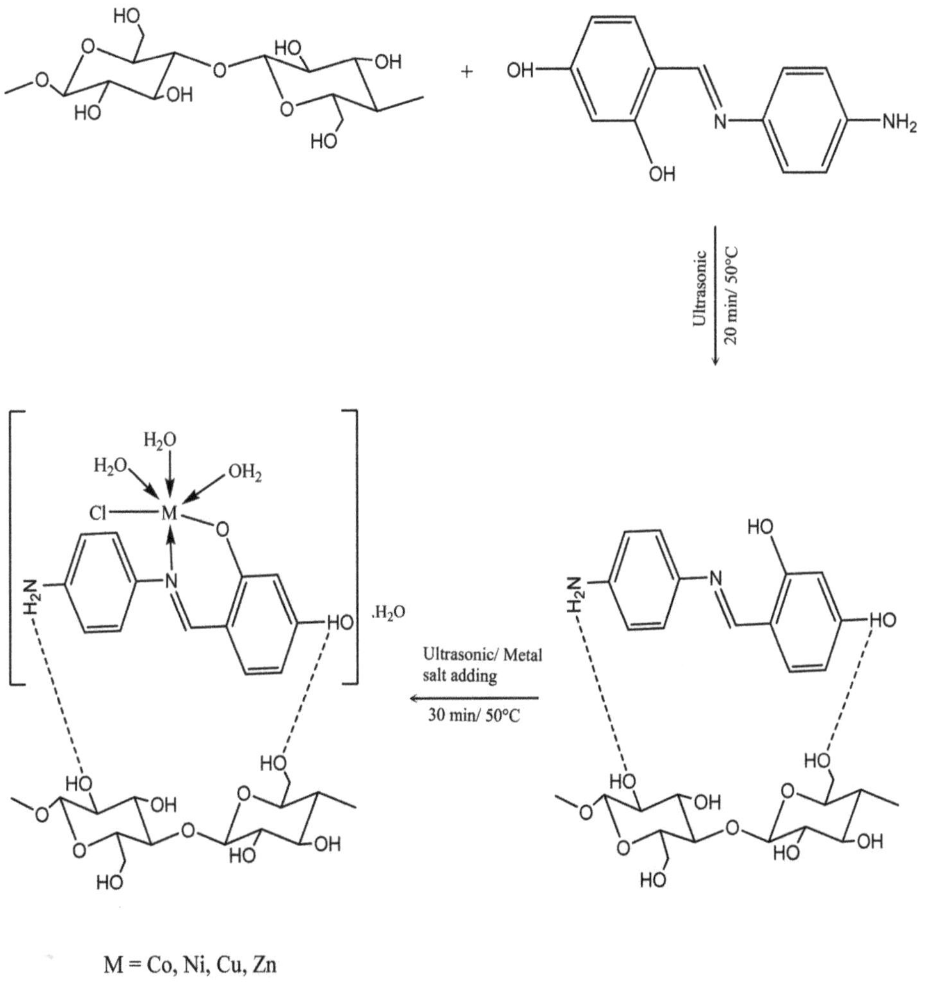

Figure 4. Schematic mechanism for deposition of Schiff base ligand metal complex on the surface of cotton fabric.

3.2.2. XRD of the Modified Cotton

The crystalline structure of the unmodified/modified cotton fabrics was measured by XRD diffractometer. The XRD of fabric samples of the two techniques are represented in Figure 4. The diffraction peaks are detected at 2θ values of 15.2°, 16.7°, 23.1°, and 34.7° related to the cellulose crystalline structure of all the samples [41].

XRD of the modified cotton fabric in the case of in situ formation (T2), and the fabric coated with the previously synthesized complexes (T1) was investigated. The appearance of the new peaks at 2θ value around 21°. As shown in Figure 5, the 2θ peak values in situ (L = 21°, Co = 20.6°, Ni = 20.8°, Cu = 20.7°, and Zn = 20.8°). The missing of this peak in the blank cotton fabric can be an indication of the successful interaction between the cellulose chain of cotton fabric with the ligand and its metal complexes.

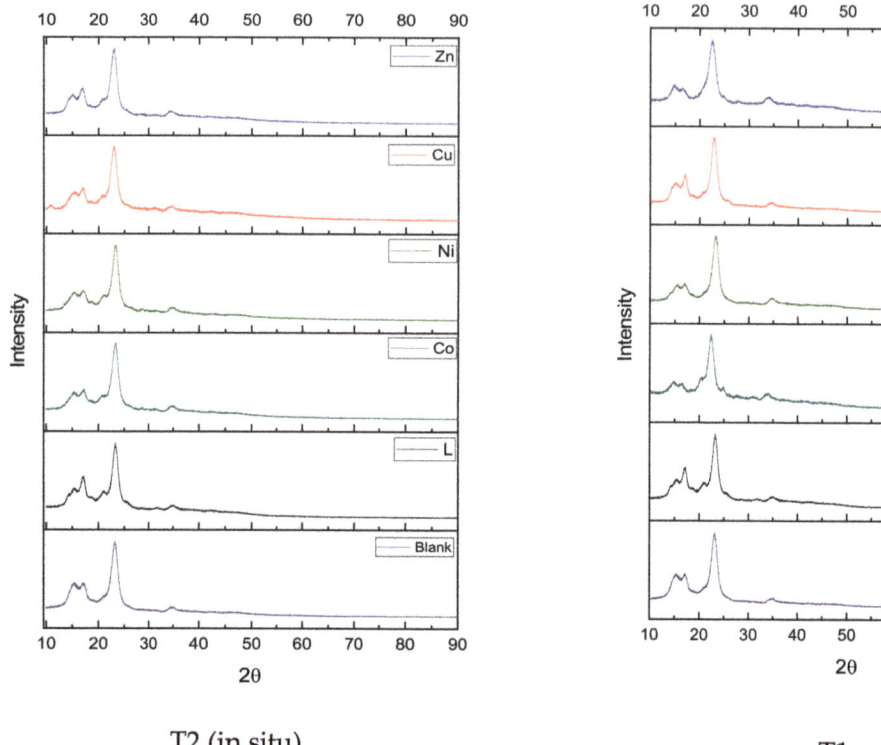

Figure 5. XRD analysis of unmodified and modified cotton fabric with ligand and their metal complexes via, (T2) technique 2 and (T1) technique 1.

The peak of 2θ value around 21° in technique 2 (in situ) is more intense than in technique 1, and this may be due to the nano size causing more cutting of the cellulose crystalline structure. This interaction between cellulose and ligand or its metal complexes is due to the hydrogen bond formation. The interaction was formed by cutting off the cellulosic intermolecular hydrogen bond and formation of a new hydrogen bond between N and O (terminal –OH and–NH$_2$ of ligand or its metal complexes) and the hydrogen atom of the primary –OH groups of the cellulosic fabric [39,42,43].

3.2.3. SEM of Nanometal Complex Modified Cotton Fabric:

SEM analysis of the cotton fabrics was used to characterize the changes in the surface morphology of the surface of cotton fabric [44]. Figure 6a–f are the SEM of the blank cotton fabric, ligand, and nanometal complexes Schiff base treated cotton fabric, respectively. In Figure 6a, the blank cotton exhibits a smooth and flat surface structure. In contrast, the surface of cotton fabric coated with ligand in Figure 6b has fine particles and the surface represents more rough. In comparing to figures (a and b), figures (c, d, e, and f) represented the increase in surface roughness on the surface of the coated cotton fabrics. This result is evidenced by the successful deposition of all metal-complexes Schiff base on the surface of cotton fabrics.

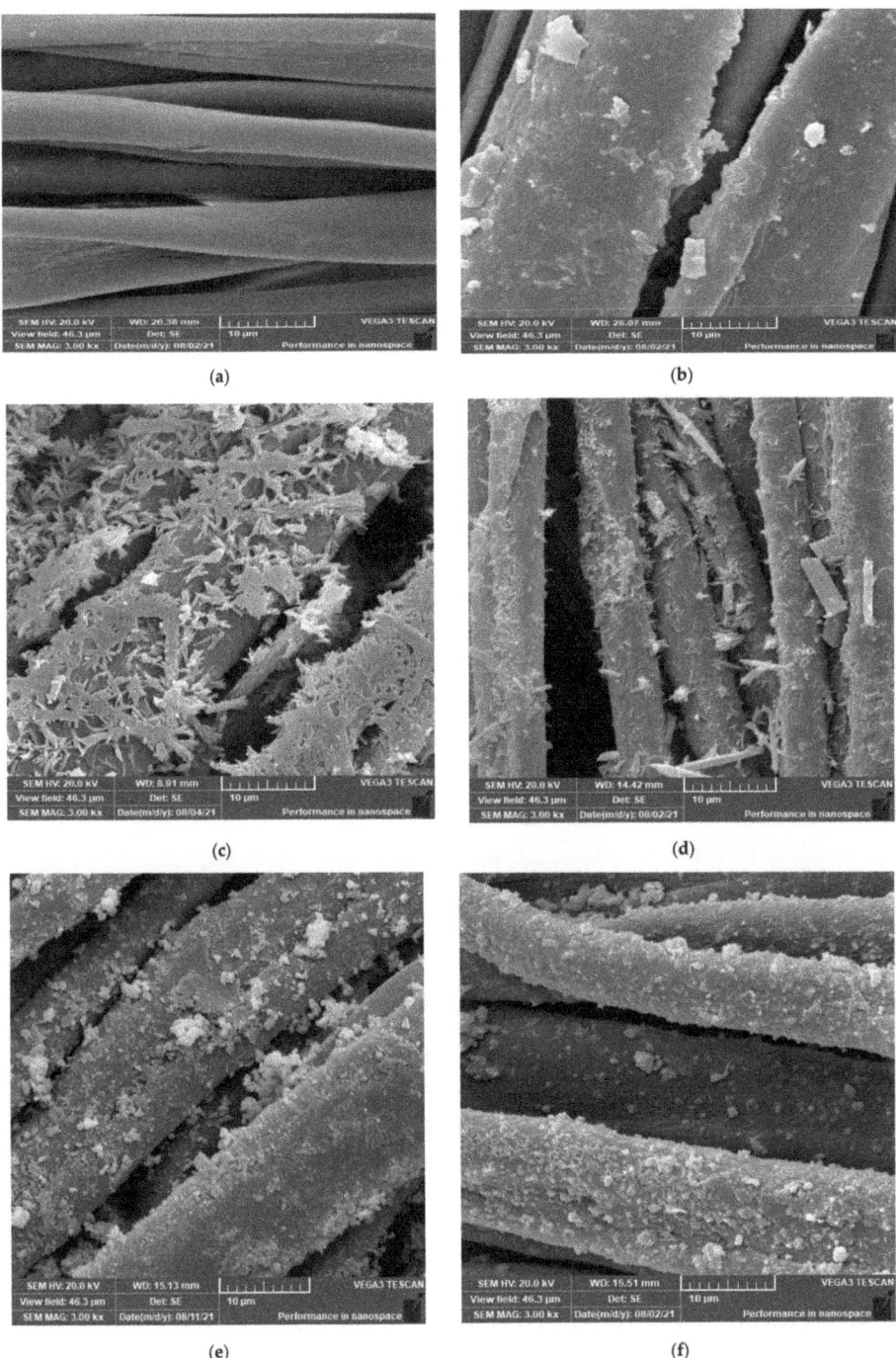

Figure 6. SEM images of (**a**) blank cotton fabric, (**b**) ligand modified cotton fabric, (**c**) Co-complex Schiff-base-modified cotton fabric, (**d**) Ni-complex Schiff-base-modified cotton fabric, (**e**) Cu-complex Schiff-base-modified cotton fabric, and (**f**) Zn-complex Schiff-base-modified cotton fabric.

3.2.4. Antimicrobial Properties

Antimicrobial activities of synthesized HL modified cotton fabric and its metal Nano complexes modified cotton fabric via technique 1 and technique 2 were studied for antibacterial and antifungal properties by disc diffusion method. The results were recorded by measuring the growth inhibition (zone of inhibition) surrounding the disc of the fabric. The result mentioned in Table 2 and Figure 7 indicated the antimicrobial effect for all treatments is ranging from 13 to 31 mm of a clear zone of inhibition depending on the type of metal complex used.

Table 2. Antimicrobial efficiency of unmodified cotton fabric, ligand, and ligand metal.

Complexes	Bacterial Species Inhibition Zone (mm)				Fungal Species Inhibition Zone (mm)			
	G+		G−					
	S. aureus		E. coil		Candida albicans		Aspergillus flavus	
	Technique 2	Technique 1	Technique 2	Technique 1	Technique 2	Technique 1	Technique 2	Technique 1
Unmodified Cotton Fabric	0	0	0	0	0	0	0	0
Ligand Modified Cotton Fabric	10	10	9	9	0	0	0	0
Co Complex- Modified Cotton Fabric	26	13	31	0.0	17	0	0	0
Ni Complex Modified Cotton Fabric	28	16	21	15	27	19	0	0
Cu Complex Modified Cotton Fabric	23	18	19	13	13	0	0	0
Zn Complex Modified Cotton Fabric	22	18	22	18	-	0	0	0

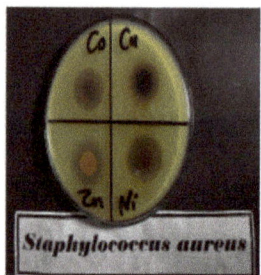

Cotton modified of Co(II), Ni(II), Cu(II), and Zn(II) complexes activity against S. aureus. (technique 2)

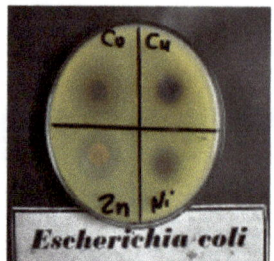

Cotton modified of Co(II), Ni(II), Cu(II), and Zn(II) complexes activity against E. coli. (technique 2).

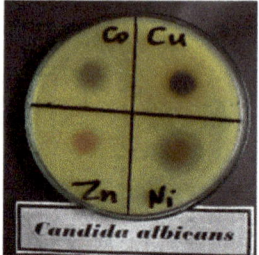

Cotton modified of Co(II), Ni(II), Cu(II), and Zn(II) complexes activity against Candida albicans. (technique 2)

(a)

Figure 7. Cont.

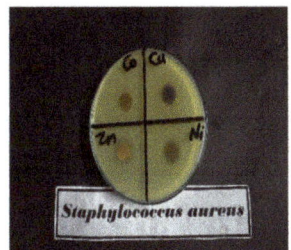 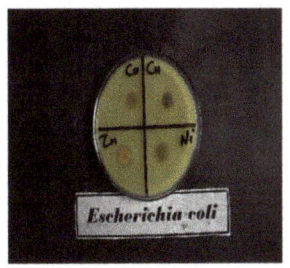 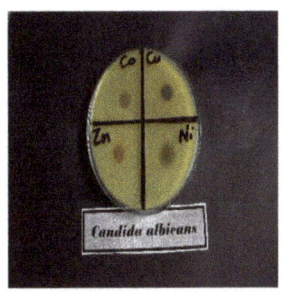

Cotton modified of Co(II), Ni(II), Cu(II), and Zn(II) complexes activity against *S. aureus*. (technique 1)

Cotton modified of Co(II), Ni(II), Cu(II), and Zn(II) complexes activity against *E. coli*. (technique 1)

Cotton modified of Co(II), Ni(II), Cu(II), and Zn(II) complexes activity *against Candida albicans*. (technique 1)

(**b**)

Figure 7. Images of antimicrobial results of modified cotton fabric by nanometal complexes of Co(II), Ni(II), Cu(II), and Zn(II) (**a**) Technique 2 and (**b**) Technique 1.

Cotton modified ligand (HL) has a weak effect on bacteria species and is not effective on fungus species. Ligand metal complexes modified cotton fabric (from technique 1) show moderate effect against *S. aureus* and *E. coli* bacteria except for Co (II) complex modified cotton fabric has no effect. This may be due to variations in bacterial cell wall organization structure. On the other hand, all the metal complexes modified cotton fabrics have no effect against both *Candida albicans* and *Aspergillus flavus* except Ni (II) complex modified cotton fabric.

In comparing with technique 1, nanometal complexes modified cotton fabric (from technique 2) show high activity against both bacteria *S. aureus* and *E. coli*. In addition, Co (II) and Ni (II) nanometal complexes modified cotton fabric show good efficiency against *Candida albicans* but Cu (II) nanometal complexes give weak activity and Zn (II) has no effect. All nanometal complexes modified cotton fabric show no effect against *Aspergillus flavus*.

The higher inhibition zone of ligand metal complex and nanometal complexes modified cotton fabric may be due to using the metal chloride and nanosize effect. It also can be described according to the chelation theory and Overton's concept. Overton's concept of cell permeability described that the lipid membrane surrounding the cell favors the passage of only lipid-soluble materials due to which liposolubility is an important factor that controls the antimicrobial activity.

According to chelation theory, when metal ion chelates with its ligand, the polarity will be lowered to a higher extent because of the ligand orbital overlapping with the partial sharing of the metal ion positive charge with the donor groups. This also increases the delocalization of p-electrons across the whole chelate ring, which increases the lipophilicity of metal complexes. This increased lipophilicity enhances the penetration of complexes into lipid membrane and restricts the further multiplicity of microorganisms. The metal complexes also affect the respiration procedure of the cell; hence, they block the synthesis of proteins, which prevents the further growth of organisms [45,46].

The effectiveness variation of different compounds against different microorganisms is determined by the impermeability of microorganisms' cells or by differences in ribosome of microbial cells. The results also depict that nanometal complexes modified cotton fabric give higher activity against bacteria and fungus compared with metal–ligand complexes coated cotton fabric. As illustrated from XRD and TEM analysis, all the complexes' crystal sizes are in the nano-domain. This nano character of the prepared complexes increased the antimicrobial activity via facilitating penetration of nano-complexes into the microbial cell [47]. The results of antibacterial demonstrated that the modified cotton fabric with nanometal complexes will have potential applications in biomaterial and textile fields.

3.2.5. UV Blocking

UPF values were measured to determine the UV-radiation protection characteristics of untreated cotton fabrics and nano metal complexes modified fabrics. Three types of protection can be found in textile materials, according to BS EN 13758-2: 2003: the excellent protection (UPF range > 40), very good (UPF range 30–40), and good protection (UPF range 20–29) [37]. According to the results in Table 3, the calculated UPF value of unmodified cotton fabric is 4.5. The calculated UPF value for nano complexes modified cotton fabric is varied from 93.5 to 507.5, which is higher than for the unmodified fabric.

Table 3. UPF values of cotton fabric modified with ligand and its nanometal complexes.

Treatment	UPF Value		UV-A		UV-B		UV Protection	
No. of washing cycle	1	10	1	10	1	10	1	10
Unmodified cotton fabric	4.5	4.1	26	28	18.8	20	Non-ratable	Non-ratable
Ligand modified cotton fabric	117.5	115.3	0.6	0.64	0.9	0.95	Excellent	Excellent
Co-complex modified cotton fabric	128.5	125.01	0.7	0.73	0.8	0.86	Excellent	Excellent
Ni-complex modified cotton fabric	217.5	211.6	0.4	0.45	0.5	0.55	Excellent	Excellent
Cu-complex modified cotton fabric	507.5	495	0.2	0.22	0.2	0.21	Excellent	Excellent
Zn-complex modified cotton fabric	93.5	88	0.9	0.97	1.1	1.5	Excellent	Excellent

The ligand modified cotton fabric has a 117.5 UPF value. On the other hand, there is a significant increase in UPF values after the formation of nanometal complexes on cotton fabric. It was noticed from the results in Table 3 that the value of UPF is varied according to the type of metal used. The values of UPF of nanometal complexes modified cotton fabric follow the order: Cu (II) > Ni (II) > Co (II) > Zn (II). The improving performance of ligand modified cotton fabric and its nanometal complexes modified fabric could be attributed to $\pi-\pi^*$ and $n-\pi^*$ transitions of the conjugated system [26].

The data of Table 3 shows the durability of the product during washing cycles. According to the obtained results, raising the number of washing cycles to 10 caused a small decrease in the UPF values of the washed modified fabrics, and the fabric maintained excellent rating. By this, we still have the protective property after several washing cycles. This confirms the strong bending of nanometal complexes based on Schiff-base-modified cotton fabric.

3.2.6. The Add-On (%) Loading and Tensile Strength

Table 4 shows the percentage of the values for add-on measurements and the mechanical properties of chemically modified cotton fabric via two techniques used. The amount of chemicals deposited on the cotton fabric during modification is indicated by the add-on values. As presented in Table 4, the add-on values via technique 2 are higher than the values via technique 1. The results present that the add–on values for modified cotton fabric with the ligand is 2.01%, whereas a significant increase in add-on values varied from 3.71% to 6.99% for nanometal complexes modified fabrics via technique 2. Table 4 also demonstrated the add-on values varied from 0.94% to 2.63% for metal complexes modified cotton fabrics via technique 1. In contrast, Table 4 shows that there is a significant decrease in the values of tensile strength for both techniques are used. This may be attributed to the varied modification of cotton fabric.

Table 4. Add–on measurements and tensile strength of the treated cotton fabric.

Treatment	Add On (%)		Tensile Strength (Newton)	
	Technique 2	Technique 1	Technique 2	Technique 1
Unmodified cotton fabric	0	0	539 ± 1.04	539 ± 1.04
Ligand modified cotton fabric	2.01 ± 0.05	-	508 ± 0.2	-
Co-complex modified cotton fabric	4.02 ± 0.2	0.94 ± 0.2	484 ± 0.2	498 ± 0.3
Ni-complex modified cotton fabric	6.87 ± 0.3	1.59 ± 0.3	471 ± 0.3	493 ± 0.3
Cu-complex modified cotton fabric	6.99 ± 0.2	1.34 ± 0.2	469 ± 0.2	521 ± 0.2
Zn-complex modified cotton fabric	3.71 ± 0.3	2.63 ± 0.3	485 ± 0.3	494 ± 0.2

4. Conclusions

Co(II), Ni(II), Cu(II), and Zn(II) complexes from Schiff base (E)-4-(((4-aminophenyl)imino)methyl)benzene-1,3-diol (HL) were successfully prepared. The structural features of metal complexes have been proven. The ligand acts as monobasic bidentate. The stoichiometry of the metal complexes is a 1:1 ratio with the general formula [M L Cl (H$_2$O)$_3$] H$_2$O. The magnetic susceptibility and UV-visible spectroscopy results support that Co(II), Ni(II), Cu(II), and Zn(II) complexes have octahedral geometry. The synthesized complexes from the Schiff base were also successfully deposited on the surface of cotton fabric by the different two techniques of modification. Technique 2 (the in situ formation of nano complexes) showed high efficiency against antimicrobial activity and durable UV protection properties compared with technique 1.

The study revealed that the modification of the cotton with metal complexes caused enhancement in antimicrobial properties with simultaneous improvement in the UV protective properties of the fabrics. The increased antibacterial properties of the modified fabrics may be attributed to the effect of the metal chelation theory. The modified cotton fabric showed acceptable antifungal against *Candia albicans* especially in the case of Co (II) and Ni (II) complexes. The improvement of the UV protection values as a result to π–π* and n–π* transition of the ligand and its metal complexes.

After the washing cycles, all the modified cotton fabric with metal complexes maintained UV protection properties. The results revealed that Schiff base divalent Co, Ni, Cu, and Zn complexes are multifunctional textile finishes that could be promising as industrial textile products.

Supplementary Materials: The following supporting information can be downloaded at: https://www.mdpi.com/article/10.3390/coatings12081181/s1, Figure S1: ^{1}H NMR of Ligand (a) and Zn (II) complex (b); Figure S2: ^{13}C NMR of Ligand (a) and Zn (II) complex (b); Figure S3. IR analysis of the ligand and its complexes. Figure S4. S: XRD pattern of the ligand and its metal complexes. Figure S5. TEM images of nano complexes (a) Co(II), (b) Ni(II), (c) Cu(II), and (d) Zn(II). Figure S6. EDX analysis of unmodified and modified cotton fabric (a) blank cotton fabric, (b) ligand modified cotton fabric, (c) Co-complex Schiff base modified cotton fabric, (d) Ni-complex Schiff base modified cotton fabric, (e) Cu-complex Schiff base modified cotton fabric, and (f) Zn-complex Schiff base modified cotton fabric. Table S1. Analytical data of ligand and metal complexes; Table S2. FTIR spectral data of the ligand and its metal complexes; Table S3. TGA data of the metal complexes; Table S4. The unit cell parameters and crystal data of the ligand and metal complex.

Author Contributions: Conceptualization, S.E.-S.S., M.M.A.E.-H. and A.N.A.; methodology, S.E.-S.S. and M.M.A.E.-H.; software, S.E.-S.S., M.M.A.E.-H. and T.M.A.-H.; validation, S.E.-S.S., M.M.A.E.-H., T.M.A.-H. and A.N.A.; formal analysis, S.E.-S.S., M.M.A.E.-H. and T.M.A.-H.; investigation, S.E.-S.S., M.M.A.E.-H. and T.M.A.-H.; resources, S.E.-S.S., M.M.A.E.-H., T.M.A.-H. and A.N.A.; data curation, S.E.-S.S., M.M.A.E.-H. and T.M.A.-H.; writing—original draft preparation, S.E.-S.S., M.M.A.E.-H. and T.M.A.-H.; writing—review and editing, S.E.-S.S., M.M.A.E.-H. and T.M.A.-H.; visualization, S.E.-S.S., M.M.A.E.-H., T.M.A.-H. and A.N.A.; supervision, S.E.-S.S., M.M.A.E.-H. All authors have read and agreed to the published version of the manuscript.

Funding: This research received no external funding.

Institutional Review Board Statement: Not applicable.

Informed Consent Statement: Not applicable.

Data Availability Statement: The data used to support the findings of this study are included within the article.

Conflicts of Interest: The authors declare no conflict of interest.

References

1. Küçük, M.; Öveçoğlu, M.L. Fabrication of SiO$_2$–ZnO NP/ZnO NR hybrid coated cotton fabrics: The effect of ZnO NR growth time on structural and UV protection characteristics. *Cellulose* **2020**, *27*, 1773–1793. [CrossRef]
2. Chen, W.; Zhu, Y.; Zhang, Z.; Gao, Y.; Liu, W.; Borjihan, Q.; Qu, H.; Zhang, Y.; Zhang, Y.; Wang, Y.-J.; et al. Engineering a multifunctional N-halamine-based antibacterial hydrogel using a super-convenient strategy for infected skin defect therapy. *Chem. Eng. J.* **2020**, *379*, 122238. [CrossRef]
3. Liu, J.; Dong, C.; Wei, D.; Zhang, Z.; Xie, W.; Li, Q.; Lu, Z. Multifunctional Antibacterial and Hydrophobic Cotton Fabrics Treated with Cyclic Polysiloxane Quaternary Ammonium Salt. *Fibers Polym.* **2019**, *20*, 1368–1374. [CrossRef]
4. Shirvan, A.R.; Kordjazi, S.; Bashari, A. Environmentally friendly Finishing of Cotton Fabric via Star-like Silver micro/nano Particles Synthesized with Neem/Salep. *J. Nat. Fibers* **2021**, *18*, 1472–1480. [CrossRef]
5. Sharaf, S.; Farouk, A.; El-Hady, M. Novel conductive textile fabric based on polyaniline and CuO nanoparticles. *Int. J. PharmTech Res.* **2016**, *9*, 461–472.
6. Abd El-Hady, M.M.; Farouk, A.; Sharaf, S. Multiwalled-Carbon-Nanotubes (MWCNTs)–GPTMS/Tannic-Acid-Nanocomposite-Coated Cotton Fabric for Sustainable Antibacterial Properties and Electrical Conductivity. *Coatings* **2022**, *12*, 178. [CrossRef]
7. Abd El-Hady, M.M.; Saeed, S.E.S. Antibacterial properties and pH sensitive swelling of insitu formed silver-curcumin nanocomposite based chitosan hydrogel. *Polymers* **2020**, *12*, 2451. [CrossRef] [PubMed]
8. Farouk, A.; Saeed, S.E.-S.; Sharaf, S.; El-Hady, M.M.A. Photocatalytic activity and antibacterial properties of linen fabric using reduced graphene oxide/silver nanocomposite. *RSC Adv.* **2020**, *10*, 41600–41611. [CrossRef] [PubMed]
9. Abd El-Hady, M.M.; Farouk, A.; Saeed, S.E.S.; Zaghloul, S. Antibacterial and UV Protection Properties of Modified Cotton Fabric Using a Curcumin/TiO$_2$ Nanocomposite for Medical Textile Applications. *Polymers* **2021**, *13*, 4027. [CrossRef] [PubMed]
10. Laddha, P.R.; Biyani, K.R. Synthesis and Biological Evaluation of Novel Schiff Bases of Aryloxy Moiety. *J. Drug Deliv. Ther.* **2019**, *9*, 44–49. [CrossRef]
11. Raczuk, E.; Dmochowska, B.; Samaszko-Fiertek, J.; Madaj, J. Different Schiff Bases—Structure, Importance and Classification. *Molecules* **2022**, *27*, 787. [CrossRef] [PubMed]
12. Thangadurai, T.D.; Gowri, M.; Natarajan, K. Synthesis and characterization of ruthenium (III) complexes containing monobasic bidentate Schiff bases and their biological activities. *Synth. React. Inorg. Met. Org. Chem.* **2002**, *32*, 329–344. [CrossRef]
13. Ramesh, R.; Sivagamasundari, M. Synthesis, spectral and antifungal activity of Ru (II) mixed-ligand complexes. *Synth. React. Inorg. Met. Org. Chem.* **2003**, *33*, 899–910. [CrossRef]
14. Shahid, K.; Ali, S.; Shahzadi, S.; Badshah, A.; Khan, K.M.; Maharvi, G.M. Organotin (IV) complexes of aniline derivatives. I. Synthesis, spectral and antibacterial studies of di-and triorganotin (IV) derivatives of 4-bromomaleanilic acid. *Synth. React. Inorg. Met. Org. Chem.* **2003**, *33*, 1221–1235. [CrossRef]
15. Rauf, A.; Shah, A.; Munawar, K.S.; Ali, S.; Tahir, M.N.; Javed, M.; Khan, A.M. Synthesis, physicochemical elucidation, biological screening and molecular docking studies of a Schiff base and its metal (II) complexes. *Arab. J. Chem.* **2020**, *13*, 1130–1141. [CrossRef]
16. Alorini, T.A.; Al-Hakimi, A.N.; Saeed, S.E.-S.; Alhamzi, E.H.L.; Albadri, A.E. Synthesis, characterization, and anticancer activity of some metal complexes with a new Schiff base ligand. *Arab. J. Chem.* **2020**, *15*, 103559. [CrossRef]
17. Saeed, S.E.S.; Alhakimi, A.N. Synthesis, characterization of Lanthanum mixed ligand complexes based on benzimidazole derivative and the effect of the added ligand on the antimicrobial, and anticancer activities. *J. Nat. Sci. Math. (JNSM)* **2022**, *15*, 35–53.
18. da Silva, C.M.; da Silva, D.L.; Modolo, L.; Alves, R.B.; de Resende, M.A.; Martins, C.V.; de Fátima, Â. Schiff bases: A short review of their antimicrobial activities. *J. Adv. Res.* **2011**, *2*, 1–8. [CrossRef]
19. Naureen, B.; Miana, G.A.; Shahid, K.; Asghar, M.; Tanveer, S.; Sarwar, A. Iron (III) and zinc (II) monodentate Schiff base metal complexes: Synthesis, characterisation and biological activities. *J. Mol. Struct.* **2021**, *1231*, 129946. [CrossRef]
20. Fouda, M.F.R.; Abd-Elzaher, M.M.; Shakdofa, M.M.E.; El Saied, F.A.; Ayad, M.I.; El Tabl, A.S. Synthesis and characterization of transition metal complexes of N'-[(1,5-dimethyl-3-oxo-2-phenyl-2,3-dihydro-1H-pyrazol-4-yl) methylene] thiophene-2-carbohydrazide. *Transit. Met. Chem.* **2008**, *33*, 219–228. [CrossRef]
21. El-Molla, M.M.; Shama, S.A.; Saeed, S.E.-S. Preparation of Disappearing Inks and Studying the Fading Time on Different Paper Surfaces. *J. Forensic Sci.* **2013**, *58*, 188–194. [CrossRef] [PubMed]
22. Saeed, S.E.S.; El-Molla, M.M.; Hassan, M.L.; Bakir, E.; Abdel-Mottaleb, M.M.; Abdel-Mottaleb, M.S. Novel chitosan-ZnO based nanocomposites as luminescent tags for cellulosic materials. *Carbohydr. Polym.* **2014**, *99*, 817–824. [CrossRef]

23. Shama, S.; El-Molla, M.; Basalah, R.F.; Saeed, S.E.-S. Fading Time Study on Prepared Thymolphthalein, Phenolphthalein and their Mixture Disappearing Ink. *Res. J. Text. Appar.* **2008**, *12*, 9–18. [CrossRef]
24. Abdel-Rahman, L.H.; El-Khatib, R.M.; Nassr, L.A.; Abu-Dief, A.M. Synthesis, physicochemical studies, embryos toxicity and DNA interaction of some new Iron (II) Schiff base amino acid complexes. *J. Mol. Struct.* **2013**, *1040*, 9–18. [CrossRef]
25. Abu-Dief, A.M.; Mohamed, I.M. A review on versatile applications of transition metal complexes incorporating Schiff bases. *Beni-Suef Univ. J. Basic Appl. Sci.* **2015**, *4*, 119–133. [CrossRef] [PubMed]
26. Hou, A.; Zhang, C.; Wang, Y. Preparation and UV-protective properties of functional cellulose fabrics based on reactive azobenzene Schiff base derivative. *Carbohydr. Polym.* **2012**, *87*, 284–288. [CrossRef]
27. Sharma, V.; Ali, S.W. A greener approach to impart multiple functionalities on polyester fabric using Schiff base of vanillin and benzyl amine. *Sustain. Chem. Pharm.* **2022**, *27*, 100645. [CrossRef]
28. Wen, W.; Zhang, Z.; Jing, L.; Zhang, T. Highly Antibacterial Efficacy of a Cotton Fabric Treated with Piperazinyl Schiff Base. *Fibers Polym.* **2021**, *22*, 3298–3308. [CrossRef]
29. Valarmathy, G.; Subbalakshmi, R.; Sumathi, R.; Renganathan, R. Synthesis of Schiff base ligand from N-substituted benzenesulfonamide and its complexes: Spectral, thermal, electrochemical behavior, fluorescence quenching, in vitro-biological and in-vitro cytotoxic studies. *J. Mol. Struct.* **2020**, *1199*, 127029. [CrossRef]
30. Daravath, S.; Vamsikrishna, N.; Ganji, N.; Venkateswarlu, K. Synthesis, characterization, DNA binding ability, nuclease efficacy and biological evaluation studies of Co (II), Ni (II) and Cu (II) complexes with benzothiazole Schiff base. *Chem. Data Collect.* **2018**, *17*, 159–168. [CrossRef]
31. El-Saied, F.A.; Shakdofa, M.M.; Abdou, S.; Abd-Elzaher, M.M.; Morsy, N. Coordination versatility of N₂O₄ polydentate hydrazonic ligand in Zn (II), Cu (II), Ni (II), Co (II), Mn (II) and Pd (II) complexes and antimicrobial evaluation. *Beni-Suef Univ. J. Basic Appl. Sci.* **2017**, *6*, 310–320.
32. Elseman, A.; Shalan, A.E.; Rashad, M.; Hassan, A.M.; Ibrahim, N.M.; Nassar, A. Easily attainable new approach to mass yield ferrocenyl Schiff base and different metal complexes of ferrocenyl Schiff base through convenient ultrasonication-solvothermal method. *J. Phys. Org. Chem.* **2017**, *30*, e3639. [CrossRef]
33. Al-Hamdani, A.A.S.; Balkhi, A.; Falah, A.; Shaker, S.A. New Azo-Schiff base derived with Ni (II), Co (II), Cu (II), Pd (II) and Pt (II) Complexes: Preparation, spectroscopic investigation, structural studies and biological activity. *J. Chil. Chem. Soc.* **2015**, *60*, 2774–2785. [CrossRef]
34. Mohammed, A.; Taha, N.I. Microwave Preparation and Spectroscopic Investigation of Binuclear Schiff Base Metal Complexes Derived from 2, 6-Diaminopyridine with Salicylaldehyde. *Int. J. Org. Chem.* **2019**, *7*, 412–419. [CrossRef]
35. Mounika, K.; Pragathi, A.; Gyanakumari, C. Synthesis Characterization and Biological Activity of a Schiff Base Derived from 3-Ethoxy Salicylaldehyde and 2-Amino Benzoic acid and its Transition Metal Complexes. *J. Sci. Res.* **2010**, *2*, 513. [CrossRef]
36. Meenarathi, B.; Siva, P.; Palanikumar, S.; Kannammal, L.; Anbarasan, R. Synthesis, characterization and drug release activity of poly (ε-caprolactone)/Fe₃O₄–alizarinred nanocomposites. *Nanocomposites* **2016**, *2*, 98–107. [CrossRef]
37. Zhang, Z.; Wang, H.; Sun, J.; Guo, K. Cotton fabrics modified with Si@ hyperbranched poly (amidoamine): Their salt-free dyeing properties and thermal behaviors. *Cellulose* **2021**, *28*, 565–579. [CrossRef]
38. Samanta, B.; Chakraborty, J.; Shit, S.; Batten, S.R.; Jensen, P.; Masuda, J.D.; Mitra, S. Synthesis, characterisation and crystal structures of a few coordination complexes of nickel (II), cobalt (III) and zinc (II) with N'-[(2-pyridyl) methylene] salicyloylhydrazone Schiff base. *Inorg. Chim. Acta* **2007**, *360*, 2471–2484. [CrossRef]
39. Li, N.; Ming, J.; Yuan, R.; Fan, S.; Liu, L.; Li, F.; Wang, X.; Yu, J.; Wu, D. Novel Eco-Friendly Flame Retardants Based on Nitrogen–Silicone Schiff Base and Application in Cellulose. *ACS Sustain. Chem. Eng.* **2019**, *8*, 290–301. [CrossRef]
40. Barbosa, H.F.G.; Cavalheiro, T.G. The influence of reaction parameters on complexation of Zn(II) complexes with biopolymeric Schiff base prepared from chitosan and salicylaldehyde. *Int. J. Biol. Macromol.* **2019**, *121*, 1179–1185. [CrossRef]
41. Su, Y.; Wang, S.; Zhang, N.; Cui, P.; Gao, Y.; Bao, T. Zr-MOF modified cotton fiber for pipette tip solid-phase extraction of four phenoxy herbicides in complex samples. *Ecotoxicol. Environ. Saf.* **2020**, *201*, 110764. [CrossRef]
42. Harmon, K.M.; Akin, A.C.; Keefer, P.K.; Snider, B.L. Hydrogen bonding Part 45. Thermodynamic and IR study of the hydrates of N-methylmorpholine oxide and quinuclidine oxide. Effect of hydrate stoichiometry on strength of H–O–H··O–N hydrogen bonds; implications for the dissolution of cellulose in amine oxide solvents. *J. Mol. Struct.* **1992**, *269*, 109–121.
43. French, A.D. Glucose, not cellobiose, is the repeating unit of cellulose and why that is important. *Cellulose* **2017**, *24*, 4605–4609. [CrossRef]
44. Gashti, M.P.; Alimohammadi, F.; Song, G.; Kiumarsi, A. Characterization of nanocomposite coatings on textiles: A brief review on microscopic technology. *Curr. Microsc. Contrib. Adv. Sci. Technol.* **2012**, *2*, 1424–1437.
45. Imran, M.; Liviu, M.; Latif, S.; Mahmood, Z.; Naimat, I.; Zaman, S.S.; Fatima, S. Antibacterial Co (II), Ni (II), Cu (II) and Zn (II) complexes with biacetyl-derived Schiff bases. *J. Serb. Chem. Soc.* **2010**, *75*, 1075–1084. [CrossRef]
46. Joseyphus, R.S.; Nair, M.S. Antibacterial and Antifungal Studies on Some Schiff Base Complexes of Zinc(II). *Mycobiology* **2008**, *36*, 93–98. [CrossRef]
47. Saif, M.; El-Shafiy, H.F.; Mashaly, M.M.; Eid, M.F.; Nabeel, A.I.; Fouad, R. Synthesis, characterization, and antioxidant/cytotoxic activity of new chromone Schiff base nano-complexes of Zn (II), Cu (II), Ni (II) and Co (II). *J. Mol. Struct.* **2016**, *1118*, 75–82. [CrossRef]

Article

Biopaper Based on Ultralong Hydroxyapatite Nanowires and Cellulose Fibers Promotes Skin Wound Healing by Inducing Angiogenesis

Jing Gao [1,†], Liang-Shi Hao [2,†], Bing-Bing Ning [1], Yuan-Kang Zhu [1], Ju-Bo Guan [2], Hui-Wen Ren [3], Han-Ping Yu [4,*], Ying-Jie Zhu [4,*] and Jun-Li Duan [1,*]

1. Department of Gerontology, Xinhua Hospital Affiliated to Shanghai Jiaotong University School of Medicine, Kongjiang Road 1665, Shanghai 200092, China; gaojingsjtu@163.com (J.G.); ningbingbing1986@163.com (B.-B.N.); zyk3336@163.com (Y.-K.Z.)
2. The Second School of Medicine, Wenzhou Medical University, Wenzhou 325027, China; haoliangshi@126.com (L.-S.H.); guanjubo@126.com (J.-B.G.)
3. Advanced Institute for Medical Sciences, Dalian Medical University, Dalian 116044, China; rhwcmu@163.com
4. State Key Laboratory of High Performance Ceramics and Superfine Microstructure, Shanghai Institute of Ceramics, Chinese Academy of Sciences, Shanghai 200050, China
* Correspondence: yuhanping@mail.sic.ac.cn (H.-P.Y.); y.j.zhu@mail.sic.ac.cn (Y.-J.Z.); duanjunli@xinhuamed.com.cn (J.-L.D.)
† These authors contributed equally to this work.

Abstract: Skin injury that is difficult to heal caused by various factors remains a major clinical challenge. Hydroxyapatite (HAP) has high potential for wound healing owing to its high biocompatibility and adequate angiogenic ability, while traditional HAP materials are not suitable for wound dressing due to their high brittleness and poor mechanical properties. To address this challenge, we developed a novel wound dressing made of flexible ultralong HAP nanowire-based biopaper. This biopaper is flexible and superhydrophilic, with suitable tensile strength (2.57 MPa), high porosity (77%), and adequate specific surface area (36.84 m$^2 \cdot$g^{-1}) and can continuously release Ca^{2+} ions to promote the healing of skin wounds. Experiments in vitro and in vivo show that the ultralong HAP nanowire-based biopaper can effectively induce human umbilical vein endothelial cells (HUVECs) treated with hypoxia and rat skin tissue to produce more angiogenic factors. The as-prepared biopaper can also enhance the proliferation, migration, and in vitro angiogenesis of HUVECs. In addition, the biopaper can promote the rat skin to achieve thicker skin re-epithelialization and the formation of new blood vessels, and thus promote the healing of the wound. Therefore, the ultralong HAP nanowire-based biopaper has the potential to be a safe and effective wound dressing and has significant clinical application prospects.

Keywords: nanomaterials; nanowires; hydroxyapatite; wound healing; angiogenesis

1. Introduction

Trauma refers to human tissue damage and dysfunction caused by various injury factors. Skin wound injury is the most common trauma, and its treatment has become an important research area. Some pathological conditions caused by acute and chronic diseases, such as burns, diabetes, senile diseases, etc., can cause skin damage that is difficult to heal [1]. As the first barrier of the human body, the integrity of the skin is essential for various life activities. Moreover, skin is the major temperature perception and regulation organ of the human body. The occurrence of skin trauma seriously disrupts the stable state and functions of the body. Therefore, promoting the rapid healing of skin wounds and ensuring the integrity of the skin are key in maintaining a stable internal environment for all life activities. Delayed skin wound healing is at risk of infection, chronic wounds, and various complications, and even worse, serious skin wound repair is usually costly, with long-term care and expensive treatment products [2]. It is bound to bring a heavy burden to the medical system, the patient's family, and society [3,4]. Therefore, exploring the way to promote rapid skin wound healing has become an important research topic in the biomedical field.

Wound healing, as one of the most complex biological processes in the body, is affected by many factors. The most important factor in wound healing is the condition of blood vessels at the site of injury [5]. Vascularization and ischemia–reperfusion are signs of early tissue repair [6] and cornerstones of wound healing [7]. Mature endothelial cells are in a static and stable state under physiological conditions with a balance of pro- and anti-angiogenic factors. Only when the body has a pathological condition such as trauma are endothelial cells activated, migrated, and differentiated and form new blood vessels [8]. Angiogenesis is highly dependent on the interactions between cells, and it is mainly achieved by the expressions of cytokines such as endothelial nitric oxide synthase (eNOS) and the activation of various signal pathways [9]. The newly formed blood vessels are essential for transmitting oxygen, nutrients, and small molecules to the wound site as well as providing circulating stem cells and active factors critical to wound repair, thus facilitating wound healing [10]. Moreover, angiogenesis provides a suitable local microenvironment for wound healing and promotes the growth and reconstruction of tissues around the wound [11]. Therefore, the induction of angiogenesis at the wound site is of great significance for wound healing.

Clinically, most wounds are disinfected with 75% alcohol or iodine, and then covered with gauze [12,13]. This has been successful in preventing deterioration of the chronic wound; however, some disadvantages should not be ignored, including the lack of ability to induce angiogenesis, simple function, low bioactivity, and unsatisfactory biocompatibility. Therefore, researchers have been dedicated to developing safer and more effective wound dressings. At present, hydrogel has become popular for medical wound dressings owing to its unique structure and physicochemical properties [14–16], but the high cost seriously hinders its widespread application. In addition, dressings should be easy to remove and change frequently to prevent wound infection for chronic wounds, but most hydrogel dressings are difficult to remove because of their excellent bonding effect with wound tissues [17]. To this end, it is urgent to develop an inexpensive, replaceable, bioactive, and biocompatible wound dressing with the ability to induce angiogenesis for wound closure and rapid healing.

HAP, the main inorganic component of hard tissues, has high biocompatibility, bioactivity, and biodegradability [18–20], making it a popular biomaterial in biomedical applications [21,22]. As a common bone defect repair material, HAP-based biomaterials not only promote new bone formation, but also exhibit excellent angiogenic potential [23–27]. Our previous studies have found that highly porous and elastic aerogel consisting of pure ultralong HAP nanowires could simultaneously induce angiogenesis and osteogenesis around the bone defect area in vivo. This indicates that HAP is able to promote angiogenesis and is expected to be a safe and effective dressing material for wound healing [27,28]. In addition, the calcium ions released from HAP degradation may regulate the proliferation and differentiation of keratinocytes, and play an important role in the formation of fibroblasts and keratinocytes [29,30].

Unfortunately, traditional HAP materials show poor mechanical properties, and are usually composed of HAP particles or rods, and are difficult to form into a stable porous structure, let alone wound dressings. Based on the idea of exploring the potential use of ultralong HAP nanowires (HAPNWs) in the context of promoting chronic wound healing, in this work, we developed a novel wound dressing made of flexible ultralong HAP nanowire-based biopaper with a small quantity of cellulose fibers (CFs) as the reinforcement, and revealed its mechanism for promoting angiogenesis and skin wound healing. The as-prepared HAPNW/CF biopaper is highly biocompatible, hydrophilic, and shows suitable mechanical properties, beneficial to promoting wound healing. Moreover, the HAPNW/CF biopaper possesses a porous networked structure, which provides a favorable environment for blood vessel formation and ingrowth.

Specifically, the HAPNW/CF biopaper can continuously release calcium ions during the degradation process to promote wound healing. Therefore, the HAPNW/CF biopaper as the wound dressing can up-regulate the expression of angiogenic factors at the wound site, and promote the formation of more blood vessels, nutrition transport, and a stable internal environment, thereby accelerating wound healing. Studies in vitro reveal that the HAPNW/CF biopaper dressing can up-regulate the expression of angiogenesis-related proteins such as eNOS, AKT, and vascular endothelial growth factor A (VEGFA). That effect can be partially offset by specific eNOS inhibitors, indicating that angiogenesis is achieved through the activation of eNOS-related signaling pathways. Animal models also suggest the greater re-epithelialization ability of the HAPNW/CF biopaper than the blank control group. In brief, we developed a wound dressing made of HAPNW/CF biopaper with the advantages of angiogenesis promotion, high bioactivity, high biocompatibility, simple preparation, stable covering, and easy replacement. Our work not only suggests excellent prospects of the HAPNW/CF biopaper dressing in fast wound healing, but also proves the high potential and mechanism of ultralong HAP nanowires in promoting angiogenesis and chronic wound healing.

2. Materials and Methods

2.1. Materials and Chemicals

Calcium chloride ($CaCl_2$), sodium dihydrogen phosphate dihydrate ($NaH_2PO_4 \cdot 2H_2O$), and sodium hydroxide (NaOH) were purchased from Sinopharm Chemical Reagent Co., Ltd. (Shanghai, China). Methanol was obtained from Shanghai Lingfeng Chemical Reagent Co., Ltd. (Shanghai, China). Oleic acid was purchased from Aladdin Industrial Corporation (Shanghai, China). All chemicals were of analytical grade and used as received without further purification.

2.2. Preparation of Ultralong HAP Nanowires (HAPNWs)

HAPNWs were synthesized by the calcium oleate precursor solvothermal method previously reported [31–33]. Firstly, 135 mL of deionized water, 105 mL of oleic acid (0.33 mol), and 60 mL of methanol were mixed together under mechanical stirring. Then, NaOH (10.500 g) aqueous solution (150 mL), $CaCl_2$ (3.330 g, 0.03 mol) aqueous solution (120 mL), and $NaH_2PO_4 \cdot 2H_2O$ (9.360 g, 0.06 mol) aqueous solution (180 mL) were slowly added to the above mixture. After that, the resulting reaction system was transferred into a 1 L Teflon-lined stainless-steel autoclave, sealed, and thermally treated at 185 °C for 25 h. The solvothermal product was cooled down to room temperature, collected, and washed with ethanol twice and deionized water twice, and the obtained HAPNWs were dispersed in deionized water for further use.

2.3. Fabrication of the HAPNW/CF Biopaper Consisting of HAPNWs and CFs

CFs were added to the above aqueous suspension containing HAPNWs under stirring. The weight ratio of HAPNWs to CFs was 80:20. The HAPNW/CF biopaper was fabricated through a simple vacuum-assisted filtration method. Specifically, the aqueous suspension containing HAPNWs and CFs was poured onto a filter paper under the suction generated by a vacuum pump, and then the wet HAPNW/CF paper was formed on the filter paper. Finally, the HAPNW/CF biopaper was dried at 95 °C for 10 min. The HAPNW paper consisting of only HAPNWs or the CF paper consisting of only CFs was also prepared by the vacuum-assisted filtration method using the aqueous suspension containing HAPNWs or CFs.

2.4. Materials Characterization

2.4.1. Structural Characterization

The microstructures of the samples were observed by scanning electron microscopy (SEM, S-4800 and TM-3000, Hitachi, Tokyo, Japan) and transmission electron microscope (TEM, JEM-2100, JEOL, Tokyo, Japan). The HAPNW/CF biopaper was identified by X-ray diffraction (XRD) with a Rigaku D/max 2550 V instrument with Cu Kα radiation (λ = 1.54178 Å). Fourier transform infrared spectroscopy (FTIR) spectra were recorded on an FTIR spectrometer (FTIR-7600, Lambda Scientific, Edwardstown, Australia) in the wavenumber range of 400 to 4000 cm^{-1} at a resolution of 4 cm^{-1}. Thermogravimetric (TG) analysis was performed on a thermal analyzer (STA 409/PC, Netzsch, Selb, Germany) with a heating rate of 10 °C·min^{-1} in a flowing air atmosphere.

2.4.2. Degradation and Ca^{2+} Ions Release Test

The HAPNW/CF biopaper (80 mg) was soaked in 40 mL of normal saline at 36.5 °C under constant shaking (160 rpm) in a desk-type constant temperature oscillator (HZQ-X 160, Peiying, Suzhou, China). The supernatant solution (1 mL) was withdrawn for inductively coupled plasma (ICP) analysis (JY 2000-2, Horiba, Paris, France) to measure the concentrations of Ca^{2+} ions at different intervals and then replaced with the same volume of fresh normal saline.

2.4.3. Specific Surface Area Measurement

The Brunauer–Emmett–Teller (BET) specific surface area of the sample was obtained by analyzing N_2 adsorption–desorption through a surface area and pore size analyzer (TriStar II 3020, Micromeritics, Norcross, GA, USA).

2.4.4. Density and Porosity Test

The density of the sample was calculated by dividing the weight by the apparent volume. The porosity was determined by immersing the sample into ethanol for 30 min and calculated according to the following equation:

$$\text{Porosity (\%)} = (\Delta m/\rho)/V_0 \times 100\% \qquad (1)$$

where Δm is the weight difference of the sample after and before the absorption with ethanol, ρ is the density of ethanol (0.789 g·cm^{-3}), and V_0 is the apparent volume of the sample.

2.4.5. Water Absorption and Water Contact Angle Measurements

The water absorption capacity of the sample was calculated as the following formula: water absorption (%) = $(m - m_0)/m_0 \times 100\%$, where m_0 and m are the weight of the sample before and after immersion in deionized water, respectively.

The water contact angle was measured using an optical contact angle system (model SL200A/B/D, Solon, Shanghai, China).

2.4.6. Mechanical Tests

The tensile behaviors of the samples were measured on a universal testing machine (DRK 101A, Drick, Jinan, China) at a displacement rate of 5 mm·min^{-1}. The elastic modulus was calculated as the slope of the linear fraction of the tensile stress–strain curve. The fracture behavior of the HAPNW/CF biopaper was observed by the SEM (TM-3000, Hitachi, Tokyo, Japan) at an acceleration voltage of 15 kV.

2.5. In Vitro Cellular Studies

2.5.1. Cell Culture

HUVECs were purchased from American Type Culture Collection (Manassas, VA, USA). Cells were cultured under standard culture conditions (37 °C, 5% CO_2) and in

Dulbecco's Modified Eagle's Medium (DMEM) supplemented with 10% fetal bovine serum (Gibco, Grand Island, NE, USA) and 1% penicillin/streptomycin. The cells were cultured in a humidified air incubator (HERAcell 150i, Thermo Fisher Scientific Inc., Waltham, MA, USA). Each experiment was repeated 3 times.

2.5.2. Establishment of a Model of Cellular Ischemia and Hypoxia

According to previous studies, HUVECs were transferred to a three-gas incubator, where the original medium was replaced with a serum-free DMEM medium and incubated at 37 °C for 24 h under the conditions of 1% oxygen, 5% carbon dioxide, and 94% nitrogen [34].

2.5.3. Cell Viability

HUVECs after hypoxia treatment were seeded onto 96-well plates at a density of 2×10^5 cells·mL^{-1}. Cells were intervened with the HAPNW/CF biopaper (weight ratio was 80:20) at different doses of 0, 62.5, 125, 250, 500, and 1000 µg·mL^{-1} for 24 h. The CCK-8 kit (Beyotime Biotechnology, Shanghai, China) was used to determine cell viability. Following the manufacturer's instructions, the cells were incubated with 10 µL of CCK-8 solution at 37 °C for 2 h. The OD value was measured at 450 nm using a microplate reader (Epoch, BioTek, Winooski, VT, USA).

2.5.4. Western Blot Analysis

Total cell protein was extracted using $1 \times$ LDS sample buffer (Thermo Fisher Scientific, Waltham, MA, USA). Proteins from skin tissues were extracted in radio-immunoprecipitation assay (RIPA) buffer (Beyotime Biotechnology, Shanghai, China) containing phosphatase inhibitors and protease inhibitors (Thermo Fisher Scientific, Waltham, MA, USA). An equal amount of protein sample was loaded on a sodium dodecyl sulfate polyacrylamide gel electrophoresis (SDS-PAGE) gel, and then transferred to a polyvinylidene fluoride (PVDF) membrane for Western blot analysis. The main antibodies used were as follows: anti-eNOS (Cell Signaling Technology, Danvers, MA, USA), anti-AKT (Cell Signaling Technology, Danvers, MA, USA), and anti-VEGFA (Abcam, Waltham, MA, USA) diluted at 1:1000. Anti-β-actin (Beyotime Biotechnology, Shanghai, China) was diluted with 1:5000 as an internal reference. The ECL Western Blot Detection System (AI600, GE Healthcare, Buckinghamshire, UK) was used for detection. The results were normalized to the control group ratio.

2.5.5. Cell Migration and Tube Formation

The migration ability of HUVECs was evaluated using 8.0 µm Transwell plates (Corning, Corning, NY, USA). HUVECs with a density of 5×10^5 cell·mL^{-1} were inoculated in the upper lumen of the Transwell plate using the serum-free medium. The sample and the specific eNOS inhibitor N-nitro-L-arginine methyl ester (L-NAME) were placed in the lower lumen, and cultured in a cell incubator for 24 h. The upper compartment of the Transwell plate was removed and the cells in the upper compartment were wiped, fixed in 4% paraformaldehyde for 15 min, and stained with crystal violet (Beyotime Biotechnology, Shanghai, China) for 20 min. The number of migrating HUVECs was counted by a light microscope. The tubeforming ability of HUVECs was evaluated by the tubeforming experiments. Matrigel (BD Biosciences, San Jose, CA, USA) was pre-coated on a 96-well plate, and HUVECs with a density of 5×10^5 cell·mL^{-1} were inoculated on it. The cells were intervened with or without the sample and L-NAME, respectively. After culturing in a cell culture incubator for 6–8 h, the formation of small tubes was observed with an optical microscope. The results were processed using Image J software.

2.5.6. EdU Cell Proliferation Assay

To observe the effect of the HAPNW/CF biopaper (developed by Shanghai Institute of Ceramics, Chinese Academy of Sciences) and L-NAME intervention on cell proliferation, the EdU kit (Beyotime Biotechnology, Shanghai, China) was used. According to the manufacturer's instructions, HUVECs were intervened with or without the sample and L-NAME, incubated in the cell incubator with EdU working solution for 2 h, fixed at room temperature with immunostaining fixative for 15 min, incubated at room temperature with the permeating solution for 15 min, incubated at room temperature with click reaction solution for 30 min, and finally incubated at room temperature with Hoechst 33342 for 10 min. The sample was imaged using a fluorescence microscope.

2.6. In Vivo Experiments

2.6.1. Establishment of the Animal Model of Skin Wound

Eight-week-old male Sprague–Dawley (SD) rats were purchased from Shanghai Jihui Laboratory Animal Breeding Co., Ltd. (Shanghai, China) and placed in a temperature and humidity control room. The light:dark cycle was 12:12 h, and they could drink and eat freely. All experimental protocols were approved by the Ethics Committee of Xinhua Hospital Affiliated to Shanghai Jiao Tong University School of medicine under approval number XHEC-F-2021-079, and the experiments were in accordance with relevant institutional guidelines. After weighing, rats were anesthetized by intraperitoneal injection of 1% pentobarbital sodium at a dose of 40 mg·kg^{-1}. Two round full-thickness skin wounds with the same diameter of 1 cm were made on the back of all 5 rats after anesthesia. One side was used as the control group, and the other side was implanted with the HAPNW/CF biopaper with an equal size. Full-thickness skin samples were taken after 7 days.

2.6.2. Histological Analysis

The full-thickness skin samples of rats were taken 7 days after surgery, fixed in 10% formalin for 24 h, paraffin-embedded, sliced into 5 μm thickness according to the standard procedure, and stained with hematoxylin and eosin (H&E). Images were obtained using a microscope to measure the thickness of the re-epithelialized skin. Immunohistochemical staining was performed on paraffin sections using CD34 antibodies (Abclonal, Wuhan, China). The images were obtained by a microscope, and the positive rate of CD34 staining was calculated to evaluate the formation of blood vessels.

2.7. Statistical Analysis

All data were presented as mean ± standard deviation (SD). The differences between groups were compared multiple times by one-way analysis of variance (ANOVA) and post hoc tests using the least significant difference, and *p*-values less than 0.05 were considered statistically significant.

3. Results

3.1. Characterization of the HAPNW/CF Biopaper and Control Samples

In this work, HAPNWs were synthesized by our previously reported calcium oleate precursor solvothermal method with minor modifications using $CaCl_2$, $NaH_2PO_4·2H_2O$, and NaOH in the water/methanol/oleic acid reaction system with a Ca/P molar ratio of 0.5 and a Ca/oleic acid molar ratio of 0.09 at 185 °C for 25 h. Then, the HAPNWs and CFs were hybridized to form a porous HAPNW/CF biopaper by a simple vacuum-assisted filtration process, and the HAPNW/CF biopaper was investigated for promoting skin wound healing, as shown in Figure 1.

The morphologies of CFs, HAPNWs, and the HAPNW/CF biopaper are shown in Figure 2. In many cases, HAPNWs are self-assembled along their longitudinal direction into highly aligned nanowire bundles with larger diameters, and the diameter of a single nanowire is only about 10 nm (Figure 2A–C). The self-assembled alignment structure of HAPNWs (nanowire bundles) can enhance their mechanical strength and flexibility. It is observed that ultralong HAP nanowire bundles can naturally bend without fracture. The diameters of CFs are tens of micrometers, and a higher-resolution SEM image reveals that the CFs also consists of many approximately parallel fine fibers (Figure 2D,E).

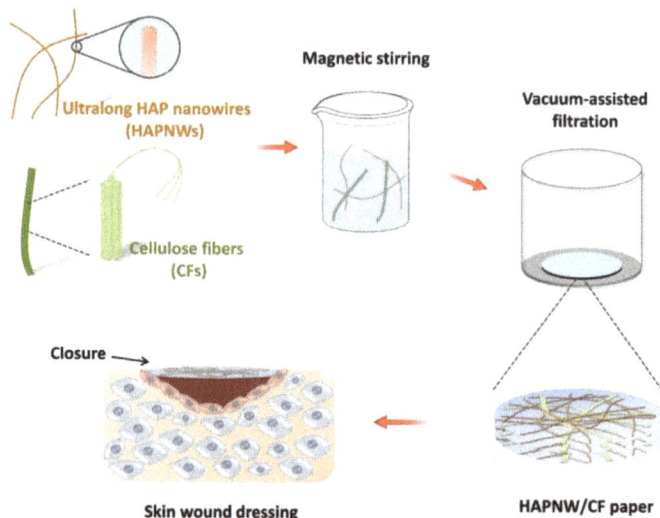

Figure 1. Schematic illustration for the structures of HAPNWs and CFs, and the preparation of the HAPNW/CF biopaper for the skin wound dressing.

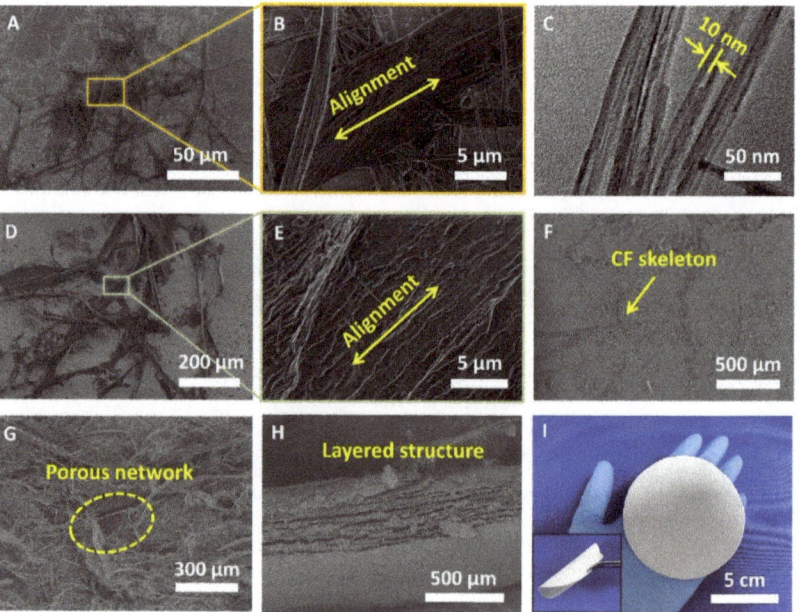

Figure 2. (**A**,**B**) Scanning electron microscopy (SEM) images of the HAPNWs with the self-assembled alignment structure. (**C**) Transmission electron microscopy (TEM) image of the HAPNWs. (**D**,**E**) SEM images of the CFs. (**F–H**) The surface morphology (**F**,**G**) and the cross-section (**H**) of the as-prepared HAPNW/CF biopaper. (**I**) A digital image of a large-sized HAPNW/CF biopaper with a diameter of 9 cm, and the insert is a digital image of the HAPNW/CF biopaper showing high flexibility.

Through a vacuum-assisted filtration process, HAPNWs and CFs were synergistically interwoven with each other to form the HAPNW/CF biopaper. The SEM image of the HAPNW/CF biopaper shows that CFs were embedded in the ultralong HAP nanowire network, which plays an important role as the reinforcement agent, similar to the steel bars in reinforced cement (Figure 2F). In a higher-magnification SEM image of the HAPNW/CF biopaper, a porous network structure constructed with HAPNWs and CFs is clearly observed (Figure 2G). Moreover, the cross-section SEM observation of the HAPNW/CF biopaper exhibits a layered structure, and the thickness of one single layer is about 10 μm (Figure 2H). Thus, the CF-reinforced ultralong HAP nanowire network as well as the layered structure in the HAPNW/CF biopaper cooperate to facilitate the high mechanical performance. Compared with the conventional brittle HAP ceramics, the as-prepared HAPNW/CF biopaper is highly flexible, and can be bent at a large angle (nearly 180°) without obvious cracking (the inset of Figure 2I). In addition, a large-sized HAPNW/CF biopaper with a diameter of 9 cm (Figure 2I) is fabricated rapidly and conveniently through the above-mentioned method for large-area skin wound dressings.

XRD patterns of the as-prepared HAPNW paper and HAPNW/CF biopaper are shown in Figure 3A, which reveals that both the HAPNW paper and the HAPNW/CF biopaper are composed of the crystal phase of hexagonal HAP (JCPDS No. 09-0432), and the peak intensity corresponding to the (0 0 2) crystal plane at 2θ = 25.8° is obviously diminished because of the preferential growth of ultralong HAP nanowires along its c-axis direction. The CFs do not exhibit strong diffraction peaks except for a broad peak locating around 2θ = 10–20° in the XRD pattern of HAPNW/CF biopaper. The TG curves of the samples are shown in Figure 3B. As expected, the weight ratio of the CFs in the HAPNW/CF biopaper is about 22 wt.% calculated by the TG analysis (Figure 3B). CFs are not thermally stable and are completely decomposed and oxidized at temperatures below 450 °C. In contrast, HAPNWs are highly stable at high temperatures, and only a small amount of weight loss (mainly due to the loss of adsorbed water) is observed.

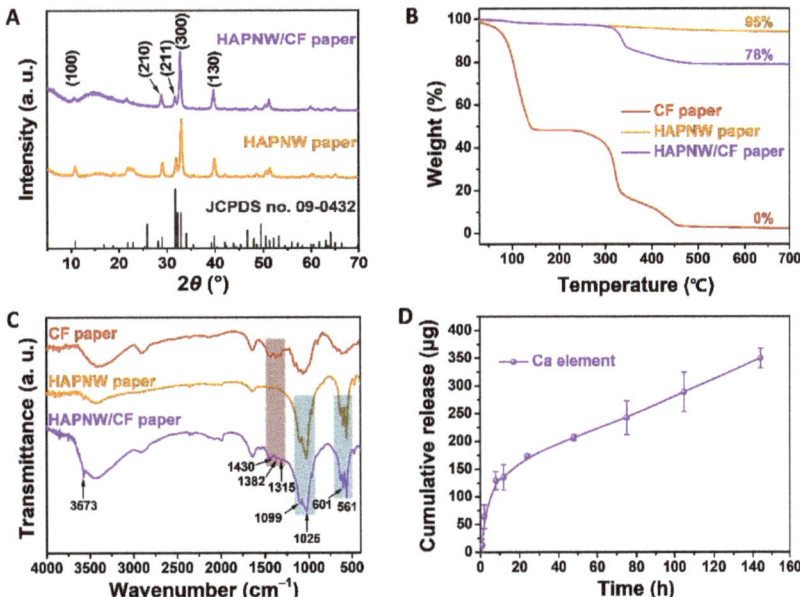

Figure 3. (**A**) X-ray diffraction (XRD) patterns of the HAPNW paper and HAPNW/CF biopaper. (**B**) Thermogravimetric (TG) curves and (**C**) Fourier transform infrared (FTIR) spectra of the HAPNW paper, the CF paper, and the HAPNW/CF biopaper. (**D**) The cumulative release curve of Ca^{2+} ions from the HAPNW/CF biopaper immersed in the normal saline for different times.

FTIR spectra of the HAPNW paper made of only HAPNWs, the CF paper made of only CFs, and the HAPNW/CF biopaper are shown in Figure 3C. The absorption peak at 3573 cm^{-1} is the characteristic peak of the OH group from HAP. The absorption peaks at 1099 and 1025 cm^{-1} correspond to the PO$_4^{3-}$ group with the stretching mode, and the absorption peaks at 601 and 561 cm^{-1} originate from the bending mode of the O–P–O bond of HAP. In addition, the absorption peaks of CFs at 1430 cm^{-1} (C=O), 1382 cm^{-1} (O–CH$_3$), and 1315 cm^{-1} (C–O–C) can be observed in the FTIR spectrum of the HAPNW/CF biopaper, confirming the presence of CFs in the biopaper. Specifically, both HAPNWs and CFs show a broad absorption peak from 3500 to 3300 cm^{-1} ascribed to the hydroxyl groups on their surface, while this peak for HAPNWs is slightly shifted from 3446 to 3436 cm^{-1} after hybridization with CFs, indicating the interactions between HAPNWs and CFs including hydrogen bonding and van der Waals forces.

In addition, the Ca^{2+} ion release and degradation behavior of the HAPNW/CF biopaper in vitro are investigated, and the experimental results are shown in Figure 3D, indicating that the cumulative released amount of Ca^{2+} ions from the HAPNW/CF biopaper in normal saline continuously increases with the increase in the release time. The continuous release of Ca^{2+} ions from the HAPNW/CF biopaper is beneficial to promoting the skin wound healing. In the degradation and Ca^{2+} ion release test, the HAPNW/CF biopaper (80 mg) was soaked in 40 mL of normal saline. The cumulative release of Ca^{2+} ions from the HAPNW/CF biopaper is about 0.35 mg at a release time of about 150 h, indicating that the percentage of the released Ca^{2+} ions from the HAPNW/CF biopaper is about 0.44 wt.%.

The Brunauer–Emmett–Teller (BET) specific surface areas are measured by the N$_2$ adsorption/desorption isotherms. As shown in Figure 4A,B, the BET specific surface area of the HAPNW paper is measured to be 47.29 m^2·g^{-1}, and that of the CF paper is very small (0.84 m^2·g^{-1}). Although the presence of CFs will decrease the specific surface area of the HAPNW/CF biopaper, the value is maintained at a relatively high level of 36.84 m^2·g^{-1}. The high specific surface area of the HAPNW/CF biopaper is beneficial for providing more sites for cell and protein adhesion and enhancing the angiogenesis in the skin wound region. The differences in the densities of the HAPNW paper (0.416 g·cm^{-3}), CF paper (0.425 g·cm^{-3}), and HAPNW/CF biopaper (0.414 g·cm^{-3}) are not significant (Figure 4C). Their low densities are ascribed to the porous structure and high porosities. The porosities of the HAPNW paper, the HAPNW/CF biopaper, and the CF paper are 80%, 77%, and 70%, respectively, as shown in Figure 4C.

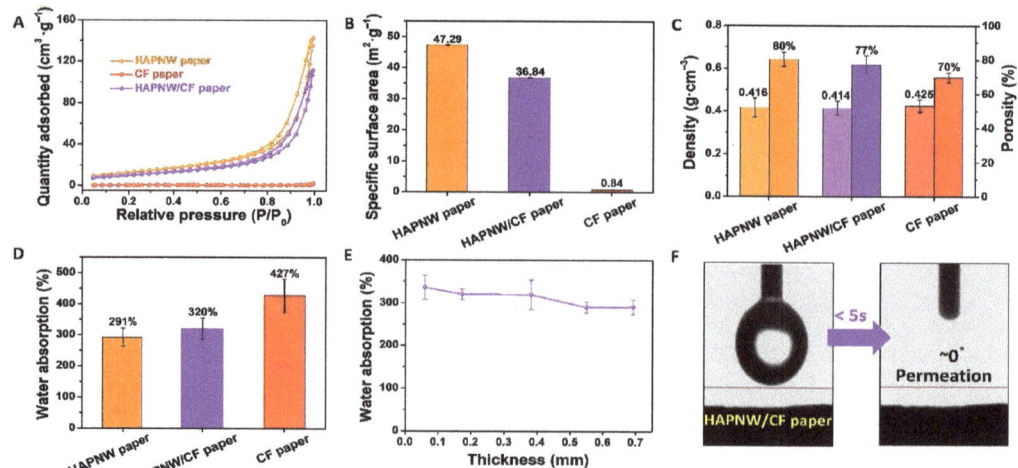

Figure 4. (A–D) Nitrogen adsorption–desorption isotherms (A), specific surface areas (B), densities and porosities (C), and water absorption percentages (D) of the HAPNW paper, the CF paper, and the HAPNW/CF biopaper. (E) Water absorption percentages of the HAPNW/CF biopaper with different thicknesses. (F) Water contact angle of the HAPNW/CF biopaper.

The as-prepared HAPNW paper, the CF paper, and the HAPNW/CF biopaper are highly hydrophilic. As shown in Figure 4D, the water uptake percentage of the HAPNW paper is about 291%, and that of the CF paper is higher (427%). The abundant hydroxyl groups on the surface of CFs are ascribed to its high water absorption capacity. Therefore, the introduction of CFs in the HAPNW/CF biopaper can improve its water absorption capacity, and the water uptake percentage of the as-prepared HAPNW/CF biopaper is about 320%. Moreover, the thickness of the HAPNW/CF biopaper has no significant effect on its water absorption capacity (Figure 4E). The water contact angle of the HAPNW/CF biopaper is 0°, and the water permeates into the biopaper within 5 s, which is consistent with the superhydrophilicity of the as-prepared HAPNW/CF biopaper (Figure 4F).

The mechanical properties of the HAPNW paper can be improved by the addition of CFs. As shown in Figure 5A,B, the tensile strength and Young's modulus of the HAPNW paper are 1.03 MPa and 465 MPa, respectively, and those of the CF paper are 7.42 MPa and 259 MPa, respectively. The tensile strength of the HAPNW paper is relatively low, and this will limit its applications, whereas the CF paper has a relatively high tensile strength but low Young's modulus, ascribed to the intrinsic soft structure of CFs. Thus, the addition of CFs in the HAPNW paper can improve its mechanical strength and toughness. Compared to the HAPNW paper, the tensile strength of the HAPNW/CF biopaper reaches 2.57 MPa, and the Young's modulus increases to 581 MPa. These experimental results are also consistent with the high flexibility of the HAPNW/CF biopaper (Figure 2I).

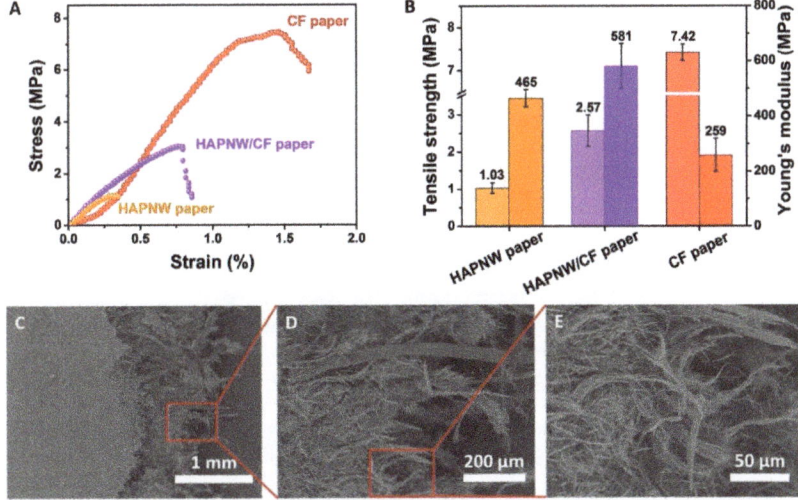

Figure 5. (**A**) Tensile stress–strain curves of the HAPNW paper, the CF paper, and the HAPNW/CF biopaper. (**B**) Tensile strengths and Young's moduli of the HAPNW paper, the CF paper, and the HAPNW/CF biopaper. (**C**–**E**) SEM images of the fractured HAPNW/CF biopaper.

To further demonstrate the mechanical enhancement mechanism, the fracture behaviors of the HAPNW/CF biopaper have been investigated by the SEM observation. As shown in Figure 5C–E, the breakage and pull out of HAPNWs are observed in the broken region of the HAPNW/CF biopaper. Moreover, the twisting and deformation of HAPNWs as well as the fractured CFs are clearly revealed in the fracture area. The obvious tearing surfaces, accompanied by the twisting, deformation, breakage, and pull-out of HAPNWs and CFs, indicate the stretching of HAPNWs and CFs under the external force.

3.2. Safety Evaluation of the HAPNW/CF Biopaper

The CCK8 kit was used to evaluate the cell viability of the HAPNW/CF biopaper (weight ratio 80:20) with different doses (62.5, 125, 250, 500, 1000 µg·mL^{-1}) using the CCK8 kit and HUVECs cultured for 24 h, as shown in Figure 6. We found that the cell viabilities of the 500 and 1000 µg·mL^{-1} HAPNW/CF treatment groups are significantly higher than that of the control group, and the difference is statistically significant ($p < 0.01$). There is no statistical difference between the 500 and 1000 µg·mL^{-1} HAPNW/CF treatment groups. Compared with the control group, the remaining treatment groups have no statistical difference. These experimental results indicate the high biocompatibility and safety of the HAPNW/CF biopaper.

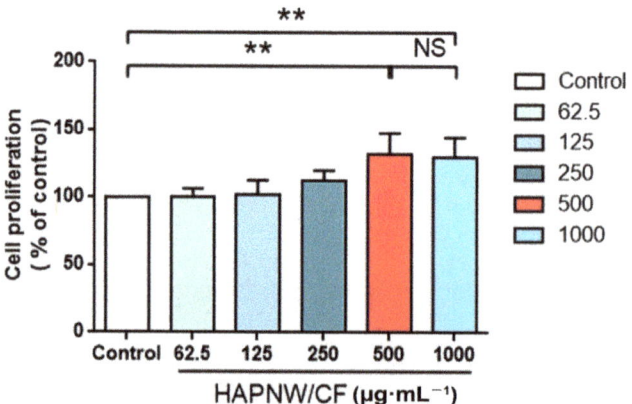

Figure 6. Safety analysis of HAPNW/CF (weight ratio 80:20) with different doses (62.5, 125, 250, 500, 1000 µg·mL^{-1}) using the CCK8 kit and HUVECs cultured for 24 h (** $p < 0.01$).

3.3. Evaluation of the Effect of the HAPNW/CF Biopaper on HUVECs

Western blot analysis was used to analyze the expressions of angiogenesis-related proteins eNOS, AKT, and VEGFA after 24 h intervention with different doses (62.5, 125, 250, 500, 1000 µg·mL^{-1}) of HAPNW/CF (weight ratio 80:20) (Figure 7). Compared with the control group, the expressions of eNOS in the 250 and 500 µg·mL^{-1} HAPNW/CF treatment groups significantly increased, which are statistically significant ($p < 0.05$ and $p < 0.01$). There is no statistical difference between 250 and 500 µg·mL^{-1} HAPNW/CF treatment groups (Figure 7B). Compared with the control group, the expressions of AKT in the 500 and 1000 µg·mL^{-1} HAPNW/CF treatment groups are significantly higher, which are statistically significant ($p < 0.05$). There is no statistical difference between the 500 and 1000 µg·mL^{-1} HAPNW/CF treatment groups (Figure 7C). Compared with the control group, the expression of VEGFA in the 500 µg·mL^{-1} HAPNW/CF treatment group is significantly increased, which is statistically significant ($p < 0.05$). There is no statistical difference between the remaining treatment groups and the control group (Figure 7D). These experimental results indicate that the HAPNW/CF biopaper can significantly up-regulate the expressions of angiogenesis-related proteins eNOS, AKT, and VEGFA, and this effect is related to the concentration of the material.

3.4. Effect of L-NAME on the Ability of the HAPNW/CF Biopaper to Promote Angiogenesis

Angiogenesis is realized through multiple signaling pathways. We used specific eNOS inhibitor L-NAME to observe its effect on the ability of the HAPNW/CF biopaper to promote angiogenesis. Western blot analysis (Figure 8A,B) shows that there are significant differences between the control group and the HAPNW/CF treatment group, the HAPNW/CF treatment group and the HAPNW/CF + L-NAME treatment group, and the control group and the HAPNW/CF + L-NAME treatment group ($p < 0.001$, $p < 0.01$, $p < 0.05$). These experimental results indicate that HAPNW/CF can promote the expression of the pro-angiogenic protein eNOS, and L-NAME can partially counteract this effect. Transwell migration experiments (Figure 8C,D) show that compared with the control group and the HAPNW/CF + L-NAME treatment group, the migration ability of cells in the HAPNW/CF treatment group is stronger, and the experimental results have statistical differences ($p < 0.01$), but there was no statistical difference between the control group and the HAPNW/CF + L-NAME treatment group. These experimental results indicate that HAPNW/CF can enhance the migration ability of HUVECs, and L-NAME can counteract this effect.

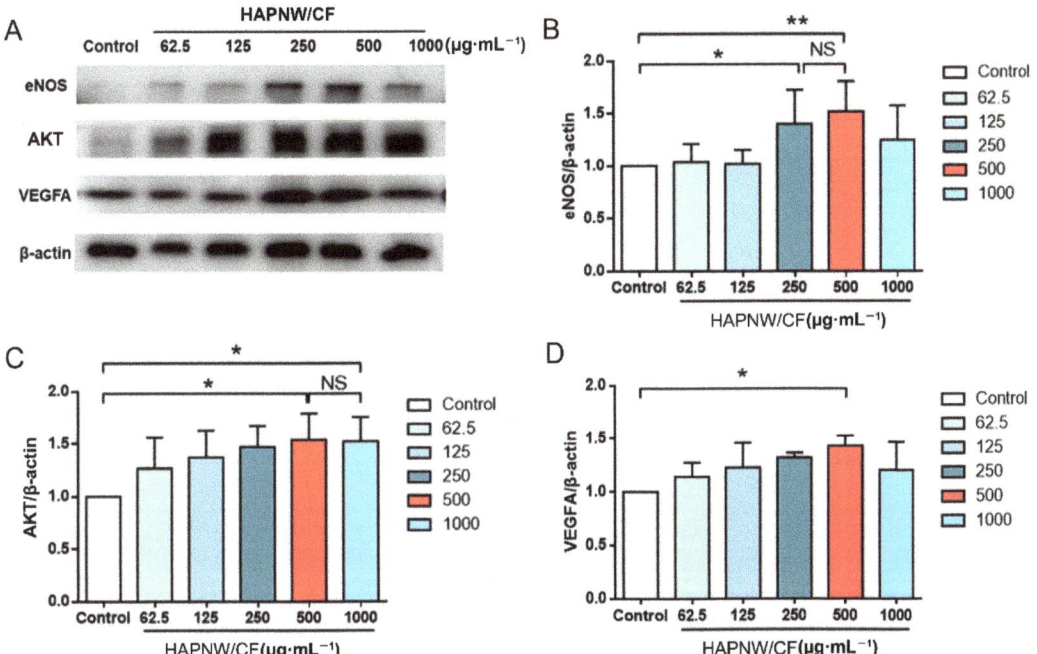

Figure 7. Effect of different doses (62.5, 125, 250, 500, 1000 µg·mL^{-1}) of HAPNW/CF (weight ratio 80:20) in HUVECs intervention. (**A–D**) Western blot analysis of protein expressions of eNOS, AKT, and VEGFA after 24 h intervention with different doses of HAPNW/CF (weight ratio 80:20; 62.5, 125, 250, 500, 1000 µg·mL^{-1}) (* $p < 0.05$; ** $p < 0.01$).

Tubeforming experiments (Figure 8E,F) indicate that compared with the control group and the HAPNW/CF + L-NAME treatment group, the HAPNW/CF treatment group has a stronger cell tubeforming ability, and the experimental results are statistically different ($p < 0.01$). There is also a statistical difference between the control group and the HAPNW/CF + L-NAME treatment group ($p < 0.05$). This means that HAPNW/CF can enhance the tubeforming ability of HUVECs, while L-NAME partially offsets this effect. The EdU proliferation experiments (Figure 8G,H) show that compared with the control group and the HAPNW/CF + L-NAME treatment group, the HAPNW/CF treatment group has a higher cell proliferation, and the experimental results are statistically different ($p < 0.01$). However, there is no statistical difference between the control group and the HAPNW/CF + L-NAME treatment group. The experimental results indicate that HAPNW/CF can enhance the proliferation ability of HUVECs, and L-NAME can counteract this effect.

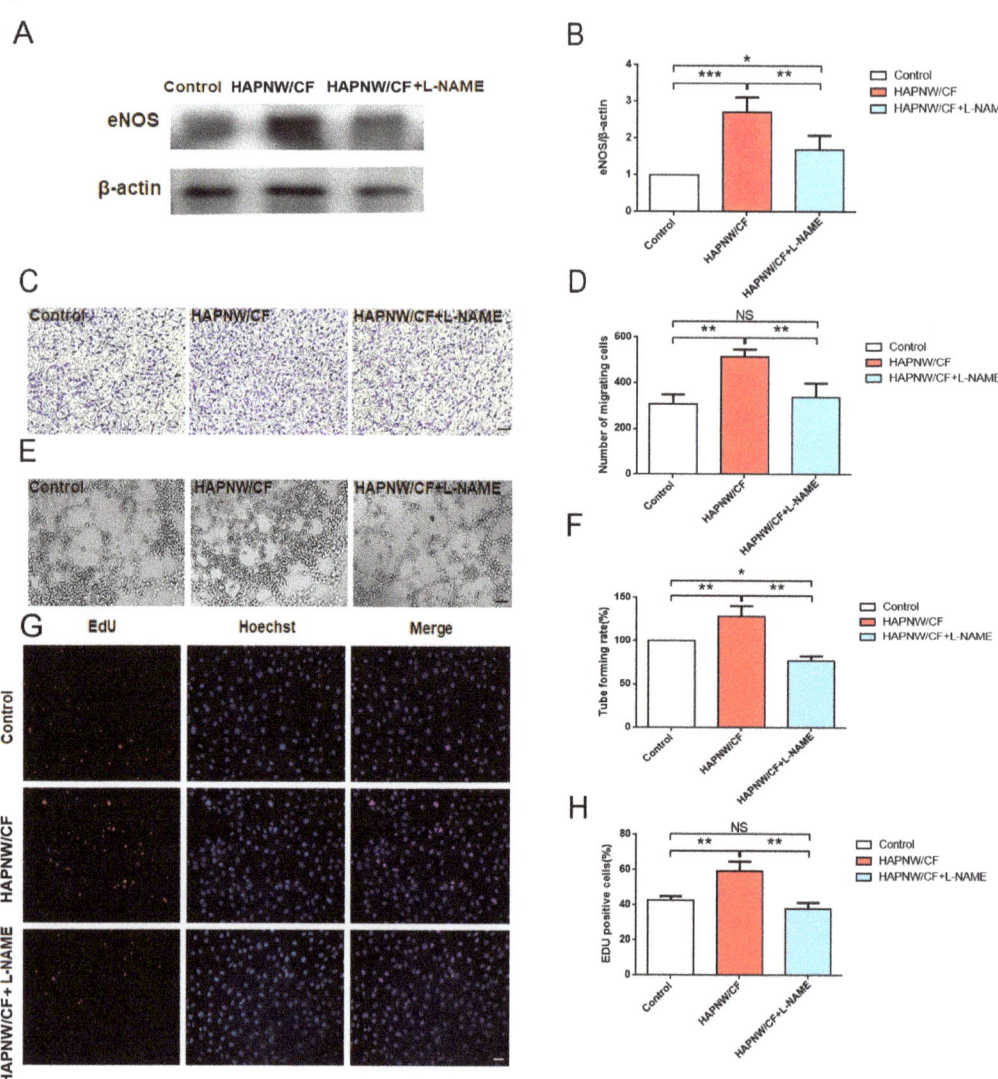

Figure 8. The effect of L-NAME on the ability of the HAPNW/CF biopaper to promote angiogenesis. (**A,B**) Western blot analysis of eNOS protein expressions in the control group, the HAPNW/CF (500 μg·mL^{-1}) group, and the HAPNW/CF (500 μg·mL^{-1}) + L-NAME (100 μM) group. (**C,D**) Transwell migration experiments, scale bar = 100 μm. (**E,F**) Tube-forming experiments, scale bar = 100 μm. (**G,H**) EdU proliferation tests, scale bar = 50 μm (* $p < 0.05$; ** $p < 0.01$; *** $p < 0.001$).

3.5. Effect of the HAPNW/CF Biopaper on the Expression of Pro-Angiogenic Protein in a Full-Thickness Skin Trauma Model

Western blot analysis was used to analyze the expression of angiogenesis-related proteins eNOS, AKT, and VEGFA on the 7th day after the HAPNW/CF biopaper intervention in the full-thickness skin trauma model (Figure 9). Compared with the control group, the expressions of eNOS, AKT, and VEGFA in the HAPNW/CF biopaper treatment group are significantly higher with statistical significance ($p < 0.05$). The experimental results indicate that the HAPNW/CF biopaper can significantly up-regulate the expression of angiogenesis-related proteins eNOS, AKT, and VEGFA in full-thickness skin trauma model of SD rats.

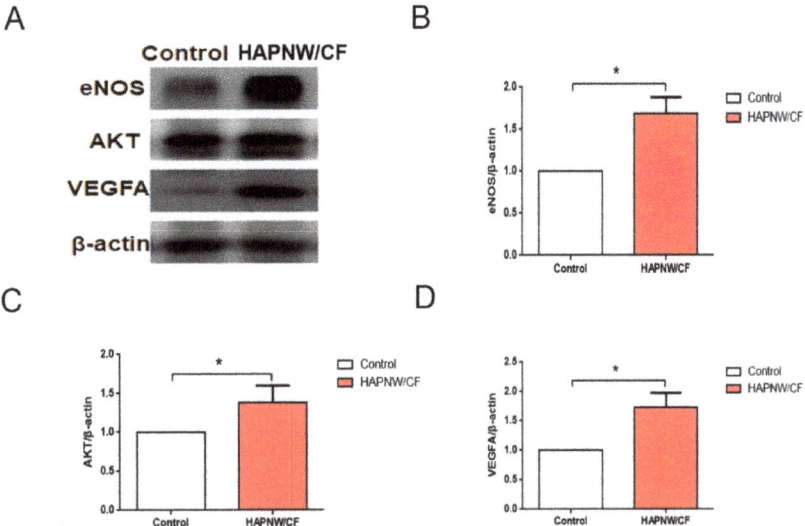

Figure 9. The effect of the HAPNW/CF biopaper on the expression of pro-angiogenic protein in a full-thickness skin trauma model. (**A–D**) Western blot analysis of eNOS, AKT, and VEGFA protein expressions in the control group and the HAPNW/CF biopaper group after intervention in the SD rat back skin full-thickness trauma model for 7 days (* $p < 0.05$).

3.6. Histological Analysis of Angiogenesis in Full-Thickness Skin Trauma Model

The H&E staining results after 7 days of the HAPNW/CF biopaper intervention in the full-thickness skin trauma model (Figure 10A,B) show that compared with the control group, the thickness of wound healing in the HAPNW/CF biopaper treatment group significantly increases, and the difference is statistically significant (* $p < 0.05$), indicating that the HAPNW/CF biopaper can promote skin wound healing, and the wound healing depends more on promoting the re-epithelialization of the skin wound rather than shrinking the skin around the wound. CD34 immunohistochemical staining (Figure 10C,D) shows that compared with the control group, the percentage of CD34-positive cells in the HAPNW/CF biopaper treatment group is higher, and the difference is statistically significant (* $p < 0.05$). The experimental results indicate that the HAPNW/CF biopaper can promote angiogenesis at the wound area, and thus promote skin wound healing.

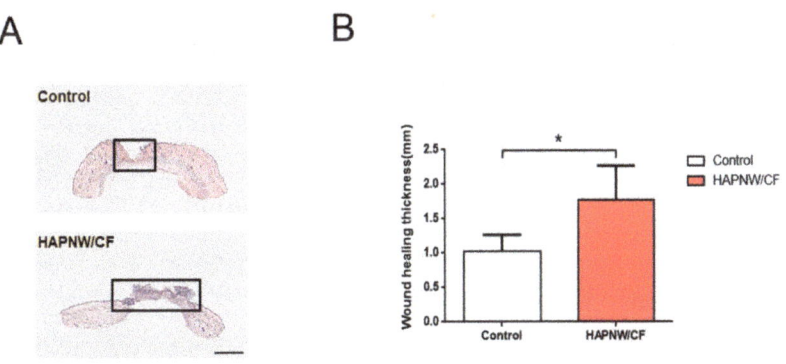

Figure 10. *Cont.*

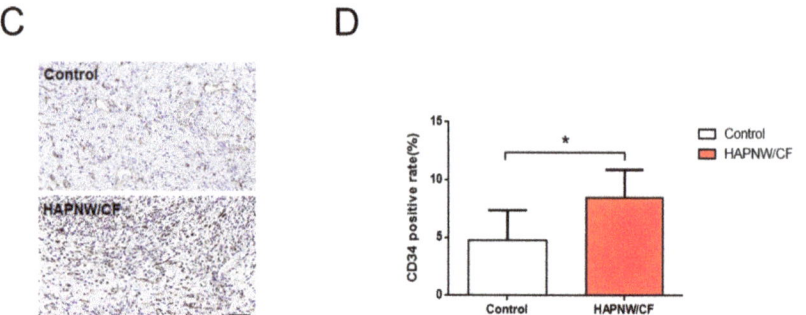

Figure 10. Histological analysis of angiogenesis in the presence of the HAPNW/CF biopaper in a full-thickness skin trauma model. (**A,B**) H&E staining of trauma 7 days after operation, scale bar = 2.5 mm. (**C,D**) CD34 immunohistochemical staining of trauma 7 days after operation, scale bar = 50 μm (* $p < 0.05$).

4. Discussion

The effective healing of skin wounds is essential to ensure the important barrier function of the skin. Skin wound healing is a very complex process that includes hemostasis, inflammation, angiogenesis, migration of various types of cells to the wound, and formation of extracellular matrix [35]. HAP, as a popular kind of bone defect repair biomaterial, can simultaneously induce angiogenesis and osteogenesis around the bone defect area, and may be a potential wound dressing. To this end, we fabricated a biopaper comprising ultralong HAPNWs and CFs by a simple vacuum-assisted filtration method. The as-prepared HAPNW/CF biopaper possesses a layered and porous structure for the ingrowth of new vessels. Both ultralong HAPNWs and CFs cooperate to improve the mechanical performance of the biopaper, which can also be bent and recover to its original state without obvious cracks. A large-sized HAPNW/CF biopaper (with a diameter of 9 cm) was also fabricated rapidly and conveniently for large-area skin wound dressing. The continuous release of Ca^{2+} ions from the degradation of the HAPNW/CF biopaper is beneficial to promoting skin wound healing. Moreover, the HAPNW/CF biopaper is highly hydrophilic and has high specific surface area, favorable for cell adhesion.

Our in vitro study revealed that the HAPNW/CF intervention in HUVECs treated with hypoxia can induce more expression of eNOS, AKT, and VEGFA compared with the control group, and this effect is related to the concentration of HAPNW/CF. The CCK-8 experiments also show that HAPNW/CF has high biocompatibility with HUVECs. HAPNW/CF at appropriate doses has the highest effectiveness and safety. The expressions of these angiogenesis-related proteins are conducive to angiogenesis and are very important for the progress of skin wound healing [36]. The new blood vessels are beneficial to the improvement of the ischemic and hypoxic conditions of the skin wound, and ensure the nutrient supply of the wound. They can also promote the formation of granulation tissue and re-epithelialization of the skin wound.

Pro-angiogenic factors can promote the formation of blood vessels. After the activation of endothelial cells, the cells form new blood vessels through adhesion, migration, and angiogenesis. The basement membrane of endothelial cells is degraded, the budding appears on the edge of the skin wound, and a new vascular system is formed [37]. Our further research has found that the pro-angiogenic effect of HAPNW/CF at the cellular levels can be partially offset by L-NAME as the specific eNOS inhibitor. HAPNW/CF can induce cells to express more eNOS. However, when HAPNW/CF and L-NAME are applied at the same time, the expression of eNOS is partially suppressed. Additionally, HAPNW/CF can also enhance the proliferation, migration, and tube formation capabilities of HUVECs, and this effect can be completely or partially offset when HAPNW/CF and L-NAME are used at the same time. These results suggest that HAPNW/CF promotes angiogenesis partly through the eNOS signaling pathway. The eNOS is the major source of NO in the cardiovascular system [38]. Under ischemic and hypoxic conditions, NO can play a role through the phosphatidylinositol-3-kinase (PI3K) signaling pathway, affecting the activation, recruitment, migration, and differentiation in endothelial cells [39]. Moreover, HAPNW/CF can continuously release Ca^{2+} ions, which may be a key early trigger for skin wound healing [40]. The effects of Ca^{2+} ions on cell proliferation, differentiation, signal transduction, and gene expression were also reported [41]. This may further promote HAPNW/CF-mediated wound healing.

The effect of the HAPNW/CF biopaper on promoting wound healing has also been proven in animal models. The latest research shows that this kind of material has obvious advantages in wound healing. Compared with traditional materials, it has better hemostasis, excellent cell compatibility, and stronger skin wound healing function [42]. In the rat model of hypoxia and ischemia, we found higher levels of eNOS, AKT, and VEGFA proteins expressed in the HAPNW/CF biopaper intervention group than controls. It has been found that the HAPNW/CF biopaper could also induce more expressions of angiogenesis-related factors in the animal model and promote skin wound healing. The histological analysis further confirmed this point. The H&E staining showed that the HAPNW/CF biopaper group had a thicker re-epithelialization level than the control group. The healing of skin wounds was mainly achieved by restoring the thickness of the skin re-epithelialization, rather than the shrinkage of the wound. The CD34 staining experiments showed more angiogenesis in the HAPNW/CF biopaper group than in the control group, indicating that the HAPNW/CF biopaper could promote skin wound healing by inducing more angiogenesis. Moreover, it benefits from the material itself as an excellent wound dressing for the physical protection of wounds [43]. Some studies reported that dressings containing calcium could better promote wound healing [44]. Due to the characteristics of the HAPNW/CF biopaper, the material has the characteristics of high porosity and high specific surface area, which are conducive to the migration of cells and diffusion of nutrients. Studies show that the surface topography of nanomaterials plays an important role in guiding the orientation of microvessels and the morphogenesis of blood vessels [45]. The existence of pores in nanostructured materials provides a basis for cell infiltration and blood vessel formation. The existence of fibers in the HAPNW/CF biopaper is favorable for guiding the directional migration and proliferation of cells during angiogenesis [46]. This angiogenic effect of nanomaterials is significantly dose- and size-dependent [47]. Wound dressings with excellent biocompatibility, biodegradability, non-toxicity, and functionality have great clinical application prospects [48,49]. We hope our work can improve wound healing treatment for more patients and solve more clinical problems.

5. Conclusions

In conclusion, we developed a HAPNW/CF biopaper consisting of ultralong HAP nanowires and CFs with adequate flexibility, mechanical properties, and superhydrophilicity for promoting skin wound healing. The as-prepared HAPNW/CF biopaper has a porous structure with high porosity, which is favorable for blood vessel formation and ingrowth. In addition, the HAPNW/CF biopaper can continuously release Ca^{2+} ions to promote skin wound healing. We have proven that the HAPNW/CF biopaper has high biocompatibility and the ability to induce angiogenesis for promoting the skin wound healing. Furthermore, we have revealed that the angiogenesis promotion of the HAPNW/CF biopaper is achieved through the activation of eNOS-related signaling pathways. The animal experiments show that the HAPNW/CF biopaper can promote thicker skin re-epithelialization and the formation of new blood vessels, thus promoting the healing of the wound. It is expected that the HAPNW/CF biopaper is a promising and effective dressing for accelerating wound healing with strong potential in clinical applications.

Author Contributions: Conceptualization, J.G. and L.-S.H.; methodology, J.G., L.-S.H. and B.-B.N.; data curation, Y.-K.Z., J.-B.G. and H.-W.R.; writing—original draft preparation, J.G. and L.-S.H.; writing—review and editing, J.G., L.-S.H., B.-B.N., Y.-K.Z., J.-B.G., H.-W.R., H.-P.Y., Y.-J.Z. and J.-L.D.; project administration, H.-P.Y., Y.-J.Z. and J.-L.D.; funding acquisition, J.-L.D. All authors have read and agreed to the published version of the manuscript.

Funding: This research was funded by the National Natural Science Foundation of China (12074253, 12174289), and the Medical and Industrial Cross Research Fund of "Star of Jiaotong University" Program of Shanghai Jiao Tong University (YG2019ZDA21).

Institutional Review Board Statement: This study's experimental protocol was approved by the Ethics Committee of Xinhua Hospital Affiliated to the Shanghai Jiao Tong University School of Medicine under approval number XHEC-F-2021-079, and the experiments were in accordance with relevant institutional guidelines.

Informed Consent Statement: Not applicable.

Data Availability Statement: Not applicable.

Conflicts of Interest: The authors declare no conflict of interest.

References

1. Subramaniam, T.; Fauzi, M.B.; Lokanathan, Y.; Law, J.X. The role of calcium in wound healing. *Int. J. Mol. Sci.* **2021**, *22*, 6486. [CrossRef] [PubMed]
2. Järbrink, K.; Ni, G.; Sönnergren, H.; Schmidtchen, A.; Pang, C.; Bajpai, R.; Car, J. The humanistic and economic burden of chronic wounds: A protocol for a systematic review. *Syst. Rev.* **2017**, *6*, 15. [CrossRef]
3. Keyes, B.E.; Liu, S.; Asare, A.; Naik, S.; Levorse, J.; Polak, L.; Lu, C.P.; Nikolova, M.; Pasolli, H.A.; Fuchs, E. Impaired epidermal to dendritic T cell signaling slows wound repair in aged skin. *Cell* **2016**, *167*, 1323–1338. [CrossRef]
4. Kim, H.S.; Sun, X.; Lee, J.H.; Kim, H.W.; Fu, X.; Leong, K.W. Advanced drug delivery systems and artificial skin grafts for skin wound healing. *Adv. Drug Deliv. Rev.* **2019**, *146*, 209–239. [CrossRef] [PubMed]
5. Carmeliet, P. Angiogenesis in health and disease. *Nat. Med.* **2003**, *9*, 653–660. [CrossRef]
6. Liu, W.; Zhang, G.; Wu, J.; Zhang, Y.; Liu, J.; Luo, H.; Shao, L. Insights into the angiogenic effects of nanomaterials: Mechanisms involved and potential applications. *J. Nanobiotechnol.* **2020**, *18*, 9. [CrossRef]
7. Hendrickx, B.; Verdonck, K.; Van den Berge, S.; Dickens, S.; Eriksson, E.; Vranckx, J.J.; Luttun, A. Integration of blood outgrowth endothelial cells in dermal fibroblast sheets promotes full thickness wound healing. *Stem Cells* **2010**, *28*, 1165–1177. [CrossRef] [PubMed]
8. Risau, W. Mechanisms of angiogenesis. *Nature* **1997**, *386*, 671–674. [CrossRef]
9. Senger, D.R.; Davis, G.E. Angiogenesis. *Cold Spring Harb. Perspect. Biol.* **2011**, *3*, a005090. [CrossRef]
10. Greco Song, H.H.; Rumma, R.T.; Ozaki, C.K.; Edelman, E.R.; Chen, C.S. Vascular tissue engineering: Progress, challenges, and clinical promise. *Cell Stem Cell* **2018**, *22*, 340–354. [CrossRef]
11. Bae, H.; Puranik, A.S.; Gauvin, R.; Edalat, F.; Carrillo-Conde, B.; Peppas, N.A.; Khademhosseini, A. Building vascular networks. *Sci. Transl. Med.* **2012**, *4*, 160ps23. [CrossRef] [PubMed]
12. Yang, Z.; Huang, R.; Zheng, B.; Guo, W.; Li, C.; He, W.; Wei, Y.; Du, Y.; Wang, H.; Wu, D.; et al. Highly stretchable, adhesive, biocompatible, and antibacterial hydrogel dressings for wound healing. *Adv. Sci.* **2021**, *8*, 2003627. [CrossRef]
13. Guo, M.; Wang, Y.; Gao, B.; He, B. Shark tooth-inspired microneedle dressing for intelligent wound management. *ACS Nano* **2021**, *15*, 15316–15327. [CrossRef]
14. Kim, M.S.; Oh, G.W.; Jang, Y.M.; Ko, S.C.; Park, W.S.; Choi, I.W.; Kim, Y.M.; Jung, W.K. Antimicrobial hydrogels based on PVA and Diphlorethohydroxycarmalol (DPHC) derived from brown alga Ishige okamurae: An in vitro and in vivo study for wound dressing application. *Mater. Sci. Eng. C* **2020**, *107*, 110352. [CrossRef] [PubMed]
15. Qiu, X.; Zhang, J.; Cao, L.; Jiao, Q.; Zhou, J.; Yang, L.; Zhang, H.; Wei, Y. Antifouling antioxidant zwitterionic dextran hydrogels as wound dressing materials with excellent healing activities. *ACS Appl. Mater. Interfaces* **2021**, *13*, 7060–7069. [CrossRef] [PubMed]
16. Liu, H.; Li, Z.; Zhao, Y.; Feng, Y.; Zvyagin, A.V.; Wang, J.; Yang, X.; Yang, B.; Lin, Q. Novel diabetic foot wound dressing based on multifunctional hydrogels with extensive temperature-tolerant, durable, adhesive, and intrinsic antibacterial properties. *ACS Appl. Mater. Interfaces* **2021**, *13*, 26770–26781. [CrossRef] [PubMed]
17. Zhao, X.; Liang, Y.P.; Huang, Y.; He, J.H.; Han, Y.; Guo, B.L. Physical double-network hydrogel adhesives with rapid shape adaptability, fast self-healing, antioxidant and NIR/pH stimulus-responsiveness for multidrug-resistant bacterial infection and removable wound dressing. *Adv. Funct. Mater.* **2020**, *30*, 1910748. [CrossRef]
18. Deng, C.; Xu, C.; Zhou, Q.; Cheng, Y. Advances of nanotechnology in osteochondral regeneration. *Wiley Interdiscip. Rev. Nanomed. Nanobiotechnol.* **2019**, *11*, e1576. [CrossRef]

19. Mofazzal Jahromi, M.A.; Sahandi Zangabad, P.; Moosavi Basri, S.M.; Sahandi Zangabad, K.; Ghamarypour, A.; Aref, A.R.; Karimi, M.; Hamblin, M.R. Nanomedicine and advanced technologies for burns: Preventing infection and facilitating wound healing. *Adv. Drug Deliv. Rev.* **2018**, *123*, 33–64. [CrossRef]
20. Augustine, R.; Prasad, P.; Khalaf, I.M.N. Therapeutic angiogenesis: From conventional approaches to recent nanotechnology-based interventions. *Mater. Sci. Eng. C* **2019**, *97*, 994–1008. [CrossRef]
21. Lu, B.Q.; Zhu, Y.J. One-dimensional hydroxyapatite materials: Preparation and applications. *Can. J. Chem.* **2017**, *95*, 1091–1102. [CrossRef]
22. Zhu, Y.J.; Lu, B.Q. Deformable biomaterials based on ultralong hydroxyapatite nanowires. *ACS Biomater. Sci. Eng.* **2019**, *5*, 4951–4961. [CrossRef] [PubMed]
23. Son, J.; Kim, J.; Lee, K.; Hwang, J.; Choi, Y.; Seo, Y.; Jeon, H.; Kang, H.C.; Woo, H.M.; Kang, B.J.; et al. DNA aptamer immobilized hydroxyapatite for enhancing angiogenesis and bone regeneration. *Acta Biomater.* **2019**, *99*, 469–478. [CrossRef]
24. Song, Y.; Wu, H.; Gao, Y.; Li, J.; Lin, K.; Liu, B.; Lei, X.; Cheng, P.; Zhang, S.; Wang, Y.; et al. Zinc silicate/nano-hydroxyapatite/collagen scaffolds promote angiogenesis and bone regeneration via the p38 MAPK pathway in activated monocytes. *ACS Appl. Mater. Interfaces* **2020**, *12*, 16058–16075. [CrossRef] [PubMed]
25. Ji, X.; Yuan, X.; Ma, L.; Bi, B.; Zhu, H.; Lei, Z.; Liu, W.; Pu, H.; Jiang, J.; Jiang, X.; et al. Mesenchymal stem cell-loaded thermosensitive hydroxypropyl chitin hydrogel combined with a three-dimensional-printed poly(ε-caprolactone)/nano-hydroxyapatite scaffold to repair bone defects via osteogenesis, angiogenesis and immunomodulation. *Theranostics* **2020**, *10*, 725–740. [CrossRef]
26. Dong, T.; Duan, C.; Wang, S.; Gao, X.; Yang, Q.; Yang, W.; Deng, Y. Multifunctional surface with enhanced angiogenesis for improving long-term osteogenic fixation of poly (ether ether ketone) implants. *ACS Appl. Mater. Interfaces* **2020**, *12*, 14971–14982. [CrossRef]
27. Huang, G.J.; Yu, H.P.; Wang, X.L.; Ning, B.B.; Gao, J.; Shi, Y.Q.; Zhu, Y.J.; Duan, J.L. Highly porous and elastic aerogel based on ultralong hydroxyapatite nanowires for high-performance bone regeneration and neovascularization. *J. Mater. Chem. B* **2021**, *9*, 1277–1287. [CrossRef]
28. Zhu, Y.J. Multifunctional fire-resistant paper based on ultralong hydroxyapatite nanowires. *Chin. J. Chem.* **2021**, *39*, 2296–2314. [CrossRef]
29. Salama, A. Recent progress in preparation and applications of chitosan/calcium phosphate composite materials. *Int. J. Biol. Macromol.* **2021**, *178*, 240–252. [CrossRef]
30. Kawai, K.; Larson, B.J.; Ishise, H.; Carre, A.L.; Nishimoto, S.; Longaker, M.; Lorenz, H.P. Calcium-based nanoparticles accelerate skin wound healing. *PLoS ONE* **2011**, *6*, e27106. [CrossRef]
31. Lu, B.Q.; Zhu, Y.J.; Chen, F. Highly flexible and nonflammable inorganic hydroxyapatite paper. *Chem. Eur. J.* **2014**, *20*, 1242–1246. [CrossRef] [PubMed]
32. Zhang, Y.G.; Zhu, Y.J.; Chen, F.; Wu, J. Ultralong hydroxyapatite nanowires synthesized by solvothermal treatment using a series of phosphate sodium salts. *Mater. Lett.* **2015**, *144*, 135–137. [CrossRef]
33. Jiang, Y.Y.; Zhu, Y.J.; Chen, F.; Wu, J. Solvothermal synthesis of submillimeter ultralong hydroxyapatite nanowires using a calcium oleate precursor in a series of monohydroxy alcohols. *Ceram. Int.* **2015**, *41*, 6098–6102. [CrossRef]
34. Huang, J.J.; Shi, Y.Q.; Li, R.L.; Hu, A.; Lu, Z.Y.; Weng, L.; Han, Y.P.; Wang, S.Q.; Zhang, L.; Hao, C.N.; et al. Therapeutic ultrasound protects huvecs from Ischemia/Hypoxia-induced apoptosis via the PI3K-akt pathway. *Am. J. Transl. Res.* **2017**, *9*, 1990–1999.
35. Reinke, J.M.; Sorg, H. Wound repair and regeneration. *Eur. Surg. Res.* **2012**, *49*, 35–43. [CrossRef] [PubMed]
36. Wan, X.; Liu, S.; Xin, X.; Li, P.; Dou, J.; Han, X.; Kang, I.-K.; Yuan, J.; Chi, B.; Shen, J. S-nitrosated keratin composite mats with no release capacity for wound healing. *Chem. Eng. J.* **2020**, *400*, 125964. [CrossRef]
37. Tonnesen, M.G.; Feng, X.; Clark, R.A. Angiogenesis in wound healing. *J. Investig. Dermatol. Symp. Proc.* **2000**, *5*, 40–46. [CrossRef]
38. Dudzinski, D.M.; Igarashi, J.; Greif, D.; Michel, T. The regulation and pharmacology of endothelial nitric oxide synthase. *Annu. Rev. Pharmacol. Toxicol.* **2006**, *46*, 235–276. [CrossRef]
39. Balligand, J.L.; Feron, O.; Dessy, C. eNOS Activation by physical forces: From short-term regulation of contraction to chronic remodeling of cardiovascular tissues. *Physiol. Rev.* **2009**, *89*, 481–534. [CrossRef]
40. Wood, W. Wound healing: Calcium flashes illuminate early events. *Curr. Biol.* **2012**, *22*, R14–R16. [CrossRef]
41. Pastar, I.; Stojadinovic, O.; Yin, N.C.; Ramirez, H.; Nusbaum, A.G.; Sawaya, A.; Patel, S.B.; Khalid, L.; Isseroff, R.R.; Tomic-Canic, M. Epithelialization in wound healing: A comprehensive review. *Adv. Wound Care* **2014**, *3*, 445–464. [CrossRef] [PubMed]
42. Zheng, Y.; Ma, W.; Yang, Z.; Zhang, H.; Ma, J.; Li, T.; Niu, H.; Zhou, Y.; Yao, Q.; Chang, J.; et al. An ultralong hydroxyapatite nanowire aerogel for rapid hemostasis and wound healing. *Chem. Eng. J.* **2022**, *430*, 132912. [CrossRef]
43. Dhand, C.; Venkatesh, M.; Barathi, V.A.; Harini, S.; Bairagi, S.; Goh Tze Leng, E.; Muruganandham, N.; Low, K.Z.W.; Fazil, M.; Loh, X.J.; et al. Bio-inspired crosslinking and matrix-drug interactions for advanced wound dressings with long-term antimicrobial activity. *Biomaterials* **2017**, *138*, 153–168. [CrossRef]
44. Zhao, W.Y.; Fang, Q.Q.; Wang, X.F.; Wang, X.W.; Zhang, T.; Shi, B.H.; Zheng, B.; Zhang, D.D.; Hu, Y.Y.; Ma, L.; et al. Chitosan-calcium alginate dressing promotes wound healing: A preliminary study. *Wound Repair Regen.* **2020**, *28*, 326–337. [CrossRef]
45. Sharma, D.; Ross, D.; Wang, G.; Jia, W.; Kirkpatrick, S.J.; Zhao, F. Upgrading prevascularization in tissue engineering: A review of strategies for promoting highly organized microvascular network formation. *Acta Biomater.* **2019**, *95*, 112–130. [CrossRef] [PubMed]

46. Formentín, P.; Catalán, Ú.; Fernández-Castillejo, S.; Alba, M.; Baranowska, M.; Solà, R.; Pallarès, J.; Marsal, L.F. Human aortic endothelial cell morphology influenced by topography of porous silicon substrates. *J. Biomater. Appl.* **2015**, *30*, 398–408. [CrossRef] [PubMed]
47. Yu, H.; Chen, X.; Cai, J.; Ye, D.; Wu, Y.; Fan, L.; Liu, P. Novel porous three-dimensional nanofibrous scaffolds for accelerating wound healing. *Chem. Eng. J.* **2019**, *369*, 253–262. [CrossRef]
48. Bibire, T.; Yilmaz, O.; Ghiciuc, C.M.; Bibire, N.; Dănilă, R. Biopolymers for surgical applications. *Coatings* **2022**, *12*, 211. [CrossRef]
49. Hamza, R.Z.; Al-Motaani, S.E.; Al-Talhi, T. Therapeutic and ameliorative effects of active compounds of combretum molle in the treatment and relief from wounds in a diabetes mellitus experimental model. *Coatings* **2021**, *11*, 324. [CrossRef]

Article

Preparation of Graphene Oxide Composites and Assessment of Their Adsorption Properties for Lanthanum (III)

Jie Zhou [1,2], Xiaosan Song [1,2,*], Boyang Shui [1,2] and Sanfan Wang [1,2]

1 School of Environment and Municipal Engineering, Lanzhou Jiaotong University, No. 88, Anning West Road, Lanzhou 730070, China; 0619100@stu.lzjtu.edu.cn (J.Z.); 201701131@stu.lzjtu.edu.cn (B.S.); wsf1612@mail.lzjtu.cn (S.W.)
2 Engineering Research Center of Comprehensive Utilization of Water Resources in Cold and Drought Areas, Ministry of Education, No. 88, Anning West Road, Lanzhou 730070, China
* Correspondence: songxs@mail.lzjtu.cn

Abstract: In this study, graphene oxide (GO) was prepared using the improved Hummers' method, and GO was carboxylated and modified into hydroxylated graphene oxide (GOH). Diatomaceous earth (DE), which exhibits stable chemical properties, a large specific surface area, and high porosity, as well as chitosan/magnetic chitosan, was loaded by solution blending. Subsequently, carboxylated graphene oxide/diatomite/chitosan (GOH/DCS) and carboxylated graphene oxide/diatomite/magnetic chitosan (GOH/DMCS) composites were prepared through simple solid–liquid separation. The results showed that the modified GOH/DCS and GOH/DMCS composites could be used to remove lanthanum La(III)), which is a rare earth element. Different factors, such as initial solution concentration, pH of the solution, adsorbent dosage, adsorption contact time, and adsorption reaction temperature, on adsorption, were studied, and the adsorption mechanism was explored. An adsorption–desorption recycling experiment was also used to evaluate the recycling performance of the composite material. The results show that at the initial solution concentration of 50 mg·g^{-1}, pH = 8.0, 3 g·L^{-1} adsorbent dosage, reaction temperature of 45 °C, and adsorption time of 50 min, the adsorption effect is the best. The adsorption process is more in line with the pseudo-second-order kinetic model and Langmuir model, and the internal diffusion is not the only controlling effect. The adsorption process is an endothermic and spontaneous chemical adsorption process. The maximum adsorption capacity of GOH/DMCS for La(III) at 308K is 302.51 mg/g through model simulation. After four adsorption–desorption cycles, the adsorption capacity of the GOH/DMCS composite for La(III) initially exceeded 74%. So, GOH/DMCS can be used as a reusable and efficient adsorbent.

Keywords: graphene oxide; composite materials; lanthanum; adsorption

Citation: Zhou, J.; Song, X.; Shui, B.; Wang, S. Preparation of Graphene Oxide Composites and Assessment of Their Adsorption Properties for Lanthanum (III). *Coatings* **2021**, *11*, 1040. https://doi.org/10.3390/coatings11091040

Academic Editor: Philippe Evon

Received: 8 August 2021
Accepted: 24 August 2021
Published: 29 August 2021

Publisher's Note: MDPI stays neutral with regard to jurisdictional claims in published maps and institutional affiliations.

Copyright: © 2021 by the authors. Licensee MDPI, Basel, Switzerland. This article is an open access article distributed under the terms and conditions of the Creative Commons Attribution (CC BY) license (https://creativecommons.org/licenses/by/4.0/).

1. Introduction

Lanthanum (La) is a metal used in optical glasses, alloys, catalysts, and ceramics [1]. To extract trace elements of this rare earth metal, significant quantities of water and chemical reagents are needed, through a variety of chemical procedures. As a result, chemical reagents, rare earth elements, and radioactive substances end up in wastewater, which destroys surrounding vegetation, causing serious environmental pollution and affecting the lives of nearby residents [2].

Treating wastewater that contains rare earth elements is a complex and universal ecological problem. Membrane separation, photocatalysis, electrolysis, ion exchange, and adsorption are commonly used for wastewater treatment. The adsorption method is widely used because of its many advantages, such as low cost, a wide range of applications, ease of operation, and effectiveness in removing pollutants from the water [3]. For instance, Iannicelli-Zubiani et al. [4] developed an effective adsorbent for aqueous rare earths recovery, in which activated carbon (AC) was modified with pentaethylenehexamine. The strong improvement in the efficiency values detected by using modified carbons (uptake

100% until initial concentrations of about 2600 ppm and release over 95%) demonstrated that the coordination mechanism due to the modifying agent is effective. Abdel-Magied et al. [5] studied the adsorption properties of rare earth lanthanum by hierarchical porous zeolitic imidazolate frameworks nanoparticles (ZIF-8 NPs) prepared by a template-free method at room temperature using water. Adsorption equilibrium was reached after 7 h and moderate adsorption capacity was obtained for lanthanum (28.8 mg·g^{-1}) at a pH of 7.0. Kusrini et al. [6] investigated a biosorbent derived from the inner part of durian (Durio zibethinus) rinds. La ion was efficiently adsorbed by the biosorbent with optimum adsorption capacity as high as 71 mg La per gram biosorbent. The removal of lanthanum is due to the favorable chelation and strong chemical interactions between the functional groups on the surface of the biosorbent and the metal ions. If the REE concentration in wastewater is high, biological adsorption, extraction, and traditional zeolite, clay, and activated carbon adsorption can be used. Among the above methods, adsorption has been widely concerned and applied because of its simple operation, low cost, and high efficiency. Nanomaterials (NMs) provide a promising technology for wastewater with very low REE concentration because of its potential high adsorption efficiency as an adsorbent. Thus, research on the preparation and modification of efficient and reusable nanomaterials adsorbents is an essential part of realizing sustainable industrial development, protecting the environment and reducing pollution.

Nanomaterial adsorbents have many excellent physical and chemical properties, but there are limits to their applications [7]. Graphene oxide (GO), a graphene derivative, easily aggregates or accumulates between layers, with a decreased adsorption specific surface area that can weaken its adsorption effects [8,9]. Thus, modified GO materials have become a hot research topic. Nanocomposites are sometimes known as free radical substitutes in traditionally filled or blended polymers [10]. However, graphene oxide sheets can only be dispersed in aqueous media that are incompatible with most organic polymers due to their hydrophilic properties.

Presently, there are only two methods to modify the hydrophilicity of graphene oxide: magnetic modification [10] and chemical functionalization. In chemical functionalization, the exfoliation behavior of graphite oxide can be changed by modifying its surface properties, and functional groups such as amide and carbamate bonds will form on the carboxyl and hydroxyl groups of graphite oxide, respectively. As a result, this reduces the hydrophilicity of graphene oxide sheets [11].

Nanomaterials have a paradoxical combination of excellent performance with restricting limitations [12]. This makes them a vital topic for developing and preparing high-efficiency and recyclable green adsorbents to improve the adsorption and removal of water pollutants by the modification of graphene oxide-based materials. Specifically, one area of concern is that graphene oxide-based composite materials are difficult to separate from water [13]. Therefore, this study provides a reference for the removal of heavy metals from wastewater, to alleviate water pollution and facilitate ecological construction.

In this work, magnetic modified chitosan (CS) was compounded with hydroxylated graphene oxide (GOH) and diatomaceous earth (DE). This study focused on enhancing the adsorption capacity and effectively reducing the hydrophilicity of graphene oxide sheets, as well as creating graphene oxide-based composite materials that are easily separated in water to realize material recycling. We assessed the adsorption behavior of La(III) by diatomite and magnetic chitosan-modified graphene oxide-based adsorbents. In addition, scanning electron microscopy (SEM), Fourier transform infrared spectroscopy (FTIR), X-ray diffraction (XRD), zeta potential analysis (Zeta potential), and a vibrating sample magnetometer (PPMS-VSM) were used to characterize the materials, surface morphology, chemical structures, compositions, and their physical and magnetic properties. The modified composite materials (GOH/DCS and GOH/DMCS) were used for static adsorption experiments on the target pollutant, La(III), and the effects of five factors on the adsorption effect were investigated. Specifically, the initial concentration of the solution, pH value of the solution, adsorbent dosage, adsorption contact time, and the adsorption reaction temperature were assessed. The adsorption process was fitted and analyzed by adsorption kinetics, isotherms, and thermodynamics. Then, regeneration experiments to assess the adsorption saturated GOH/DMCS were carried out to evaluate the regeneration performance of the composites.

2. Materials and Methods

2.1. Chemicals and Reagents

All chemical reagents used for testing were analytically pure or of a higher grade, including graphite powder, H_2SO_4, H_3PO_4, $KMnO_4$, H_2O_2, HCl, absolute ethanol, NaOH, $C_2H_3ClO_2$, diatomite, chitosan (CS, deacetylation degree > 90%), CH_3COOH, nano Fe_3O_4, and $C_5H_8O_2$. All aqueous solutions were prepared with deionized water.

2.2. Preparation of Materials

Graphene oxide was prepared using the improved Hummers' method [14]. The prepared graphite oxide was ultrasonically treated in deionized water for 30 min to fully disperse and exfoliate the graphite oxide, and the GO hydrosol concentration was 2 g/L. Then, 2.0 g of sodium hydroxide was added and stirred until completely dissolved, which was followed by the addition of 3.0 g of chloroacetic acid, which was ultrasonically dispersing for 1 h. Then, the solution was magnetically stirred for 24 h under constant heating at a temperature of 25 °C, and the mixed solution was centrifuged and washed with anhydrous ethanol three times. Afterward, the mixture was washed with ample amounts of deionized water several times until the pH was close to 6.0. Then, the mixed solution was dried in an oven at 60 °C and grounded and sieved with a 200-mesh sieve to obtain the carboxylated graphite oxide. Finally, the carboxylated graphite oxide was uniformly dispersed in quantitative distilled water by ultrasonic stripping to prepare the carboxylated graphene oxide hydrosol (GOH) at a concentration of 2 g/L.

To prepare the magnetic chitosan, 0.3 g of chitosan was dissolved in 50 mL of 2% glacial acetic acid solution and heated and stirred until completely dissolved. Then, 0.15 g of nano Fe_3O_4 was added to the above solution and mechanically stirred until it was thoroughly dispersed in the chitosan solution. Then, 2.0 mL of glutaraldehyde was added to the mixed solution and mechanically stirred at 60 °C for 4 h. The obtained product was washed with ethanol and deionized water several times until it reached a pH of 6, and then, it was dried in a vacuum drying oven at 60 °C for 12 h. Finally, magnetic chitosan (MCS) was prepared by grinding and sieving with a 200-mesh sieve.

The prepared carboxylated graphite oxide was ultrasonically treated in deionized water for 30 min, and the GOH hydrosol concentration was 2 g/L. First, 1.0 g of chitosan was weighed and added to 50 mL of 3% glacial acetic acid solution, heated, and stirred until it was fully dissolved. Then, 3 g of diatomite was added into the GO-COOH suspension, which was returned to room temperature. Then, it was ultrasonicated for 1 h, magnetically stirred for 12 h, and vacuum filtrated after washing with water. Then, the material was dried to achieve a constant weight in a drying oven at 60 °C, after which it was removed, ground, and placed through a 200-mesh sieve to prepare the carboxylated graphene oxide/diatomite/chitosan (GOH/DCS) material.

The prepared carboxylated graphite oxide was ultrasonically treated in deionized water for 30 min. The concentration of the GO hydrosol was 2 g/L. First, 1.0 g of magnetic chitosan and 3 g of diatomite were weighed and added to the suspension of GO-COOH in turn, ultrasonicated for 1 h, and magnetically stirred for 12 h. The obtained product was washed with ethanol and deionized water several times until the pH was 6 and then dried in a vacuum drying oven at 60 °C for 12 h. Finally, the product was grounded and screened with a 200-mesh sieve to prepare the carboxylated graphene oxide/diatomite/magnetic chitosan (GOH/DMCS) composite material.

2.3. Testing and Characterization

The materials were characterized by scanning electron microscope (SEM), Fourier transform infrared spectroscopy (FT-IR), X-ray diffraction (XRD), zeta potential analysis, and vibrating sample magnetometer (PPMS-VSM). The basic information of the instruments is shown in Table 1.

Table 1. Main characterization instruments.

Instrument	Model	Manufacturer
SEM	TESCAN MIRA3	Tesken Trading Co., Ltd., Shanghai, China
FT-IR	VERTEX70	Swiss Brook
XRD	D/MAX-2500/PC	Nippon Neotoku Corporation
Zeta potential	Zetasizer Nano S90	Malvern Instrument Co. Ltd., Worcestershire, UK
PPMS-VSM	PPMS-9	Quantum Design, San Diego, CA, USA

2.4. Adsorption Experiment

To study the adsorption behavior of GOH/DCS and GOH/DMCS, several groups of 25 mL and 25 mg/L of rare earth La(III) solutions were taken, and the effects of diatomite dosage, adsorption time, pH value, adsorption temperature, adsorbent dosage, and initial concentration of La(III) solution on the adsorption of La(III) by composite materials were studied. Next, a filter with a 0.45-µm pore size and the absorbance of the filtered supernatant with arsenazo III spectrophotometer was measured, the adsorption efficiency (w) and adsorption capacity (q_e) were calculated, and the pH value of each suspension with hydrochloric acid and sodium hydroxide was controlled.

The equilibrium adsorption capacity (q_e) and adsorption rate (w) of La(III) by GOH/DCS and GOH/DMCS were calculated according to Equation (1) and Equation (2), respectively.

$$q_e = \frac{(C_0 - C_e)V}{m} \tag{1}$$

$$w = \frac{(C_0 - C_e)}{C_0} \times 100\% \tag{2}$$

where q_e is the adsorption capacity (mg/g); C_0 is the initial concentration of the solution, mg/L; C_e is the concentration of La(III) in the solution after adsorption equilibrium, mg/L; V is the liquid volume, mL; m is the amount of adsorbent (mg); and w is the adsorption rate (%).

Magnetic separation was carried out on the adsorption saturated GOH/DMCS composite material from the solution. Then, it was placed in 0.2 mol/L HNO$_3$ desorption solution, at room temperature of 25 °C, oscillating frequency of 250 r/min, oscillating for t min to desorb until desorption equilibrium. Then, the absorbance of filtered supernatant was measured by arsenazo III spectrophotometer, and the desorption rate was calculated. The GOH/DMCS composite material after magnetic separation is cleaned to neutrality by deionized water many times, dried in an oven at 60 °C, and the above adsorption–desorption experimental operation is continuously circulated.

According to Equation (3), the desorption rate of rare earth element La(III) by GOH/DMCS was calculated.

$$R_d = \frac{V C_e'}{M q_e} \times 100\% \tag{3}$$

where R_d is the desorption rate, %; C_e' is the concentration of La(III) in the desorbed solution, mg/L; V is the volume of liquid, L; M is the amount of adsorbent (g); and q_e is the adsorption capacity (mg/g).

2.5. Adsorption Kinetics, Adsorption Isotherms, and Adsorption Thermodynamics Experiments

2.5.1. Adsorption Kinetics

In order to quantify the change law of adsorption with time, describe the solute absorption rate, and reveal the mechanism involved in the adsorption process, the experimental data are processed by pseudo-first order kinetics, pseudo-second order kinetics, and internal diffusion model equations, which are as follows [15]:

$$\ln(q_e - q_t) = \ln q_e - k_1 t \tag{4}$$

$$\frac{t}{q_t} = \frac{1}{q_e^2 k_2} + \frac{t}{q_e} \tag{5}$$

$$q_t = k_d t^{1/2} + C \tag{6}$$

where t (min) is the adsorption time, q_e (mg·g^{-1}) and q_t (mg·g^{-1}) are the equilibrium adsorption capacity and adsorption capacity at time t, respectively. k_1 (min^{-1}), k_2 (g·(mg·min)$^{-1}$), and k_d (mg·g^{-1}·min$^{-1/2}$) are the pseudo-first order kinetics, pseudo-second order kinetics, and diffusion rate constants in particles, respectively. C is the boundary layer constant.

2.5.2. Adsorption Isotherm

Adsorption isotherm study is of great significance to determine the adsorption capacity of La(III) on modified GOH/DMCS composites. Langmuir and Freundlich isotherm models are used to analyze adsorption data [16].

The linear equations of Langmuir and Freundlich isothermal models are as follows:

$$\frac{C_e}{q_e} = \frac{C_e}{q_m} + \frac{1}{q_m \cdot K_L} \tag{7}$$

$$\ln q_e = \ln K_F + \frac{1}{n} \ln C_e \tag{8}$$

where C_e (mg·L^{-1}) is the concentration of La(III) at adsorption equilibrium, q_e (mg·g^{-1}) is the adsorption capacity at adsorption equilibrium, q_m is the maximum adsorption capacity (mg·g^{-1}), K_L is the Langmuir constant (L·mg^{-1}) and K_F is the Freundlich constant (mg·g^{-1})

The dimensionless parameter R_L can further analyze the adsorption process. When $R_L = 0$, the adsorption process is irreversible. When $R_L = 0\sim 1$, the adsorption process is favorable. When $R_L > 1$, it is unfavorable for adsorption. The calculation formula is as follows:

$$R_L = \frac{1}{1 + C_0 K_L} \tag{9}$$

where C_0 is the initial concentration of solution, mg/L.

2.5.3. Adsorption Thermodynamics

Adsorption thermodynamics uses three calculated thermodynamic parameters, namely Gibbs free energy change (ΔG^0), enthalpy change (ΔH^0), and entropy change (ΔS^0), to study the degree and driving force of adsorption process. Gibbs energy change (ΔG^0) indicates the degree of spontaneity in the adsorption process, while the higher negative value reflects that it is more beneficial to adsorption in the adsorption process. Enthalpy change (ΔH^0), which is the change of enthalpy of an object, is a thermodynamic energy state function, which indicates that the adsorption process belongs to endothermic or exothermic reaction. Entropy change (ΔS^0) reflects the disorder or disorder degree of the interface between solid and solution during adsorption [17].

ΔH^0 and ΔS^0 are calculated by the van't Hoff Equation:

$$\ln K_d = -\frac{\Delta H^0}{RT} + \frac{\Delta S^0}{R} \tag{10}$$

where K_d is dispersion coefficient, L/g; ΔH^0 is enthalpy change, KJ·mol^{-1}; R is gas constant, 8.31×10^{-3} KJ·mol^{-1}·k^{-1}; T is absolute temperature, K; and ΔS^0 is entropy change, J·K^{-1}·mol^{-1}.

ΔG^0 is calculated using the following Equation:

$$\Delta G^0 = -RT \ln K_d \tag{11}$$

$$K_d = \frac{q_e}{C_e} \tag{12}$$

Thereinto, the change of Gibbs free energy of ΔG^0 is KJ·mol^{-1}.

3. Results and Discussion

3.1. Material Characterization

As shown in Figure 1, the surface of GO (a) prepared by the improved Hummers' method was smooth, with an accumulated flaky structure and numerous wrinkles. Compared with the GO, the surface of GOH (b) was relatively rough, with numerous irregular wrinkles and cracks, which were possibly due to the activation of the epoxide and ester groups, and the conversion of hydroxyl groups (−OH) into carboxyl groups (−COOH) [18,19]. As observed in the SEM image (c) of the pretreated DE, DE was disc-shaped with a smooth surface and regular porosity [20]. Many impurities were observed on the diatomite surface. Acidification pre-treatment was used to remove impurities and to better prepare adsorbents for integrating with the composite materials. Thus, the internal structure of CS (d) was solid and filamentous, with a relatively rough surface and porous structure. The SEM image of MCS (e) showed that the chitosan was grainy after magnetic modification, with agglomerated white round particles, and the diameter of the magnetic particles was about 30 nm [21]. In the SEM image (f) of GOH/DCS, the interior of the modified adsorption material still showed the appearance of regular and large pores, thus increasing the specific surface area of the sheet structure. The increase in pore structure and the specific surface area provided a greater possibility for the adsorption of the rare earth La elements [22]. Furthermore, GOH/DMCS (g) had a fold on the porous disk structure, which differed from those on the DE surface, which was covered with a thick layer of folds, increasing the dispersity. This was attributed to the successful compounding of the carboxylated graphene oxide and modified materials. In addition, the apparent surface roughness of the intact DE structure, and the irregular adhesion of magnetic chitosan particles on the surface of the modified materials, confirmed the successful preparation of the magnetic chitosan/diatomite-modified carboxylated graphene oxide.

The infrared spectrograms of GO, GOH, DE, CS, GCS, GOH/DCS, and GOH/DMCS are shown in Figure 2. As shown in the FTIR spectrogram for GO in Figure 2a, the prominent peak at 3406 cm^{-1} corresponds to the asymmetric stretching vibration peak of −OH, while the vibration at 1650 cm^{-1} was the stretching vibration peak of C=O in the carboxyl group of GO, and C=O was grafted on the edge of GO. The peak at 1507 cm^{-1} reflected the vibration absorption of the C=C skeleton and the characteristic bending band of the O−H bond in the undamaged aromatic region. The C−O stretching vibration band and C−O−C vibration absorption band appeared at 1260 cm^{-1} [23]. Many oxygen-containing functional groups appeared in the interlayer and edge of GO, and the above conclusions indicated that GO was successfully prepared by the modified Hummers' method. For GOH, the FTIR spectrum shows characteristic bands for GO, and a characteristic peak at 1171 cm^{-1} was attributed to the antisymmetric tensile vibration of C−O−C. Compared with GO, the characteristic bending of the O−H bond and the stretching vibration of C−O for the carboxyl group also appeared at 1600 cm^{-1}. The prominent stretching vibration of −OH in GO shifted from 3406 cm^{-1}, and a less obvious peak appeared at 3419 cm^{-1} in the FTIR spectrum of GOH, indicating that the carboxyl groups in the alkaline environment replaced some hydroxyl groups in GO. Thus, the FTIR results confirmed the successful preparation of GOH. In the FTIR spectrum of DE Figure 2b, the peak at 3434 cm^{-1} was attributed to the stretching vibration of the free silanol group (Si−OH). In addition, the bands at 1613 cm^{-1}

and 1525 cm^{-1} were related to the bending vibrations of water molecules (H−O−H) [24]. The band at 1090 cm^{-1} reflected the stretching of the siloxane (Si−O−Si) group, while 849 and 792 cm^{-1} were associated with Si−O tension and SiO−H vibrations. The peak at 572 cm^{-1} reflected the stretching vibration of Si−O−Si [25,26]. In the FTIR spectrum of CS (c), there were apparent −OH asymmetric stretching vibrations at 3437 cm^{-1} and N-H stretching vibrations for the absorption peak on −NHCO−. A C−H stretching vibration appeared at 2992 cm^{-1}, and at 1603 cm^{-1}, the stretching vibration of the alcohol hydroxyl group C=O from -NHCO− occurred. The peak at 1514 cm^{-1} was related to the N-H stretching vibration of −NH$_2$, and a C−H absorption peak appeared at 1428 cm^{-1}. The absorption peak at 1078 cm^{-1} was attributed to the stretching vibration of C−O. [27] The FTIR spectrum of MCS showed that the characteristic peak at 580 cm^{-1} was Fe$_3$O$_4$, which indicated that the magnetic particles were successfully loaded. The hydroxyl peak intensity at 3437 cm^{-1} on CS weakened and moved to the left to 3429 cm^{-1}, which was possibly caused by the interactions between nano-ferric oxide particles and chitosan functional groups during magnetization.

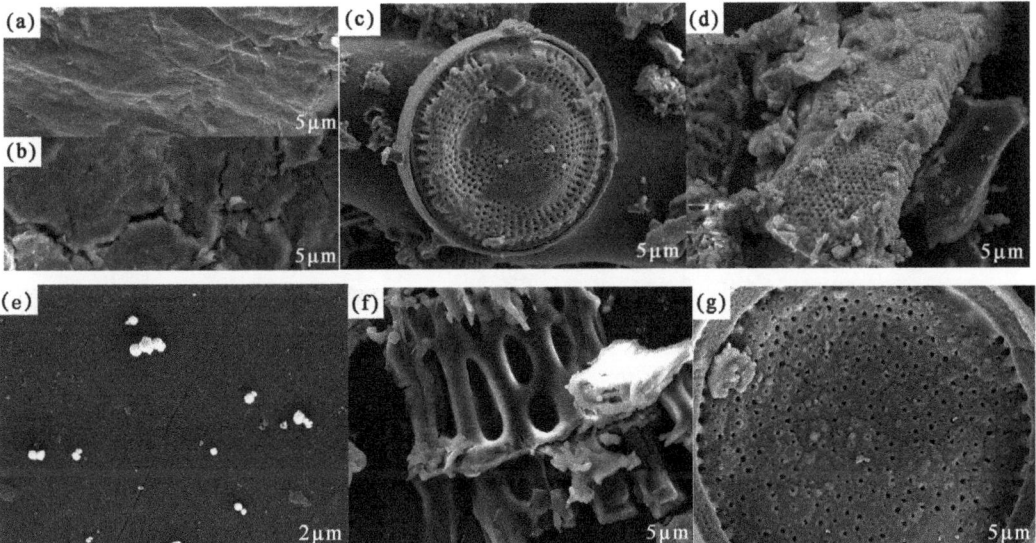

Figure 1. SEM images of (a) GO, (b) GOH, (c) DE, (d) CS, (e) MCS, (f) GOH/DCS, and (g) GOH/DMCS.

In the FTIR spectrum of GOH/DCS (d), the characteristic peaks of DE and CS appeared beside the characteristic peaks of the carboxyl group. The appearance of NH stretching and the characteristic peak of the amide group at 1612 cm^{-1} indicated that chitosan was successfully grafted onto the GOH molecule [28]. The typical band of the chitosan polymer (OH) appeared at 3410 cm^{-1}. In addition, 2813 cm^{-1} corresponded to the C-H tensile absorption band. The modified infrared band was wider and stronger, which indicated that GOH/DCS carried more chitosan. The band at 1090 cm^{-1} reflected the stretching of the siloxane (Si−Si) group in DE. Thus, an SiO−H vibration peak appeared at 786 cm^{-1}.

In summary, the GOH/DCS materials were successfully modified and prepared. In the FTIR spectrum of GOH/DMCS, a shift in the OH stretching vibration peak of the carboxyl group appeared at 3421 cm^{-1}, which indicated that hydrogen bonding occurred between the hydroxyl groups in each material. The other peaks at 786, 1090, and 1612 cm^{-1} corresponded to the FTIR spectra of DE. Thus, the modification of the carboxylated graphene oxide/diatomite composite did not change the functional group structure of DE, and GOH covered the surface of DE through covalent bonding. Additionally, the carboxyl groups in the carboxylated graphene oxide and original graphene oxide were successfully compounded with GCS through covalent bonding. The electron micrographs also showed that the composite modified by GOH/DMCS retained the DE structure, which was consistent with the obtained FTIR results. Therefore, the excellent properties of DE, such as numerous pores, large specific surface area, and strong adsorption capacity, were retained when the composite was successfully prepared.

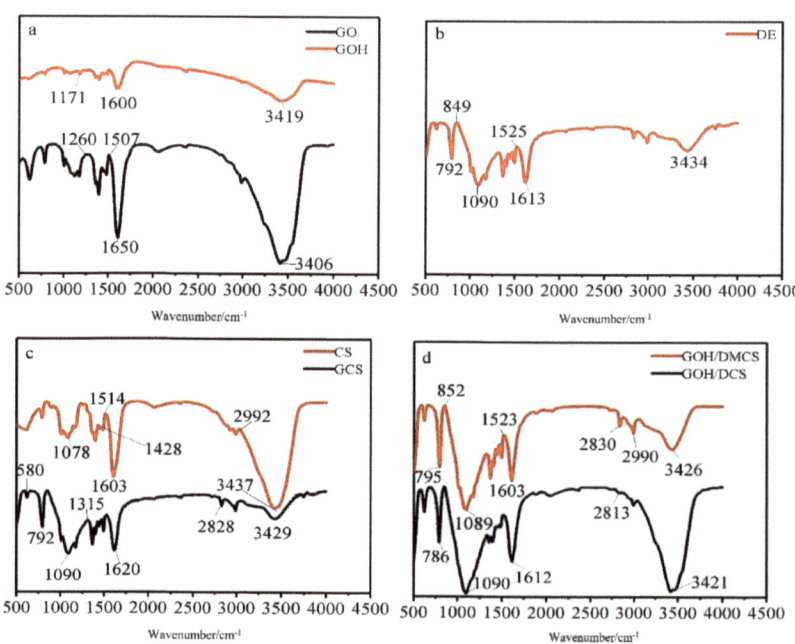

Figure 2. FTIR spectra of (a) GO and (GOH), (b) DE, (c) CS and MCS, and (d) GOH/DCS and GOH/DCS.

The crystal structure changes in each sample were analyzed by XRD, and the XRD analysis results of GO, GOH, CS, GOH/DCS, DE, and GOH/DMCS are shown in Figure 3. The diffraction peak of the (d001) crystal plane appeared in the GO (a), which was prepared by the improved Hummers' method at $2\theta = 10.9$, according to Bragg formula ($2d\sin\theta = n\lambda$), with an interlayer spacing of 0.81 nm. In the XRD results for GOH, the diffraction peak appeared at $2\theta = 10.6$, and the interlayer spacing was 0.83 nm, which indicated that the diffraction peak of the same crystal plane shifted to a low angle. The calculated results showed that the introduction of chloroacetic acid increased the interlayer spacing of GOH, and a dense layered graphene oxide structure was layered into the disordered sheets of GOH. In addition, carboxylation treatment increased the number of -COOH groups at the edge and interlayers of GO. In the XRD results of CS, (b) the characteristic peak of the amorphous chitosan structure appeared at $2\theta = 19.92$, and the diffraction peak of the GOH/DCS composite decreased to $2\theta = 21.81$, indicating that the CS crystal phase content declined. Then, the diffraction peak of GOH appeared again, and the diatomite diffraction peak appeared at $2\theta = 36.09$. The analysis results showed that GOH was related to DE and CS. In DE(c) and GOH/DMCS, the diffraction peaks were very similar, indicating that silicon dioxide accounted for a large proportion of DE. The diffraction peak of magnetic ferric oxide simultaneously appeared, verifying the successful preparation of the magnetic graphene oxide-based composites.

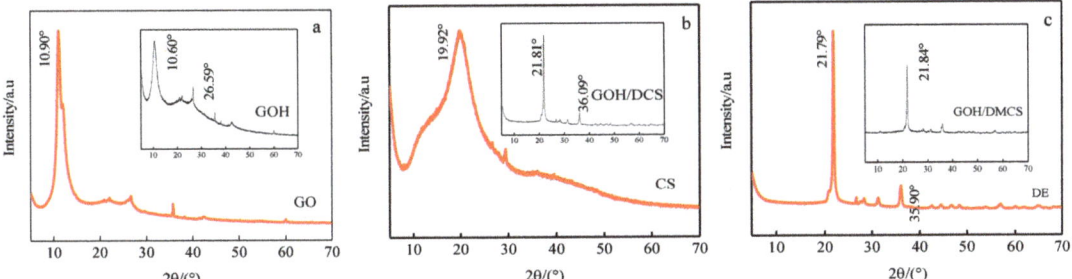

Figure 3. XRD patterns of the different materials: (a) GO and GOH, (b) CS and GOH/DCS, and (c) DE and GOH/DMCS.

In this work, the zeta potential of GOH/DMCS in the water phase was measured using a Zetasizer Nano S90 potential analyzer from the Malvern company, in a 2–12 pH environment. The zeta potential results of GOH/DMCS are shown in Figure 4. The figure shows that the electronegativity of the GOH/DMCS surface gradually increased from acidic to alkaline with changes in pH, from −11.9 mV at pH = 2 to −30.4 mV at pH = 12. In addition, compared to a potential value of −9.5 mV when the pH of graphene oxide was 2, the electronegativity of the GOH/DMCS surface was enhanced by the oxygen-containing functional groups, which were ionized as they were exposed to electricity in the aqueous solution.

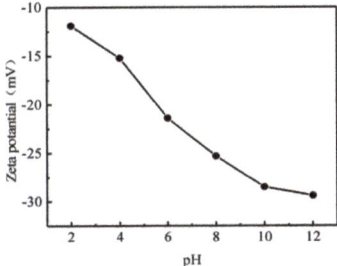

Figure 4. Zeta potential diagram of GOH/DMCS.

As shown from the negative zeta potential results under acid and alkaline conditions, the entire surface of the material became negatively charged, which was beneficial for the adsorption of positively charged rare earth La^{3+} elements. The negative charge density on the GOH/DMCS surface also increased with increased pH, and the sol stability of the composite material also increased. This did not cause aggregation when in solution form.

As observed in the hysteresis curve (Figure 5), the hysteresis loop of the GOH/DMCS composite was S-shaped with an asymmetrical origin, and there was no obvious hysteresis loop. The maximum magnetization of the GOH/DMCS composite was 13.1 emu/g, which was lower than the magnetization of Fe_3O_4 by 52.70 emu/g [29]. This was possibly due to the combination of GOH, DE, and CS non-magnetic materials, and the nanostructure of Fe_3O_4 during the preparation of the modified material, which weakened the magnetization of the composite material. However, the experimental diagram also showed that the ultrasonically dispersed uniform GOH/DMCS composite water solvent could be effectively separated in the aqueous solution in less than one minute, under an applied magnetic field. Thus, the above characterization results indicated that the GOH/DMCS composite was successfully prepared and could be used as an adsorbent that can be quickly separated and recovered.

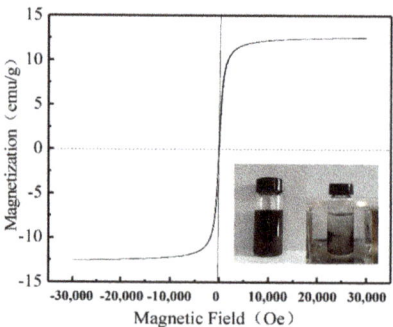

Figure 5. Magnetic hysteresis loops of GOH/DMCS.

3.2. Adsorption and Desorption Properties of La (III) by GOH-Based Composites

In this paper, compounding MCS not only changes the hydrophilicity of GO but also increases the number of functional groups (−COOH, −OH, −O−, −NH$_2$, etc.) on the surface of composite materials. There are many oxygen-containing functional groups, hydroxyl bridges, and carboxyl groups on the surface of GOH/DCS and GOH/DMCS composites, which make the whole composites have a lot of negative charges, which can interact electrostatically with positively charged rare earth elements, and they may also have complexation and chelation to achieve the removal effect. The adsorption and desorption properties of composite materials for lanthanum will be discussed below.

3.2.1. Diatomite Content and the Influence of Different Materials on Adsorption

DE content in composite materials can directly affect the adsorption effects of La(III). Therefore, GOH/DCS and GOH/DMCS composite materials were prepared with different diatomite addition amounts. Specifically, 1, 2, 3, 4, and 5 g adsorption experiments were carried out on the target pollutant La(III). Additionally, adsorption experiments for La(III) were carried out with GO and GOH matrix materials. The adsorption effects of the different materials were studied, and the results are shown in Figure 6.

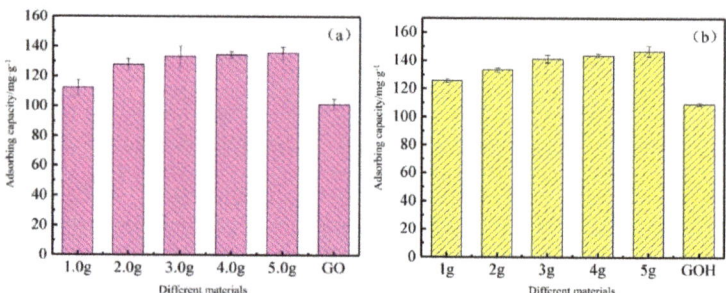

Figure 6. DE content and influence of different materials on adsorption: (**a**) GOH/DCS and GO adsorbing La(III), and (**b**) GOH/DMCS and GOH adsorbing La(III). (pH = 8, t = 50 min, T = 45 °C, La(III) concentration = 25 mg·L^{-1}, Dosage = 1 g·L^{-1})

As shown in Figure 6, with increased addition of DE, the adsorption trend of the target pollutant La(III) by the composite GOH/DMCS and GOH/DCS material was roughly the same, with an apparent initial increase, which later slowed. The reason for this was due to GOH, which played a significant role in La(III) adsorption with small amounts of DE. With increasing DE content, its specific surface area, porous structure, and functional groups all participated in the reaction process and increased the number of adsorption sites for the adsorbent; thus, the adsorption capacity increased significantly. Therefore, the composite adsorbents used in the follow-up research were all prepared using 3 g of DE. Meanwhile, as observed from the adsorption capacity in the Table 2, the adsorption capacities of La(III) by GOH/DMCS, GOH/DCS, GOH, and GO were 146.58, 135.53, 109.23, and 101.37 mg/g, respectively. The adsorption capacities of the prepared graphene oxide composites were higher than those of GO and GOH. Thus, for the adsorption results, the modified composites had a better adsorption effect on La(III).

Table 2. Adsorption capacity of different materials under the same conditions.

Material	GO	GOH	GOH/DCS	GOH/DMCS
Adsorption capacity (mg·g^{-1})	101.37	109.23	135.53	146.58

3.2.2. Influence of Adsorption Time on Adsorption and Kinetics

As shown in Figure 7, the adsorption trends of the target pollutant La(III) by GOH/DMCS and GOH/DCS were similar. The adsorption of La (III) by GOH/DMCS reached adsorption equilibrium within 60 min when the adsorption capacity was 141.19 mg/g. The adsorption of La (III) by GOH/DCS occurred within 50 min when the adsorption capacity was 134.23 mg/g. Within 60 min, the adsorption capacity increased rapidly; then, it reached adsorption equilibrium. The adsorption capacity of GOH/DMCS for La(III) was significantly higher than that of GOH/DCS, as during the initial period of adsorption, there were a sufficient number of exposed adsorption sites on the surface of the adsorbent. The negative charge on the surface of the adsorbent and the positively charged metal ions caused electrostatic interactions, and La(III) began to occupy the adsorption sites on the surface of the adsorbent. Thus, the adsorption capacity showed a rapid upward trend during the initial stage. As the contact time increased, the adsorption sites on the composite adsorbent gradually filled, and the adsorption gradually reached equilibrium. Thus, the GOH/DMCS composite material with added magnetic chitosan had a higher adsorption capacity than GOH/DCS for La(III) due to its magnetic separation. Based on the above conclusions, 50 min was chosen as the best equilibrium reaction time for the follow-up study.

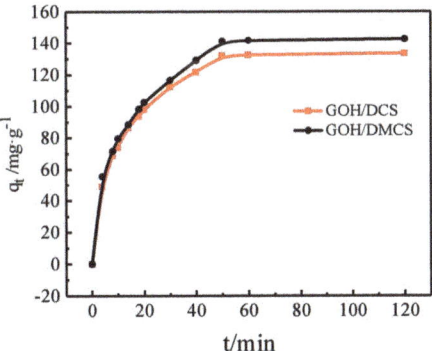

Figure 7. Influence of adsorption time on adsorption effect. (pH = 8, T = 45 °C, La(III) concentration = 25 mg·L^{-1}, Dosage = 1 g·L^{-1})

After studying the effect of adsorption time on the adsorption, we found it necessary to further explore the adsorption kinetics of GOH/DMCS on La(III). The fitting results of the kinetic model for GOH/DMCS adsorption of La(III) are shown in Figure 8, showing that more data points fell on the curve fitted by the pseudo-second-order kinetic model. As shown by the experimental parameters in Table 3, the correlation coefficient (R^2) for La(III) adsorption described by the pseudo-second-order kinetic model was 0.9854, which was higher than the correlation coefficient value of 0.9115 obtained by pseudo-first-order kinetics. Based on the above analysis results, we determined that the adsorption of La(III) on the GOH/DMCS composite material was best described by the pseudo-second-order kinetic model. The adsorption mechanism depended on the physical and chemical properties of the adsorbent and the chemical interactions of the adsorbent. Some scholars have found that the pseudo-second-order kinetic model indicates that most adsorption reactions are chemical rather than physical adsorption [30]. In addition, chemical adsorption was related to the sharing and transferring of electrons between the GOH composite material of GOH/DMCS and La(III). Thus, the initial reaction rate was calculated according to the following formula: $h = k_2 \cdot q_e^2$, where the initial reaction rate was 13.4590 mg/g·min.

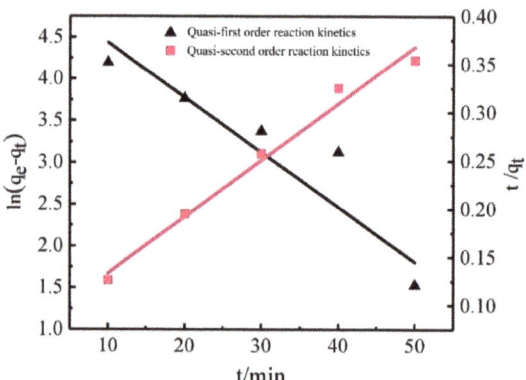

Figure 8. Kinetic model of La(III) adsorption by GOH/DMCS.

Table 3. Kinetic parameters of La(III) adsorption by GOH/DMCS.

Quasi-First-Order Dynamics Model			Quasi-Second-Order Dynamics Model			
q_e mg/g	k_1 min^{-1}	R^2	h mg/(g·min)	q_e mg/g	k_2 g/(mg·min)	R^2
162.77	0.0657	0.9115	13.4590	170.36	4.637 × 10^{-4}	0.9854

To describe the diffusion of the system in a specific range, an internal diffusion model was used to fit the adsorption process, and the results are shown in Figure 9 and Table 4. As depicted in Figure 9, the straight line fitted by q_t to $t^{0.5}$ was divided into two sections, and the straight line did not pass through the origin, C≠0, indicating that the adsorption rate was not solely controlled by internal diffusion. Thus, for the adsorption process, the adsorbate underwent two main processes. In the first stage, the external diffusion of the adsorbate La(III) in the aqueous solution was adsorbed on the surface of the composite material by GOH/DMCS. In the second stage, internal diffusion of La(III) in GOH/DMCS occurred on the surface of the composite material or between the layers. From the adsorption parameters of the internal diffusion model, k_{d1} of the first stage was greater than k_{d2} of the second stage, which proved that the adsorption rate of GOH/DMCS for La(III) was higher in the initial period.

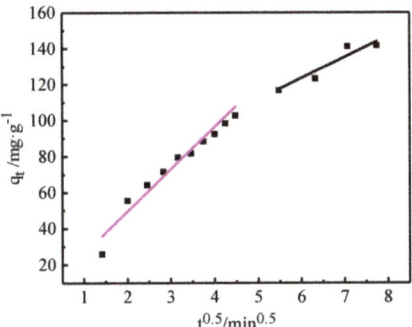

Figure 9. Intra-particle diffusion model of La(III) adsorption by GOH/DMCS.

Table 4. Intra-particle diffusion parameters of La(III) adsorption by GOH/DMCS.

First Stage			Second Stage		
k_{d1} mg·g^{-1}·min$^{-1/2}$	C_1	R^2	k_{d2} mg·g^{-1}·min$^{-1/2}$	C_2	R^2
23.4681	2.5361	0.9436	11.7543	52.9058	0.8288

3.2.3. Influence of pH Value of the Solution on Adsorption Effect

In this portion of the study, the experiments were carried out at pH 2–12, and the pH values were adjusted to 2, 4, 6, 8, 10, and 12 with NaOH or HCl. As shown in Figure 10 that when the pH was 2, the adsorption capacity of GOH/DCS for La(III) was 98.97 mg/g, and the adsorption rate was 65.98%. The adsorption capacity of GOH/DMCS for La(III) was 109.18 mg/g, and the adsorption rate was 72.78%. The figure shows that with increased pH, the adsorption rate rose, and the adsorption capacity of GOH/DMCS and GOH/DCS for the target La(III) pollutant initially increased. After the pH reached 8, the adsorption capacity was unchanged, and adsorption was optimal when pH was 8. During this time, the adsorption capacities were 133.30 and 141.19 mg/g, and the adsorption rates were 89 and 94%.

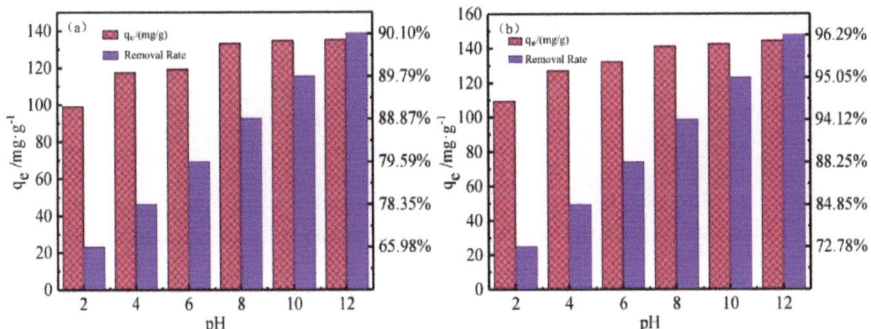

Figure 10. Influence of pH on the adsorption effects of (a) GOH/DCS and (b) GOH/DMCS. (t = 50 min, T = 45 °C, La(III) concentration = 25 mg·L^{-1}, Dosage = 1 g·L^{-1})

The pH affected the positive and negative charges on the surface of the adsorbent and affected the presence of La(III) ions in the aqueous solution. When pH was 2, the zeta potential value of GOH/DMCS in the water phase was −11.9 mV (Figure 4). The surface of the composite adsorbent was negatively charged, and the La(III) ions were electrostatically attracted for removal. However, a large number of positively charged H+ in the solution competed with the La(III) ions for adsorption sites on the adsorbent surface. Thus, due to the distribution of adsorption sites, the adsorption capacity was the lowest during this time. In addition, the GOH composite material had a high specific surface area and pore structure, which was conducive to adsorption.

Between pH 2 and 8, both the adsorption capacity and the adsorption rates trended upward, as shown in Figure 4, because the introduced oxygen-containing functional groups, −COOH and −OH, ionized into −COO- and −O in the aqueous solution. Thus, the surface of the material became negatively charged, the electronegativity of the GOH/DMCS surface gradually increased with pH from acidic to alkaline, the zeta potential values of the GOH/DMCS surface accelerated in the negative direction, and the adsorption capacity of La(III) increased. Thus, both the capacity and the adsorption rate gradually increased, and GOH/DMCS adsorbed La(III) through complexation.

At pH > 8, GOH/DMCS and GOH/DCS adsorbed and removed La (III) by forming a carboxylate complex. More OH− ions were present in the solution, and the rare earth cations and OH− ions combined, gradually becoming insoluble oxide and hydroxide precipitates. The precipitates hindered the adsorption of La(III), and the adsorption capacity was unchanged. According to the above analysis, the pH of the subsequent experiments was 8.

3.2.4. Influence of Adsorbent Dosage on Adsorption Rate

The adsorbent dosage curves are shown in Figure 11, indicating that the dosage of the adsorbent directly affected the adsorption of La(III) in the composite materials. When the dosage increased from 0.5 to 5 g/L, the adsorption rate increased, while the adsorption capacity declined. With increased adsorbent dosage, the number of active sites and functional groups in adsorbent gradually increased; therefore, there was a large contact area between the composite adsorbent and La(III), and the adsorption rate rose. However, the dosage effect of 5 g/L GOH/DCS and GOH/DMCS was smaller than 0.5 g/L of GOH/DCS and GOH/DMCS, and the higher adsorbent dosage caused an agglomeration of GOH/DCS and GOH/DMCS, resulting in a decrease in specific surface area. Thus, the adsorption capacity of GOH/DCS and GOH/DMCS for La(III) decreased with increased dosage.

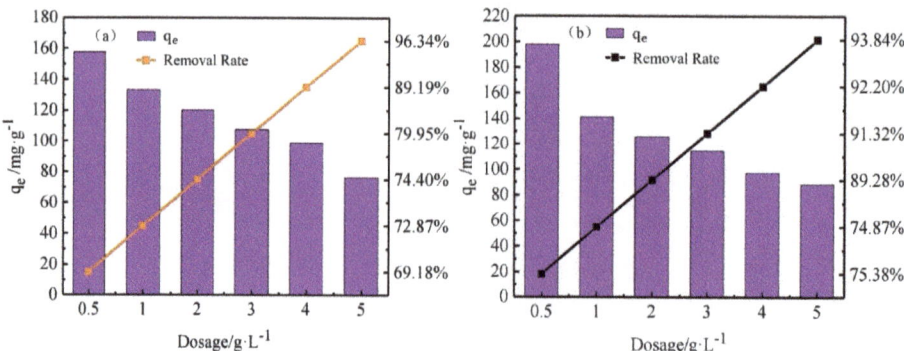

Figure 11. Influence of adsorbent dosage on adsorption for (a) GOH/DCS and (b) GOH/DMCS. (pH = 8, t = 50 min, T = 45 °C, La(III) concentration = 25 mg·L^{-1})

According to the analyzed data, the adsorption effect of GOH/DMCS on La(III) was better than that of GOH/DCS under the same dosage amount. GOH/DMCS had the highest adsorption rate of La(III), at 93.84%, and the adsorption capacity was 197.94 mg/g. The highest adsorption rate of La(III) by GOH/DCS was 96.34%, and the adsorption capacity was 157.53 mg/g. The adsorption capacity of GOH/DCS for La (III) was low, which was possibly because the −COOH at the edge of the carboxylated graphene oxide was more accessible and could more easily react and bond with -NH$_2$ in the chitosan without magnetic modification. This effect may have affected the roles of these two functional groups during adsorption, thus resulting in lower adsorption capacity compared to GOH/DMCS.

3.2.5. Influence of Temperature and Initial Concentration of the Rare Earth Ion Solution on Adsorption

We also studied the effects of different initial concentrations and temperatures on adsorption efficiency. Figure 12 shows a changing trend in the adsorption effect of La (III) by GOH/DCS and GOH/DMCS, which was similar at different temperatures. The adsorption capacities of GOH/DCS and GOH/DMCS for La (III) were 151.70 and 154.02 mg/g at 25 °C and 30 mg/L, respectively. When the temperature increased to 45 °C and the solution concentration was 30 mg/L, the adsorption capacities of GOH/DCS and GOH/DMCS for La(III) were 173.04 and 175.82 mg/g, respectively. At the same initial concentration, when the temperature increased from 25 to 45 °C, the adsorption capacity trended upward, and the adsorption process was endothermic; thus, an increase in temperature was beneficial to the reaction. When the temperature rose, the movement of the rare earth La(III) ions on the surface of the composite adsorbent accelerated, collisions intensified, and the contact probability of the adsorbent increased; therefore, the adsorption capacity increased.

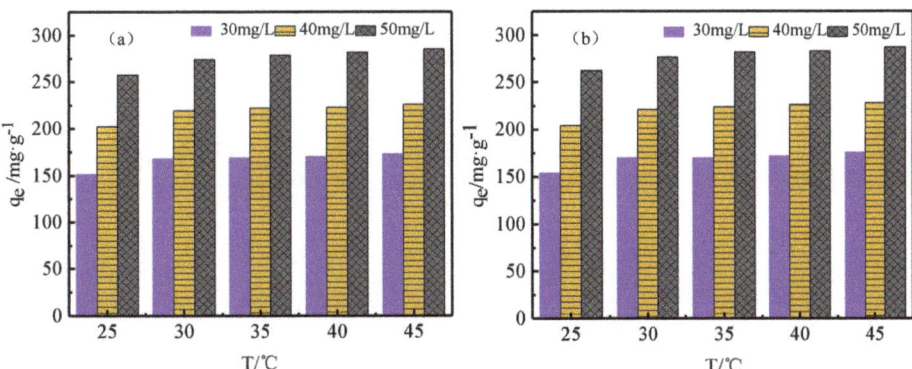

Figure 12. Influence of temperature and initial La(III) concentration on adsorption effect for (a) GOH/DCS and (b) GOH/DMCS (pH = 8, t = 50 min, Dosage = 3 g·L^{-1}).

At the same temperature, the initial concentration of the rare earth ion solution affected the adsorption and mass transfer processes, thus affecting the adsorption of La(III). As shown in Figure 12, the adsorption effects of GOH/DCS and GOH/DMCS on La(III) were the same at different initial solution concentrations. At 45 °C, when the concentration increased from 30 to 40, and then to 50 mg/L, the adsorption capacity of La(III) by GOH/DMCS increased from 175.83 to 227.94, and then to 287.01 mg/g, respectively. With increased initial concentration, the adsorption capacity exhibited a gradual upward trend. However, the adsorption capacity of GOH/DMCS was slightly higher than the GOH/DCS composite, as the GOH/DMCS composite provided more adsorption sites for La(III) during the adsorption process. Of note, this also proved that the composite was successfully modified and prepared.

3.2.6. Adsorption Isotherms

Figure 13 shows the fitting results of the adsorption data for the GOH/DMCS composites using the Langmuir (a) and Freundlich models (b), and Table 5 lists the relevant parameters of the fitting results. As observed in the figure, most of the correlation points accurately fell on the linear fitting line in the Langmuir model. As listed in Table 4, the correlation coefficient (R^2) of the Langmuir model was higher than the fitting results of the Freundlich model; thus, the adsorption process of La(III) by GOH/DMCS was more in line with the Langmuir model, indicating that the adsorption process was attributable to monolayer adsorption. The surface contained a limited number of adsorption sites with uniform adsorption energies and with no migration of adsorbate on the surface [31,32]. According to the linear fitting line, the maximum adsorption capacity at 298 and 308 K reached 294.99 and 320.51 mg/g, respectively. This was closer to the experimental values than the maximum adsorption capacity obtained by the Freundlich model. As observed from the dimensionless R_L parameter in the Langmuir model (Table 6), which was between 0 and 1 at 298 and 308 K, the adsorption of La(III) by GOH/DMCS was a favorable adsorption process, reflecting the high affinity between the adsorbent and adsorbate.

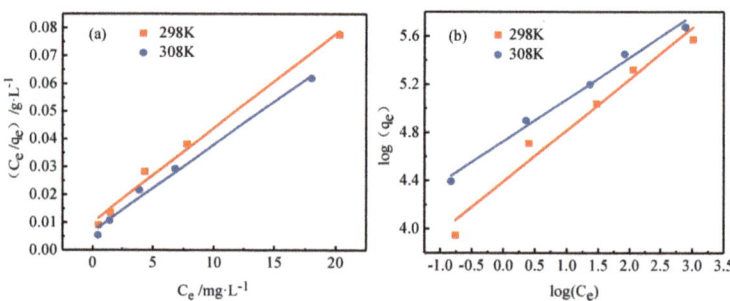

Figure 13. Isotherm models of La(III) adsorption by GOH/DMCS: (a) Langmuir model and (b) Freundlich model.

Table 5. Isothermal parameters of La(III) adsorption by GOH/DMCS.

Temperature	Langmuir Model			Freundlich Model		
	q_{max}/mg·g^{-1}	K_L/L·mg^{-1}	R^2	K_F/mg·g^{-1}·L$^{1/n}$·mg$^{-1/n}$	1/n	R^2
298 K	294.99	0.3407	0.9915	80.7784	0.4245	0.9681
308 K	320.51	0.4692	0.9912	112.9652	0.3467	0.9890

Table 6. Dimensionless parameter R_L.

Temperature	R_L
298 K	~0.0554–0.2269
308 K	~0.0409–0.1757

3.2.7. Adsorption Thermodynamics

As listed in Table 7 and Figure 14, the adsorption reaction enthalpy changes $\Delta H^0 > 0$ indicated that the adsorption process of La(III) by GOH/DMCS was an endothermic reaction. As mentioned in the above results, the adsorption capacity increased when the temperature increased from 25 to 45 °C. Thus, the adsorption process was an endothermic reaction, and the temperature increase was beneficial to the progression of the reaction. The resulting enthalpy change was the same as that of temperature. At three different temperatures, 298, 303, and 308 K, the Gibbs energy change (G^0) was less than 0. As shown in Table 6, the absolute value of G^0 increased with increased temperature, indicating that the adsorption of La(III) by GOH/DMCS proceeded spontaneously. This also verified the improved adsorption effects at higher temperatures. The entropy change value (ΔS^0) was 50.0084 J·(mol·k)$^{-1}$, indicating that the disorder of the solid–liquid interface system increased during the adsorption of La(III) by the GOH/DMCS composite. This also indicated that when discrete rare-earth ions combine with adsorbents, more ions will be released during adsorption.

Table 7. Thermodynamic parameters of La(III) adsorption by GOH/DMCS.

Temperature (K)	ΔG^0/KJ·mol^{-1}	ΔH^0/KJ·mol^{-1}	ΔS^0/J·(mol·K)$^{-1}$
298	−5.5135	10.2943	50.0084
303	−5.7433	10.2943	50.0084
308	−6.0444	10.2943	50.0084

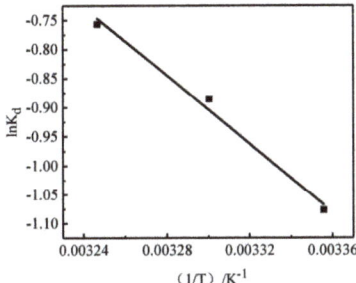

Figure 14. Thermodynamic model of La(III) adsorption by GOH/DMCS.

3.3. Analysis of Desorption Results

In practical applications, the disposal of nanomaterial (NMs) adsorbents used in adsorption processes negatively affects water quality; thus, reusability is a critical factor for creating suitable adsorbents. Desorption from adsorbents with a high affinity requires a high concentration of acid and/or chelating agents. In addition, desorption of La(III) from nanomaterials occurs in acidic solutions such as hydrochloric acid, nitric acid, or sulfuric acid. We determined that the GOH/DMCS composites prepared in this study achieved a separation effect under an external magnetic field due to their magnetism. Therefore, 0.2 mol/L of HNO$_3$ solution was used in this work for the desorption and adsorption-desorption cycling regeneration experiments.

As shown in Figure 15 that the adsorption capacities of the GOH/DMCS composites for La(III) exceeded 74% after four adsorption–desorption cycles, and the desorption percentage decreased only slightly after several adsorption–desorption cycles. This proved that the recycling of magnetic NMs can be simply and effectively achieved for adsorbent recycling, using external magnets. Of note, the gradual decrease in the desorption rate was possibly due to incomplete desorption.

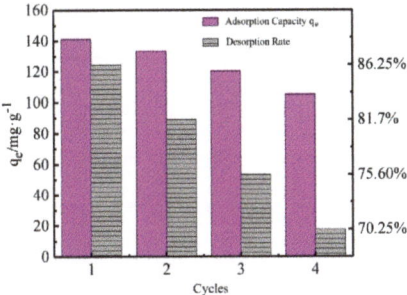

Figure 15. Cycle regeneration experiments. (pH = 8, t = 50 min, T = 45 °C, La(III) concentration = 50 mg·L^{-1}, Dosage = 3 g·L^{-1})

4. Conclusions

In this study, GO was prepared using the improved Hummers' method and was carboxylated and modified into GOH. Thus, DE and chitosan/magnetic chitosan with stable chemical properties, large specific surface areas, wear resistance, and high porosities were loaded by solution blending, and GOH/DCS and GOH/DMCS composites with suitable solid–liquid separation properties were prepared. Through La(III) adsorption experiments, various effects on adsorption were explored and fitted. The results showed that the adsorption of La(III) by the GOH/DCS and GOH/DMCS composites reached equilibrium in 50 min, and pH had the most significant effect on adsorption. When pH was 8, the adsorption capacity of La(III) by GOH/DMCS reached 141.19 mg/g. With increasing adsorbent dosage, the adsorption rate gradually increased, while the adsorption capacity gradually decreased. Therefore, the optimum adsorption effect was obtained when the addition amount of adsorbent was 3 $g \cdot L^{-1}$, and the reaction temperature and initial solution concentration were 45 °C and 50 $mg \cdot L^{-1}$, respectively. The adsorption process of La(III) by the GOH/DMCS composites followed the pseudo-second-order kinetic model, and adsorption efficiency was not solely controlled by internal diffusion. Thus, the adsorption process was more in line with the Langmuir model, which indicated that the adsorption process followed monolayer adsorption, and the surface contained a limited number of adsorption sites with uniform adsorption energies. Thus, there was no adsorbate migration on the surface. The thermodynamic fitting results showed that adsorption was an endothermic process, and an increase in temperature was beneficial to the reaction. After four adsorption–desorption cycle experiments, the adsorption capacity of the GOH/DMCS composite for La(III) initially exceeded 74%, and the desorption percentages after several adsorption–desorption cycles decreased only slightly.

Author Contributions: Conceptualization, J.Z. and X.S.; methodology, J.Z.; software, X.S.; validation, X.S., B.S. and S.W.; formal analysis, J.Z.; investigation, B.S.; resources, J.Z.; data curation, B.S.; writing—original draft preparation, X.S.; writing—review and editing, J.Z.; visualization, X.S.; supervision, S.W.; project administration, X.S.; funding acquisition, S.W. All authors have read and agreed to the published version of the manuscript. Please turn to the CRediT taxonomy for the term explanation. Authorship must be limited to those who have contributed substantially to the work reported.

Funding: This research was funded by the National Natural Science Foundation of China, grant number 21466019.

Institutional Review Board Statement: Not applicable.

Informed Consent Statement: Not applicable.

Data Availability Statement: The data presented in this study are available on request from the corresponding author. The data are not publicly available due to the author have not graduated.

Conflicts of Interest: The funders had no role in the design of the study; in the collection, analyses, or interpretation of data; in the writing of the manuscript, or in the decision to publish the results.

References

1. Wu, D.; Zhang, L.; Wang, L.; Zhu, B.; Fan, L. Adsorption of lanthanum by magnetic alginate-chitosan gel beads. *J. Chem. Technol. Biotechnol.* **2011**, *86*, 345–352. [CrossRef]
2. Jowitt, S.M.; Werner, T.T.; Weng, Z.; Mudd, G.M. Recycling of the rare earth elements. *Curr. Opin. Green Sustain. Chem.* **2018**, *13*, 1–7. [CrossRef]
3. Asadollahzadeh, M.; Torkaman, R.; Torab-Mostaedi, M. Extraction and separation of rare earth elements by adsorption approaches: Current status and future trends. *Sep. Purif. Rev.* **2021**, *50*, 417–444. [CrossRef]
4. Iannicelli-Zubiani, E.M.; Gallo Stampino, P.; Cristiani, C.; Dotelli, G. Enhanced lanthanum adsorption by amine modified activated carbon. *Chem. Eng. J.* **2018**, *341*, 75–82. [CrossRef]
5. Abdel-Magied, A.F.; Abdelhamid, H.N.; Ashour, R.M.; Zou, X.; Forsberg, K. Hierarchical porous zeolitic imidazolate frameworks nanoparticles for efficient adsorption of rare-earth elements. *Microporous Mesoporous Mater.* **2019**, *278*, 175–184. [CrossRef]
6. Kusrini, E.; Usman, A.; Sani, F.A.; Wilson, L.D.; Abdullah, M.A.A. Simultaneous adsorption of lanthanum and yttrium from aqueous solution by durian rind biosorbent. *Environ. Monit. Assess.* **2019**, *191*, 488. [CrossRef]

7. Silva, A.; Martínez-Gallegos, S.; Rosano-Ortega, G.; Schabes-Retchkiman, P.; Vega-Lebrún, C.; Albiter, V. Nanotoxicity for E. Coli and characterization of silver quantum dots produced by biosynthesis with Eichhornia crassipes. *J. Nanostructures* **2017**, *7*, 1–12.
8. Kegl, T.; Košak, A.; Lobnik, A.; Novak, Z.; Kralj, A.K.; Ban, I. Adsorption of rare earth metals from wastewater by nanomaterials: A review. *J. Hazard. Mater.* **2020**, *386*, 121632. [CrossRef]
9. Chang, S.; Zhang, Q.; Lu, Y.; Wu, S.; Wang, W. High-efficiency and selective adsorption of organic pollutants by magnetic CoFe2O4/graphene oxide adsorbents: Experimental and molecular dynamics simulation study. *Sep. Purif. Technol.* **2020**, *238*, 116400. [CrossRef]
10. Zhao, X.; Yang, M. Graphene nanocomposites. *Molecules* **2019**, *24*, 2440. [CrossRef]
11. Chen, J.; Yao, B.; Li, C.; Shi, G. An improved Hummers method for eco-friendly synthesis of graphene oxide. *Carbon* **2013**, *64*, 225–229. [CrossRef]
12. Zhang, Y.; Wang, L.; Zhang, N.; Zhou, Z. Adsorptive environmental applications of MXene nanomaterials: A review. *RSC Adv.* **2018**, *8*, 19895–19905. [CrossRef]
13. Chen, P.; Li, H.; Yi, H.; Jia, F.; Yang, L.; Song, S. Removal of graphene oxide from water by floc-flotation. *Sep. Purif. Technol.* **2018**, *202*, 27–33. [CrossRef]
14. Marcano, D.C.; Kosynkin, D.V.; Berlin, J.M.; Sinitskii, A.; Sun, Z.; Slesarev, A.; Alemany, L.B.; Lu, W.; Tour, J.M. Improved synthesis of graphene oxide. *ACS Nano* **2010**, *4*, 4806–4814. [CrossRef]
15. Molavi, H.; Zamani, M.; Aghajanzadeh, M.; Kheiri Manjili, H.; Danafar, H.; Shojaei, A. Evaluation of UiO-66 metal organic framework as an effective sorbent for Curcumin's overdose. *Appl. Organomet. Chem.* **2018**, *32*, e4221–e4231. [CrossRef]
16. Liu, Y.; Shen, L. From Langmuir Kinetics to First-and Second-Order Rate Equations for Adsorption. *Langmuir* **2008**, *24*, 11625–11630. [CrossRef] [PubMed]
17. Liu, Y. Is the Free Energy Change of Adsorption Correctly Calculated. *J. Chem. Eng. Data* **2009**, *54*, 1981–1985. [CrossRef]
18. Ma, F.; Nian, J.; Bi, C.; Yang, M.; Zhang, C.; Liu, L.; Dong, H.; Zhu, M.; Dong, B. Preparation of carboxylated graphene oxide for enhanced adsorption of U(VI). *J. Solid State Chem.* **2019**, *277*, 9–16. [CrossRef]
19. Park, K.-W. Carboxylated graphene oxide–Mn2O3 nanorod composites for their electrochemical characteristics. *J. Mater. Chem. A* **2014**, *2*, 4292–4298. [CrossRef]
20. Sun, Y.; Yao, G.; Liu, M.; Zheng, S. In situ synthesis of magnetic MnFe$_2$O$_4$/diatomite nanocomposite adsorbent and its efficient removal of cationic dyes. *J. Taiwan Inst. Chem. Eng.* **2017**, *71*, 501–509. [CrossRef]
21. Fan, L.; Luo, C.; Lv, Z.; Lu, F.; Qiu, H. Preparation of magnetic modified chitosan and adsorption of Zn(2)(+) from aqueous solutions. *Colloids Surf. B Biointerfaces* **2011**, *88*, 574–581. [CrossRef] [PubMed]
22. Khraisheh, M.; Al-Ghouti, M.; Allen, S.; Ahmad, M. The effect of pH, temperature, and molecular size on the removal of dyes from textile effluent using manganese oxides-modified diatomite. *Water Environ. Res.* **2004**, *76*, 2655–2663. [CrossRef] [PubMed]
23. Nethravathi, C.; Rajamathi, M. Chemically modified graphene sheets produced by the solvothermal reduction of colloidal dispersions of graphite oxide. *Carbon* **2008**, *46*, 1994–1998. [CrossRef]
24. Ghosh, B.; Agrawal, D.C.; Bhatia, S. Synthesis of zeolite A from calcined diatomaceous clay: Optimization studies. *Ind. Eng. Chem. Res.* **1994**, *33*, 2107–2110. [CrossRef]
25. Ma, L.L.; Xie, Q.L.; Chen, N.C.; Xu, H.; Zhou, H.M.; Yu, Q.F. In Research on Lead (II) Adsorption Mechanism from Aqueous Solution by Calcium Carbonate Modified Diatomite Absorbent. *Mater. Sci. Forum* **2018**, *921*, 21–28. [CrossRef]
26. Bahramian, B.; Ardejani, F.D.; Mirkhani, V.; Badii, K. Diatomite-supported manganese Schiff base: An efficient catalyst for oxidation of hydrocarbons. *Appl. Catal. A Gen.* **2008**, *345*, 97–103. [CrossRef]
27. Roy, J.C.; Ferri, A.; Giraud, S.; Jinping, G.; Salaün, F. Chitosan–carboxymethylcellulose-based polyelectrolyte complexation and microcapsule shell formulation. *Int. J. Mol. Sci.* **2018**, *19*, 2521. [CrossRef]
28. Li, W.; Xiao, L.; Qin, C. The characterization and thermal investigation of chitosan-Fe3O4 nanoparticles synthesized via a novel one-step modifying process. *J. Macromol. Sci. Part A* **2010**, *48*, 57–64. [CrossRef]
29. Bei, Z. Removal Performance and Mechanism of Phenolic Pollutants by Go Based Water Treatment Materials. Ph.D. Thesis, Shandong University, Jinan, China, 2019.
30. Yang, Q.; Wang, X.; Luo, W.; Sun, J.; Xu, Q.; Chen, F.; Zhao, J.; Wang, S.; Yao, F.; Wang, D. Effectiveness and mechanisms of phosphate adsorption on iron-modified biochars derived from waste activated sludge. *Bioresour. Technol.* **2018**, *247*, 537–544. [CrossRef]
31. Dai, Y.; Lv, R.; Fan, J.; Zhang, X.; Tao, Q. Adsorption of cesium using supermolecular impregnated XAD-7 composite: Isotherms, kinetics and thermodynamics. *J. Radioanal. Nucl. Chem.* **2019**, *321*, 473–480. [CrossRef]
32. Awwad, N.; Gad, H.; Ahmad, M.; Aly, H. Sorption of lanthanum and erbium from aqueous solution by activated carbon prepared from rice husk. *Colloids Surf. B Biointerfaces* **2010**, *81*, 593–599. [CrossRef] [PubMed]

MDPI
St. Alban-Anlage 66
4052 Basel
Switzerland
www.mdpi.com

Coatings Editorial Office
E-mail: coatings@mdpi.com
www.mdpi.com/journal/coatings

Disclaimer/Publisher's Note: The statements, opinions and data contained in all publications are solely those of the individual author(s) and contributor(s) and not of MDPI and/or the editor(s). MDPI and/or the editor(s) disclaim responsibility for any injury to people or property resulting from any ideas, methods, instructions or products referred to in the content.

www.ingramcontent.com/pod-product-compliance
Lightning Source LLC
LaVergne TN
LVHW070745100526
838202LV00013B/1311